1st International Conference, 'Resonance': on Cognitive Approach, Social Ethics and Sustainability

First published 2024
by Routledge
4 Park Square, Milton Park, Abingdon, Oxon OX14 4RN

and by Routledge
605 Third Avenue, New York, NY 10158

Routledge is an imprint of the Taylor & Francis Group, an informa business

British Library Cataloguing-in-Publication Data
A catalogue record for this book is available from the British Library

ISBN: 9781032501680 (pbk)
ISBN: 9781003397175 (ebk)

DOI: 10.4324/9781003397175

Typeset in Sabon LT Std
by Ozone Publishing Services

1st International Conference, 'Resonance': on Cognitive Approach, Social Ethics and Sustainability

23 and 24th November, 2022 School Of Liberal Arts and Humanities, Woxsen University, India

Edited by

Dr Raul V. Rodriguez
Dr Hemachandran K
Dr Anindita Majumdar
Dr Ranita Basu

Editors

Raul V. Rodriguez
Professor, School of Business, Woxsen University

Dr. Raul Villamarin Rodriguez is the Vice President, Woxsen University and holds the Steven Pinker Professor of Cognitive Psychology and Classavo Chair Professorship in Integrative Research and Digital Learning. Dr. Rodríguez is an Adjunct Professor at Universidad del Externado, Colombia and member of the International Advisory Board at IBS Ranepa, Russian Federation, and a member of the IAB, University of Pécs Faculty of Business and Economics. He is also part of the PRME i5 Expert Pedagogy Group - India representative. He holds a Ph.D. in Artificial Intelligence and Robotics Process Automation applications in Human Resources.

Dr. Rodriguez's specific areas of expertise and interest are Machine Learning, Deep Learning, Natural Language Processing, Computer Vision, Robotic Process Automation, Multi-agent Systems, Knowledge Engineering, and Quantum Artificial Intelligence. He has the experience and feels comfortable using Prolog, Java, C++, Python, R/RStudio, Julia, Swift, Scala, MySQL, and Spark, among others. He is a registered expert in Artificial intelligence, Intelligent Systems, and Multi-agent Systems at the European Commission, a nominee for the Forbes 30 Under 30 Europe 2020 list, and awardee in the Europe India 40 under 40 Leaders. Alongside this, he is a member of the GRLI Deans and Directors cohort. He is a regular keynote speaker and panel moderator at various national and international conferences or summits such as ML Conference (Singapore). Additionally, he is a member of the Harvard Business Review Advisory Council, the ETS Business School Advisory Council (BSAC) in India, and the Institute for Robotics Process Automation & Artificial Intelligence. He is engaged in

He has co-authored two reference books: "New Age Leadership: A Critical Insight" and "Retail Store'e" and has more than 70 publications to his credit. He is a weekly contributing writer to various magazines in the field of analytics and emerging technologies. Alongside this, he is a journal reviewer and associate editor in various publications such as IEEE.

Hemachandran K

Director of Artificial Intelligence & Business Analytics, School of Business,
Woxsen University, Hyderabad, India

Dr. Hemachandran Kannan is a Professor in the department of Artificial Intelligence & Business Analytics at School of Business, Woxsen University, India and holds the Zita Zoltay Paprika of Decision Sciences and Business Economics and Course5i- Chair Professor of Business Analytics and Machine Learning. He has been a passionate teacher with 15 years of teaching experience and 5 years of research experience. A strong educational professional with a scientific bent of mind, highly skilled in AI & Business Analytics. After receiving a PhD in embedded systems, He started focusing on Interdisciplinary research. He served as an effective resource person at various national and international scientific conferences and panel discussions. He also gave lectures on topics related to Artificial Intelligence & Business Analytics. He was bestowed as Best faculty at Woxsen University in 2021-2022 and also in Ashoka Institute of Engineering & Technology in 2019 – 2020. He is having rich working experience in Natural Language Processing, Computer Vision, Building Video recommendation systems, Building Chatbots for HR policies and Education Sector, Automatic Interview processes, and Autonomous Robots. Dr. Hemachandran is currently working on various real-time use cases and projects in collaboration with Industries such as Advertflair, Nosh Technologies, Course5i and Apstek Corp. He has organized many International Conferences, Hackathons and Ideathon. He owed four patents to his credentials. He has a life membership in estimable professional bodies. An open-ended positive person who has a stupendous peer-reviewed publication record with more than 35 journals and international conference publications. As of now, he has edited CRC Press Taylor and Francis books such as Bayesian Reasoning and Gaussian Process for Machine Learning Applications, machine learning for Business Analytics and Coded Leadership Developing Scalable Management in an AI -induced Quantum World.

Dr Anindita Majumdar,

Associate Professor, School of Liberal arts and humanities,
Woxsen University

Dr. Anindita Majumdar is an Academician and RCI registered Psychologist. She is the recipient of the Young Faculty Award at Higher Education & Research Summit 2021, conferred jointly by the International Benevolent Research Foundation and the Confederation of Indian Universities (CIU). In January 2022, she received the Swami Vivekananda Excellence Award conferred by the World Achiever's Foundation for her contribution to the field of Psychological research. She received Young Researcher Award, in April 2023 for her empirical work on body privacy training through dance movement techniques, conferred jointly by InScA, IBRF, and CIU. She has 12 years of work experience in both the Education and Corporate sectors. She is associated with the Association of Rehabilitation Psychologists & Professionals (ARPP-India) and the International Benevolent Research Foundation as a Life Member. She has extensive experience in practicing Psychometric Assessment, Counselling, Psychotherapy, Skill Training & Development, Child Development & Parenting Psycho-educational Training, and Behavioural & Educational Management in the field of Special Educational Intervention. Her areas of teaching and research inclinations include Creativity & Creative Psychotherapy, Expressive Arts, Qualitative Analysis, Clinical/ Rehabilitation Psychology, Positive Psychology, Personality Studies, Applied Social Psychology, and Developmental and Educational Psychology. She has several papers presented and published in these domains at both national and international levels.

Dr Ranita Basu

Associate Professor, School of Liberal arts and humanities,
Woxsen University

Dr Ranita Basu is an academician and has been associated with education industry for more than 15 years. She did her PhD from The University of Calcutta from the Applied Psychology Department and also qualified NET exam. Her research interest deeply is in behavioral science. Her research areas are – self efficacy, emotional intelligence, work-family conflict, digital detox, employee engagement program, work satisfaction, organizational stress, HR analytic, employee motivation and anxiety etc. She has many publications in indexed journals, presented paper in conferences, contributed book chapter and had written 2 books. Apart from teaching and research, she has worked as paper setter, moderator and examiner for many reputed Institutes/ Universities in India. She has also worked as a panelist for some public sector organizations and also served a resource person for many training programs, Faculty Development Programs and Management Development Programs organized by various private sector all over in India.

Contents

Lists of Figures

List of Tables

List of Tables

1. What is Consciousness? Could Artificial Intelligence (AI) have it?

Pinky Chauhan[1], Shiva Rohit[2], C V Sai Supraja Reddy[3]
Students, Woxsen University

ABSTRACT: Artificial intelligence (AI) has been fast growing since its evolution and experiments with various new add-on features; human efficiency is one among those and the most controversial topic. This chapter focuses on its attention towards studying human consciousness and AI independently and in conjunction. It provides theories and arguments on AI being able to adapt human-like consciousness, cognitive abilities and ethics. This chapter studies responses of more than 300 candidates of the Indian population and compares it against the literature review. Furthermore, it also discusses whether AI could attain consciousness, develop its own set of cognitive abilities (cognitive AI), ethics (AI ethics) and overcome human beings' efficiency. This chapter is a study of the Indian population's understanding of consciousness, cognitive AI and AI ethics.

KEYWORDS: Consciousness, Artificial Intelligence (AI), Cognitive Abilities, Cognitive AI, Ethics, AI Ethics

1. INTRODUCTION

In the last decade, artificial intelligence (AI) has evolved significantly and seen success in various fields and as it continues to grow, consciousness has become the new possibility (Ng and Leung, 2020). The argument on machines becoming consciousness is not a new concept, the philosophical arguments have been going on long periods (Hildt, 2019). Over the years it has been seen that machines have been designed to perform a variety of tasks that fit in all the fields of life but there still are many 'human activities' that computers or machines yet have to learn; and cognitive abilities *(cognitive AI)* are one of those. One reason why machines do not hold cognitive abilities is because humans are not completely aware of our brain's functioning and need to be enlightened on how a brain functions and how it is easy for us to visualise objects but not for machines. For instance, machines (being robots) ask humans to tick a box which says, 'I'm not a Robot', why? Google more often than not asks to select traffic lights and pedestrians to confirm that humans are not robots to a robot? That is because the robots still lack that basic knowledge of images and performing such easy tasks but how so? Does AI have the understanding of the programmes they are inserted with or just simulate the human thinking?

When AI and ethics (*data ethics* or *AI ethics*) are referred in the same sentence it most likely is understood as 'ethical concerns' to academic ethicists because there is only little knowledge and prediction as 'what will happen?' and 'how it will happen?' and chiefly thought of it as a destruction to human beings. The present AI and robotics machines have caused a 'fundamental fear' that these advanced machines will cause an end to human beings' era. Ethics come under an area which is similar to consciousness and having cognitive abilities. These need to be understood as to how these vary from one human to another and most importantly what is the right form of ethics. Today, these ethics are governed by societal norms, rules and regulations which human beings need to follow in order to have an existence outside of the prison and in an open, free world. But what will it be like for AI to have these; and will they be following the set of norms and standards that human beings follow as per government's rules, regulations and policies? Would human beings accept and trust AI ethics? Additionally, there is the fear of creating machines with such advancements that they may harm the human race (Rinesi, 2015).

2. LITERATURE REVIEW

To start with, what is *Consciousness*? In very simple terms, it means the state of being aware of our internal system as well as our external world from the perspective of ourselves (Michael, 2019).

Neuroscience researchers and cognitive scientists have stated that what humans perceive is impacted by what they see. When the retinae respond to the wavelengths of the light, the brain beings to process the raw data that is inserted (through seeing), sorts which visual information should be given more attention and how the brain will manage to retain it over the course of time span (Anon, 2015).

A woman had suffered a stroke and that had her right side of the brain damaged, which led to her being completely unaware of everything which was happening on the left side of hers. To study this state of her mind, she was examined and for that she was given two pictures of the same house to live in but one house was on fire and the other was not. She did not find any difference between the two yet, she kept selecting the one that was not on fire but did not have any reason for it. Researchers said that her brain was processing everything through her vision even though her right side of the brain was damaged and that visual input triggered her consciousness. Another experiment was that when a ball was thrown on her left side, she would move away or tend to save herself from the ball hitting her but would not be aware of the ball or why she performed that action/movement. This condition is called 'hemispatial neglect' where the brain's processing is different from the experiences that we have. This is what the scientists defined as human consciousness. We have a sense of ourselves, our surroundings and we have intensions; we are also aware of the fact that we are aware of what we are currently doing (Michael, 2019).

AI that can simulate operational intelligence of humans is considered to be higher in order to human beings on the grounds of capacity, computing speed and the accuracy levels (Dong et al., 2020). There is also a surge in the research area of AI constantly exploring forms of intelligence by reproducing the human intelligence, which includes – emotional intelligence and the learning ability (Haladjian and Montemayor, 2016).

The fear of AI surpassing human intelligence is called singularity and this is said to be fast approaching while others say that presently, it is a robotics' 'Cambrian explosion' (Pratt, 2015). Additionally, if consciousness would be identified in the machines unambiguously, there is a high likelihood that there will be different kinds of phenomenal consciousness (Kriegel, 2015).

3. RESEARCH METHODOLOGY

This research chapter talks about the plausibility of *Human-like Consciousness* being present in *Artificial Intelligence (AI)* and whether or not it can adopt human cognitive abilities and ethics.

For this research chapter, primary data has been opted, which is by studying the viewpoint of 307 participants with the help of google forms in which close-ended questions were prepared with the choice to comment their opinion. The generated responses on the three topics (human-like consciousness, cognitive abilities and ethics) were compared to the secondary data, which consisted of theories and hypothesis formed by numerous professors and scientists.

The region of the population (for the primary data) consists of south India (69%) and north India (31%); in which the age groups are – under 18 (5.9%), 18–24 (21.2%), 25–34 (19.5%), 35–44 (19.9%), 45–54 (19.5%), 55–64 (8.8%) and above 64 (5.2%).

4. RESULTS AND FINDINGS

4.1. Human-Like Consciousness in Artificial Intelligence (AI)

Consciousness is experience of everything – taste of a chocolate mousse, a tune being stuck in the head, a toothache, love for a child and the bitter understanding that all the feelings will ultimately end (Koch, 2018). Searle (2005) defined consciousness as feelings, awareness and sensations that we feel in the morning when we wake up, in the night when we go to sleep and till the day we die. In fact, dreams too are a part of the conscious self and humans cannot get rid of this consciousness as it is irresistible. He also said that consciousness is not a part of the physical world but of the spiritual world and belongs to a soul, which is not a part of the physical world (Stanford Encyclopedia of Philosophy, The Chinese Room Argument, 2020).

When asked the candidates whether AI is conscious or not, majority (78.2%) said it is not conscious (Figure 1).

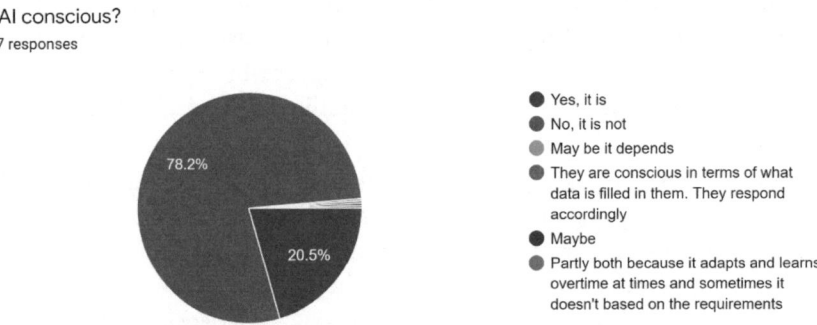

Figure 1. AI is not conscious.

Searle believes that computers do not verily think rather they just tend to manipulate symbols without knowing or understanding what they are doing. He backed this statement by his 'Chinese Room Experiment' wherein he creates a scenario where a man who does not understand Chinese language is sitting in a room which contains only a guide/manual that explains the man on how to respond in Chinese language by using a set of characters and not acknowledging the language. When someone from the outside of the room asks a question in Chinese, 'what is your favourite colour?', the man replies in Chinese, 'blue' lacking the ability to comprehend the question or the answer. Searle further said that it is what computers exactly do, they only simulate human thinking and are merely mindless automatons (Horgan, 2021).

Most of the candidates (78.5%) seem to agree with Searle, which is why 78.5% candidates think that computers simulate human thinking and do not think by their own (Figure 2).

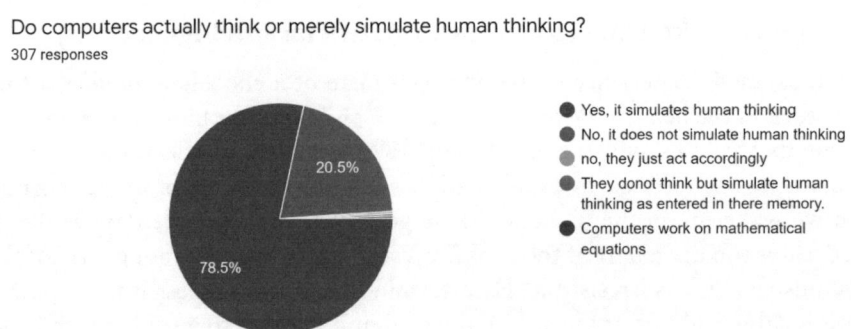

Figure 2. AI simulates human thinking.

'The bottom line is that the implemented computer program by itself is never going to be sufficient for human understanding because human understanding has more than syntax. It has semantics' (Searle, 2005). An example of the mindless automatons can be Siri or Alexa who only mimic the English language (and other languages) just like the man in the Chinese room experiment. Apart from this, 'science is objective, consciousness is subjective, therefore, there cannot be a science of consciousness'. This statement of his can spark a heated debate but he supports it by his Chinese room experiment and also the data shows that most candidates believe that computers merely simulate human thinking.

In 2015, Noam Chomsky was questioned on AI and consciousness he said, 'asking whether machines can think is like asking whether submarines can swim; if you call that swimming, it is swimming'; he calls such questions as terminological questions. What he means by this statement is that it is a personal understanding and interpretation of what is happening around the world with the use of technology. Just in a manner classifying technology as either a boon or a bane, technology being conscious or not is debatable based on the knowledge one has of it. For example, a few might debate that Siri is conscious because it says 'I'm sorry, I couldn't hear that' but the rest may debate that it is programmed in that manner and has no consciousness or awareness of what it is saying and neither what they are sorry for (Figure 3).

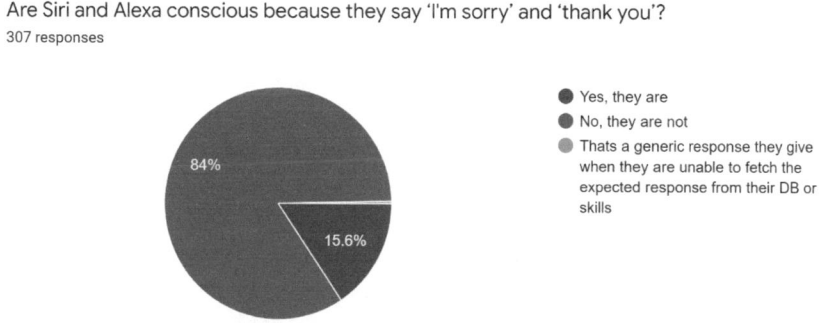

Are Siri and Alexa conscious because they say 'I'm sorry' and 'thank you'?
307 responses

- Yes, they are
- No, they are not
- Thats a generic response they give when they are unable to fetch the expected response from their DB or skills

84%

15.6%

Figure 3. Alexa and Siri are not conscious.

The issue of AI consciousness and overcoming humans is looked at objectively. If machines ever were to become conscious, they would first have to replicate human emotions and experiences and at an extent surpass human minds. However, there are authors who argue that these claims are based merely on misconceptions because in order to reach human level capabilities, the machines will have to have abstract thinking that only humans possess. There are three fundamental issues with AI consciousness, which are (a) the futurist robots do not care about what it will be like to become humans, (b) there is a materialist view to the distinctions and (c) the machines cannot be subjective about human emotions. An article on the specific subject talks about human capabilities and states that yes, robots/ machines will overtake human intelligence and as projected in science fiction movies the advancement of technology drastically overcomes humans over the years and are termed as super machines. But even today machines do surpass human beings by their speed in mathematical calculations and processing huge amounts of data in a few seconds; so, has it not already surpassed human beings? The answer to this is that computers have overcome a few of human abilities but not all of them and that is why they have not yet advanced in human capabilities (Signorelli, 2018).

When questioned the candidates whether AI could overtake human consciousness, only about 34% believe it can (Figure 4).

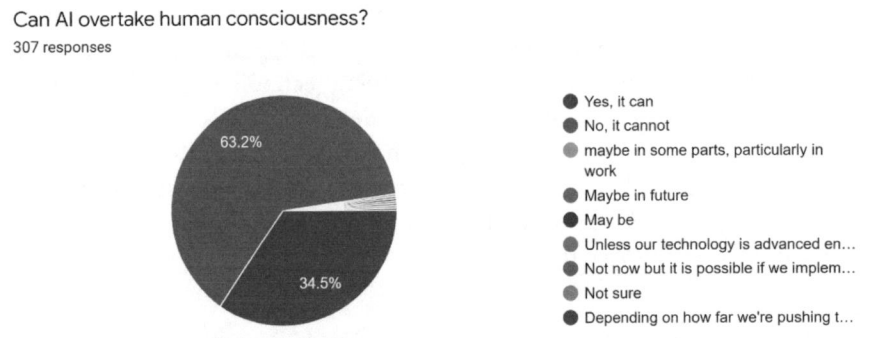

Figure 4. *AI overtaking human consciousness.*

5. COGNITIVE AI

A moral test has been proposed on the topic of AI and cognitive abilities (cognitive AI) where there is a scenario to save either a healthy, young, huge dog or a sick, slender, injured man from a shipwreck and put only one of them into the emergency boat which will take them to the shore. This is a type of a question that humans can answer (there being no right or wrong answer) but machines do not have answers to such questions because it lacks what humans have and term it 'human morality'. But what if machines are able to develop their own sense of morality? Can it be possible? For machines to develop morality they first need to acquire decision-making, imagination and empathy. The test does not provide any answer to this question but vaguely states that if machines can develop human-like consciousness, they are more likely to develop either human-like morality or their own morality and their set of other abilities which can exceed humans in every dilemma (Signorelli, 2018).

Another example of this can be robot cars; 'moral machine' provides a series of hypothetical questions arranged on whom the robot car shall kill out of two different groups having different creatures of different and/or similar characteristics and profession (Figure 5).

To sum this section, there is a prediction by an article on 'towards data science' which says, 'by 2025, humans are likely to see a categorical jump in the competencies demonstrated by AI, with machines growing markedly wiser'. But by Figures 6 and 7, it is evident that neither do the candidates believe that AI possess cognitive abilities nor would they develop it (Singer, 2021).

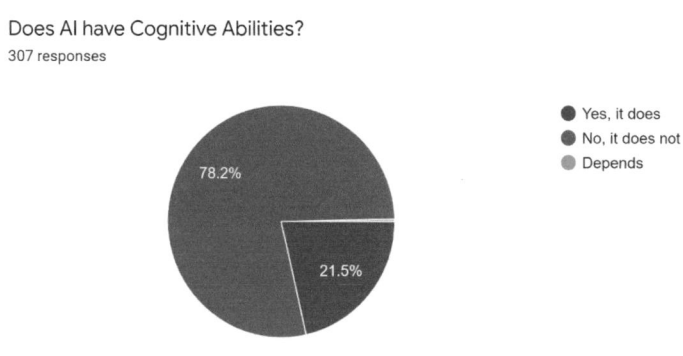

Figure 5. *AI and cognitive abilities.*

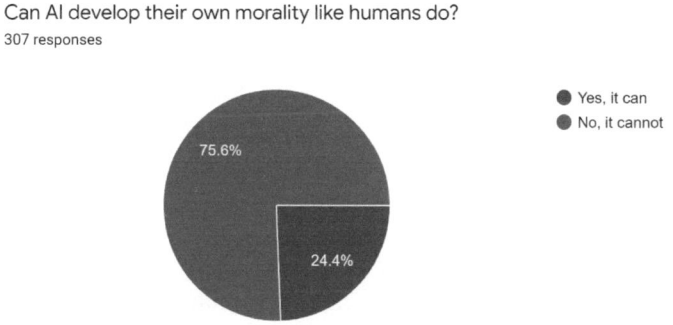

Figure 6. *AI developing its own morality.*

6. AI ETHICS

'Harvard Business Review' has an article on this AI ethics, where it presents data that firms like IBM, Optum, Goldman Sachs and a few others are being sued for unethical practices performed by their AI (without directing them to do so). Companies like Facebook, Google, Twitter and Microsoft are trying to solve these unethical practices which arise from huge amounts of data collection and analysis performed by machine learning models. These companies have in fact agreed that AI ethics is a true threat which can cost millions of losses and not just that but also reputational damage, legal suits and wasted resources (Blackman, 2020).

For example, Amazon had been working on an AI software for hiring people but dropped the finished program because they were not sure that it would treat every person equally and give out a fair chance for employment. They had their doubts that it may not hire women, discriminate on the basis of colour, caste and other characteristics. Apart from this, there would be no reasoning for selecting and/or rejecting a candidate. This plan was in progress for many years yet it was not designed ethically because AI ethics is relatively a newer concept and its execution requires data, which is not sufficient in the present time. There is another example where Toronto was to be designed as a smart city but the idea was scrapped and the organisation suffered a loss of millions of dollars, wasted time and human resources (Dastin, 2018).

Figure 7 represents that the majority of the candidates believe that AI should not be responsible for hiring backed by the examples discussed above.

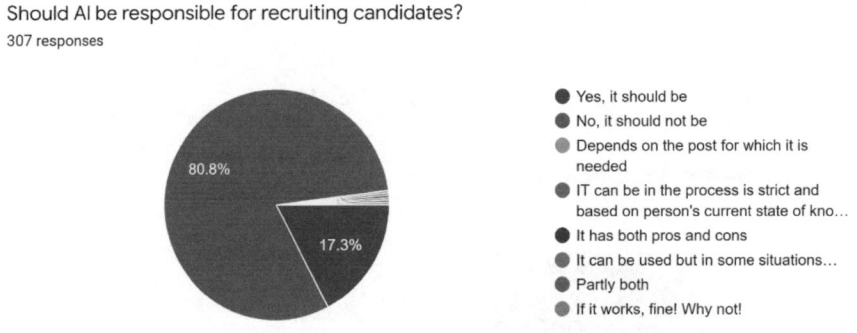

Figure 7. AI as a recruiter.

IBM stated that to earn trust in AI ethics, there are five pillars to it:

I. *Fairness* – Amazon's program was scrapped due to the doubt that their software would not be fair to all the candidates and would be prejudice against a few. This is a question that still remains as to how there is an assurance that AI will have ethics that are morally right and unbiased.

II. *Explainable* – How would the AI justify or explain itself for picking one candidate over another? The software is not planned or trained on being explainable rather only on huge amounts of data.

III. *Robustness* – What if the AI model is hacked or manipulated? This can hinder the rational choices and may benefit a party with an intention to make profits and commit crime.

IV. *Transparency* – Would the general public be aware of the activities being performed by the AI?

V. *Data Privacy* – Would the public be vulnerable to their data being exposed?

Data Ethics is still a growing field which can be the reason why majority of the candidates do not trust it to be completely fair, unbiased and treats everyone with equal integrity (Figure 8).

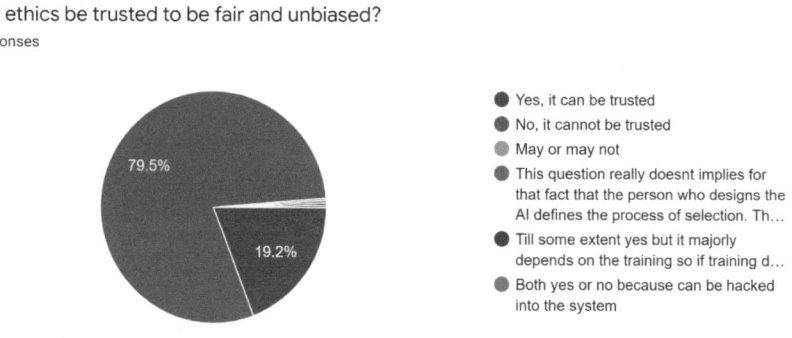

Figure 8. AI cannot be trusted.

IBM for this has proposed three principles (i) this form of AI should be controlled by humans and be specifically designed for their purposes and it should not replace humans, (ii) there must be data privacy as it will belong to the creator alone and (iii) the process of it should be explainable. These three principles provide human beings a sense of relief and assurance that AI would not be taking over humans and enslave them. Apart from IBM, Accenture proposed that there should be another application/software developed which will observe and analyse the initial/main software ethically on the basis of being fair, transparent and accountable (IBM, AI Ethics, 2021).

There are six ethical questions on the future of AI which focuses on the safety and responsibility. Those six are – Assurance, Autonomy, Agency, Indicators, Interfaces and Intentionality. Does the system have these? Will the system ever have these? If it does, then there will not be any ethical concerns cause by them to humans (Bell, 2021).

Like said earlier, it is a growing field which is why it requires further more research and understanding of it. The general public is still unaware of the topic and hence they are reluctant towards it. Figure 9 represents the data whether there should be investments in AI ethics or not. The majority is just about 53% who agree for further investment and development of this field.

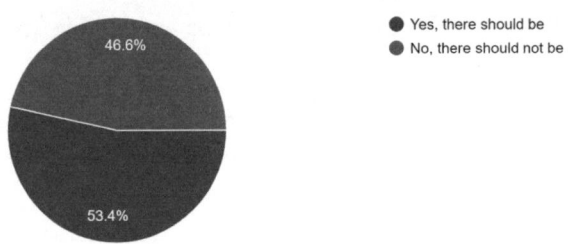

Figure 9. Investments and research in the field of AI.

It can be summed that consciousness fits in every aspect and can take its form in every field besides 'how' still remains unanswered despite of so many ongoing theories and researches.

To answer the question – can AI truly and completely develop human-like consciousness then there is no exact a yes or a no to it but through the survey it can be confidently summed that AI is not conscious as of today and will not overtake human consciousness (refer Figures 1 and 4). AI consciousness may or may not be just a hypothesis as most researchers see no limits to the possibilities to the technology but as of today, it can be confidently said that AI yet has to develop human-like consciousness because to have human-like consciousness, AI has to be inserted with this information and for that to be done, there has to be an understanding on how the human brain functions and can it ever be replicated.

While the AI developers continue to question and experiment on how data ethics or AI ethics should be ethically planned and executed there are concerns raised by academic ethicists who are unsure on the idea of AI ethics and planning such a major change when there is already enough fear of robots excelling humans and taking over. Academic Ethicists believe that by doing so, humans are exposing themselves to vulnerable ethical risks and planning mayhem; this can be backed by Figure 10.

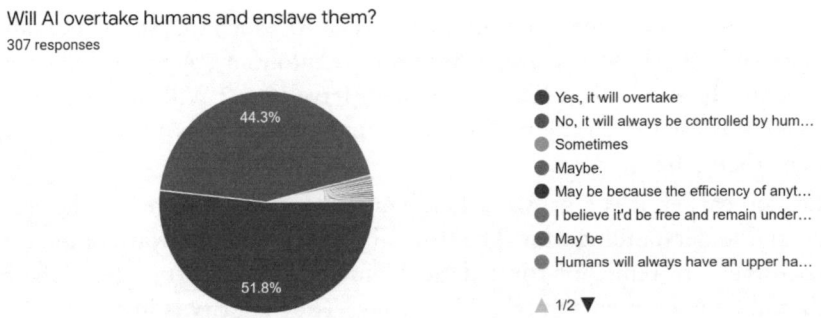

Figure 10. AI can overtake human beings and enslave them.

6. FUTURE RESEARCH

The general population is still not aware of the topics in-depth. The understanding and awareness of topics like cognitive AI and Data Ethics are relatively new, which hindered thorough study of this chapter. As AI is still considered to be in its nascent stages of growth (Financial Express, 2018), there should be further more investments for the research and innovations in the field of AI, as found out in the survey (refer to Figure 9). Future studies could focus on studying specific segments in order to have a thorough understanding of various groups; another suggestion could be a causal study to acknowledge how cultural differences and one's background affect the opinion.

7. CONCLUSION

Although there are several speculations on AI becoming more powerful than human beings and taking over, it majorly depends on how the development of the technology is executed and how careful the humans are with it. Of course, there will always be a potential risk of AI outsmarting humans but in a manner they still have. AI only lacks a few sets of human capabilities and if those could be inserted and developed in those machines earlier, developers would have but it is a demanding task due to the reasons discussed. There are countless researches and proposed hypotheses in progress to transform machines into human beings not just physically (that has been executed to an extent) but moreover mentally and emotionally (Figure 11).

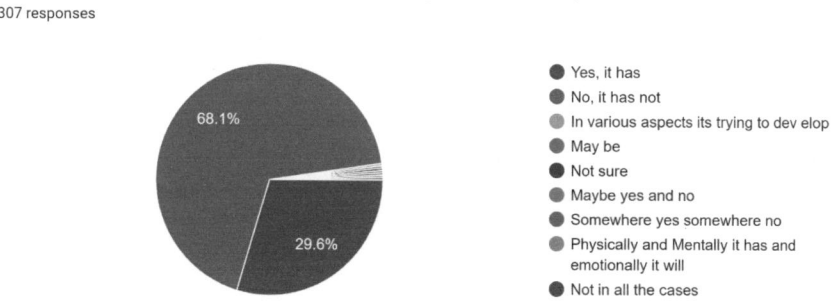

Figure 11. AI has not outsmarted human beings.

8. ACKNOWLEDGEMENT

We would like to express my gratitude to Dr. Raul V. Rodriguez for giving us this opportunity and constantly helping and guiding us in the completion of this research chapter.

REFERENCES

Birch, J., Schnell, A.K. and Clayton, N.S. (2020). Dimensions of Animal Consciousness. *Trends in Cognitive Sciences*, [online] 24(10), pp. 789–801. Available at: https://www.cell.com/trends/cognitive-sciences/fulltext/S1364-6613(20)30192-3.

Blackman, R. (2020). A Practical Guide to Building Ethical AI. *Harvard Business Review*. [online] 15 Oct. Available at: https://hbr.org/2020/10/a-practical-guide-to-building-ethical-ai.

Cole, D. (2014). *The Chinese Room Argument (Stanford Encyclopedia of Philosophy)*. [online] Stanford.edu. Available at: https://plato.stanford.edu/entries/chinese-room/.

Dastin, J. (2018). *Amazon Scraps Secret AI Recruiting Tool that Showed Bias Against Women*. [online] Reuters. Available at: https://www.reuters.com/article/us-amazon-com-jobs-automation-insight-idUSKCN1MK08G. doi:10.1016/j.concog.2016.08.011.

Dong, Y., Hou, J., Zhang, N. and Zhang, M. (2020). *Research on How Human Intelligence, Consciousness, and Cognitive Computing Affect the Development of Artificial Intelligence*. [online] Complexity. Available at: https://www.hindawi.com/journals/complexity/2020/1680845/.

Haladjian, H.H. and Montemayor, C. (2016). Artificial Consciousness and the Consciousness-Attention Dissociation. *Consciousness and Cognition*, 45, pp. 210–225.

Horgan, J. (n.d.). *Quantum Mechanics, the Chinese Room Experiment and the Limits of Understanding*. [online] Scientific American. Available at: https://www.scientificamerican.com/article/quantum-mechanics-the-chinese-room-experiment-and-the-limits-of-understanding/.

Kriegel, U. (2015). *The Varieties of Consciousness*. New York, NY: Oxford University Press.

Moral Machine. (n.d.). *Moral Machine*. [online] Available at: https://www.moralmachine.net/.

Pratt, G.A. (2015). "Is a Cambrian Explosion Coming for Robotics?" *Journal of Economic Perspectives*, 29 (3): 51-60.

Singer, G. (2021). *The Rise of Cognitive AI*. [online] Medium. Available at: https://towardsdatascience.com/the-rise-of-cognitive-ai-a29d2b724ccc.

World Economic Forum. (2015). *How Does the Brain Process What We See?* [online] Available at: https://www.weforum.org/agenda/2015/11/how-does-the-brain-process-what-we-see/.

www.ibm.com. (n.d.). *AI Ethics*. [online] Available at: https://www.ibm.com/cloud/learn/ai-ethics.

2. Exploring "Nudge" in Mental Health Research: A Systematic Review

Dipanjan Bagchi, Moupurna Mukherjee and Ishani Nag
Amity Institute of Psychology and Allied Science Kolkata, Amity University, Kolkata, West Bengal, India

ABSTRACT: Mental health stigma can play a crucial role in determining health behavior stemming from the inadequacy of awareness. This consequently feeds the already existing impediments of mental health treatment and mental health-seeking behavior. Stigma is a common problem across several societies globally and intervention is of utmost importance. With the popularity of nudge in shaping choices, the impact of the same in the field of mental health stigma still awaits extensive study. The idea of a "nudge" deals with certain cognitive predispositions that alter behavioral tendencies conducive to developing and preventing pertinent risk factors. Current literature shows that it has been implemented in mental health research as well. This paper explores the role of "choice architecture" in promoting effective mental health decision-making. The databases of CrossRef (Keywords: nudge, mental health, mental health stigma, n = 1000), PubMed (Keywords: "nudge", "mental health" OR "mental health behavior", n = 14)), Google Scholar (Keywords = "nudge", "mental health" OR "mental health behavior", n = 980) and Scopus (Keyword = TITLE-ABS-KEY ((nudge*) AND ("mental health*" OR "mental health behavior*")), n = 7) and filtered according to the inclusion and exclusion criteria, have been scoured to extract an overall understanding of the existing concepts explored so far and gauge their impact in the realm of mental health research. It was revealed that nudge-based intervention strategies could be significantly useful to avert the usual course of the stigma that precedes the decision-making that follows. We have discussed the "Nudge" strategies and techniques which have proved to be useful in this context. Also, implications, future research directions, and scope of research have been discussed in the paper.

KEYWORDS: Nudge, mental health, mental health stigma, choice architecture, mental health decision making

Chapter 2 DOI- 10.4324/9781003397175-2

Mental health problems have been identified as the second leading cause for the years lived with disability (YLDs) and 6th leading cause of disability-adjusted life years (DALYs) (*GBD Compare*, 2014). For the past two decades, the importance in identifying people suffering from mental disorders and rendering appropriate treatment has been increasing. The global estimate of the prevalence of depression is 4.3%. The National Mental Health Survey of India, 2015–16 estimates that 1 in 20 individuals suffer from depression. The data published in 2020 suggests that the suicide rate in India has been increasing steadily from 9.9% to 11.1% between 2017 and 2020 (*Crime in India 2020*, n.d.), which portrays an even more grim picture. Although a lot of initiatives have been taken by government and non-government organizations, stigma still remains one of the main barriers that keep people from availing of the services already in place.

Stigma in most cultures revolves around shame and ostracization (Boxell, 2020) because it brings down one's acceptability in society. There are two types of stress: the public stigma and the self-stigma according to Corrigan (2011). The existing literature suggests that the internalized stigma at an individual level interacts with the prevailing stigma in the community that affects the help-seeking behavior (Conner et al., 2010; Held and Owens, 2013; Vidourek & Burbage, 2019). This systemic negative attitude toward mental health-seeking behavior patterns in most communities causes a huge gap between health services and the target population that needs to be bridged. Mental health stigma is as damaging as mental illness itself as indicated by Feldman and Crandall (2007). The feeling of depression (Manos et al., 2009) can be an outcome of stigma. Other symptoms include negative attitudes toward treatment (Conner et al., 2010), poor treatment compliance (Fung & Tsang, 2010), and treatment dropouts (Wade et al, 2011).

Although efforts are being made to bring people to avail healthcare services, there seems to be a dire need to join hands with other disciplines and bring in their knowledge to tackle a complex problem like stigma. In this regard, the contribution of Daniel Kahneman (2011), Richard H Thaler and Cass R Sunstein (2008) is noteworthy. Daniel Kahneman highlighted the frailty of human rationality in favor of quick decision-making. On the other hand, Thaler and Sunstein revolutionized Behavioral Economics (BE) with the application of psychological principles in real-life problem-solving. Thaler and Sunstein (2008) highlighted the concept of "NUDGE" which was used to label all the techniques which act as the slight push which helps individuals to take up decisions in a particular avenue with favored outcomes. Many economists like Thaler and Sunstein (2008) adopt the often debated stance of libertarian paternalism. The primary assumption is that sometimes people may not act in their own best interests. Libertarian Paternalism is not coercive and often there are no other alternatives than NUDGEing people in the proper direction for their own betterment (Thaler & Sunstein, 2003). Thaler and Sunstein (2008) explain, "The first misconception is that it is possible to avoid influencing people's choices.

In many situations, some organization or agent must make a choice that will affect the behavior of some other people. There is, in those situations, no way of avoiding nudging in some direction, and whether intended or not, these nudges will affect what people choose." A case in hand is the unwillingness of people to come for treatment of mental health disorders, a decision which is clearly not in their best interest due to stigma. Hence the stance of libertarian paternalism and Nudge Theory is of utmost importance in dealing with complex problems like stigma.

According to Thaler and Sunstein (2003), "in many domains, people lack clear, stable or well-ordered preferences. What they choose is strongly influenced by details of the context in which they make their choice." The concept of nudge makes minor changes to the environment which is called choice architecture and makes the favorable choice more vivid for the decision maker to choose the same. Even the efficiency of the nudge techniques has often been accorded because of the significant impact it has on behavior which has been established with the help of randomized control trials (RCTs) (Krutsinger et al., 2020).

The present study aims to identify the techniques based on Nudge Theory that will be beneficial in bringing people into the therapeutic setting as they are not taking decisions that are best for them. As stigma prevents help-seeking behavior in people with mental illness there is a need for an interdisciplinary approach and to choose from empirically established techniques to fight this evil. Here we systematically review and identify techniques that have been studied specifically to combat the problem of stigma. We also discuss the possible concepts that researchers can explore in this context.

1. METHODOLOGY

This study used the "Preferred Reporting Items for Systematic Reviews and Meta-Analysis" (PRISMA) guidelines (Page et al., 2022; Sarkis-Onofre et al., 2021; Sohrabi et al., 2021). It is used for "planning, conducting, organizing, analyzing and reporting" the literature review. This study was conceptualized on September 26, 2022, and the data was retrieved at that point in time. Google Sheets (2016), Google Docs, and Google Meetings to search were used to organize and connect with the researchers. The method has been divided into the following parts.

1.1. Planning the Review

In addition to the PRISMA checklist, a 15-step process laid out by Pascoe et al. (2021) was considered for answering the research questions (McGinn et al., 2016; McFadden et al., 2012). The following criteria are set at the beginning of the study.

- *Context*: Application of Behavioral Economic (BE) insights (NUDGE) in mitigating stigma in mental health help-seeking behavior.

- *Study Objective*: This study's focus was to identify the BE techniques (NUDGE) used in overcoming stigma in mental health-related behavior.
- *Applications*: It is important to identify the factors related to help-seeking behavior. NUDGE techniques have already proved their efficacy in health sectors quite successfully. This study is trying to identify and evaluate the same related to bringing a person to treatment.

1.2. Research Question

RQ1: To investigate the evidence on the effectiveness of NUDGE approaches in reducing stigma in the behavior of seeking mental health assistance.

1.3. Inclusion and Exclusion Criteria

The studies were selected in a two-step process. The first one is screening the studies based on the following criteria:

- Studies focused on mental health stigma and NUDGE techniques.
- Studies published in indexed and peer-reviewed journals.
- Empirical studies using quantitative and mixed method studies.
- Studies published in the English language.
- Articles available as full text were included in the study for better understanding. Including studies only in English might seem arbitrary but it was done with the sole purpose of getting a global perspective of this topic. Moreover, a majority of the journals were published in English and it also helped eliminate the linguistic limitations of the researchers.

Studies will be excluded if:

- The information included were insufficient
- Studies were published in non-peer-reviewed journals
- Commentaries, editorial and review articles

The final stage includes filtering out and searching for duplication after the screening process. This is done collaboratively and researchers resolve any discrepancies by jointly reviewing the data. If data repetition was found only the former publication is kept.

1.4. Conducting the Review

The screening of the data will be done in compliance with the PRISMA checklist. The potential databases were identified and a systematic search was conducted (Aghaei Chadegani et al., 2013; Munn et al., 2018). Articles on mental health stigma and NUDGE techniques from four electronic databases: PubMed, Google Scholar, Scopus and CrossRef were searched. The search terms (keywords and operators) were adjusted according to the database in which they were searched (*General/keyword Search*, n.d.) . At the identification stage, a total of 2001 studies were identified for further screening. It should be noted that because of the difference in behavior of the search function the number of research papers identified is large (General/keyword Search, n.d.). Search engines where the "OR" function is not supported yield more articles.

Table 1. Search Formula and Search Strategy Employed for Each Database.

Database	Keywords	Number of Papers Identified
CrossRef	nudge, mental health, mental health stigma	1000
Google Scholar	"nudge", "mental health" OR "mental health behavior"	980
PubMed	"nudge", "mental health" OR "mental health behavior"	14
Scopus	TITLE-ABS-KEY ((nudge*) AND ("mental health*" OR "mental health behavior*"))	7
Total		2001

1.5. Organizing and Screening of the Reports

A two-step process was taken to organize and screen the data. The first step was extraction and compiling the data that was done on October 9, 2022. The second step was screening and it was conducted in consultation with the other authors over Google meet with the other authors.

Phase 1: The full papers and abstracts were considered. Relevant titles and abstracts were screened. Any discrepancies were resolved by online meeting with the researchers. The data was recorded and shared with the researchers on a Google Spreadsheet for convenience of use. Duplications were identified by alphabetically arranging the author's names.

Phase 2: Further screening of the extracted studies was done. This screening phase was guided by the inclusion and exclusion criteria stated above. Using the PRISMA flowchart (Page et al. 2022), the search overview that includes identification, screening and review is shared. Finally, thematic analysis was done on the identified documents.

2. RESULTS AND DISCUSSION

The above process yielded eight studies providing insights into the association between the application of nudge-based techniques and mental health stigma. After analysis, the results are presented chronologically, depicting a thematic analysis of the screened studies. The systematic review results are presented according to the research question.

Table 2. Prisma 2020 Flow Diagram Presenting the Process of Systematic Review.

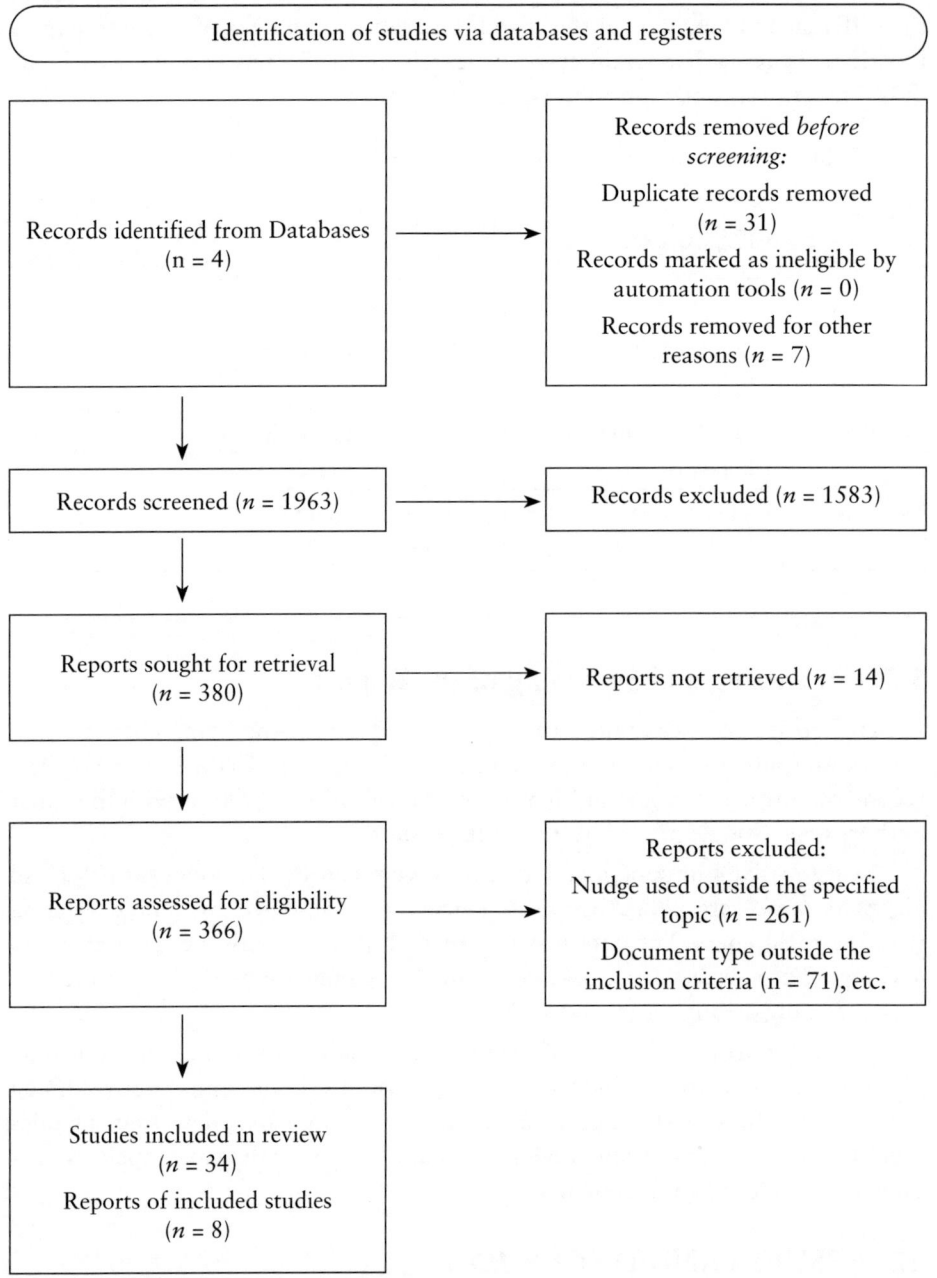

Table 3. Description of the Research Report Based on the Nudge Techniques Used and the Target Disorder.

Author	Publication Year	Type of Study	Title	Nudge Technique Used	Description of the Technique	Any Specific Disorder
Haile et al.	2020	Mixed methods, questionnaires and semi-structured interviews	Pilot Testing of a Nudge-Based Digital Intervention (Welbot) to Improve Sedentary Behavior and Well-being in the Work place	Just in time prompts	Welbot is a digital intervention that makes employees aware of their unhealthy habits. Notifications in stipulated time frames for stretch exercises and rest.	Sedentary lifestyle and mental well-being, procrastination, depression, anxiety, and stress, and work engagement.
Piao and Joo	2022	Quantitative, cross-sectional study	A Behavioral Strategy to Nudge Young Adults to Adopt In-Person Counseling: Gamification	Gamification	Gamification motivates learning by either directly using games or indirectly transferring game elements to non-game contexts	In-person counseling
Warner et al.	2021	Quantitative, cross-sectional study	Giving schools a nudge: can behavioral insights improve recruitment of schools to randomized controlled trials?	EAST	Applying behavioral insights to mail outs	Recruitment in RCT

Author	Publication Year	Type of Study	Title	Nudge Technique Used	Description of the Technique	Any Specific Disorder
Aggarwal et al.	2016	Quantitative, cross-sectional study	"Nudge" and the epidemic of missed appointments	Confirmation letter, charging of fixed percentage,	The patient receives a confirmation letter before the scheduled appointment, All patients attending for an outpatient service are asked to provide their bank details to secure the booking	Reducing missed appointments
Bauer et al.	2018	Quantitative, cross-sectional study	A Nudge in a New Direction: Integrating Behavioral Economic Strategies with Suicide Prevention Efforts	Social norm nudge, change of wordings	First nudge through emails, second nudge is in terms of statements	Suicidal thoughts and behavior (STB) Need to check the journal credibility
Liang et al.	2022	Quantitative, retrospective observational study	Effects of Behavioral Economics-Based Messaging on Appointment Scheduling Through Patient Portals and Appointment Completion: Observational Study	Behavioral economics based language	Nudge health maintenance reminders with	Rates of appointments were higher

Author	Publication Year	Type of Study	Title	Nudge Technique Used	Description of the Technique	Any Specific Disorder
Auf et al.	2021		Gamification and nudging techniques for improving user engagement in mental health and well-being apps	Gamification mechanics and nudging techniques	Creating app narrative and user interaction, in addition to adopting a non-forcible language	Improving the frequency of app daily use.
Khazanov et al.	2022	Multi-phase, mixed method, questionnaires and qualitative interviews	Leveraging behavioral economics and implementation science to engage patients at risk for suicide in mental health treatment: a pilot study protocol	EAST framework	Desired behavior should be Easy, Attractive, Social, and Timely	Effort, hopelessness, forgetting

2.1. Description of the Selected Studies

Our aim was to find the efficacy of NUDGE-based techniques in bridging the gap between stigma and treatment. A total of eight studies passed the final screening. The screened studies are peer-reviewed, published in indexed journals, and have been supported by empirical evidence.

2.2. Missed Appointments

Attempts to minimize missed healthcare appointments can be done through nudge-based language. Here we identified two studies in which two researchers used this technique to reduce missed appointments.

In one study conducted by Aggarwal et al. (2016), two different policies were employed—Policy A and Policy B. Two types of nudge-based messages were used based on loss: personal and impersonal. For Policy A, before the scheduled appointment a confirmation letter was delivered to the patient, which outlined the expenses and labor which is to be invested for the treatment and the same would go to waste in case of absenteeism. In the current nudge technique, the concept of "messenger" is used which communicates crucial information from a trusted and legitimate source (Freed et al., 2011). This leads to the active involvement of the patient in the process which makes them more insightful of the social and economic costs each treatment has to bear. The phenomenon of reinforcement again comes into play as the patients hold themselves accountable for the social welfare losses to be incurred on nonattendance. The current ordeal also happened to influence the ease with which doctors and physicians recommend medical tests to patients. This study found that there was a 31.1% reduction in cost incurred when the providers were priorly informed about the price of tests (Cummings et al., 1982). This suggests that humans have a tendency to avert the potential losses that they are likely to cause. In this case, the perceived loss seemingly caused by the patient for not attending the health checkup will elicit a proactive loss-averting behavior more readily than simply attending the checkup in the first place.

For Policy B, there is a financial penalty that is incurred by the patient for not attending the health checkup. The nudge technique gains commitment from the client by influencing the system 1 processing, which will inevitably try to avoid financial loss rather than the inconvenience caused due to attending (Kahneman and Tversky, 1979).

In another study by Liang et al (2022) two types of messages were used as reminders using the nudge techniques of—curiosity (email A) and exclusivity (email B). The control group consisted of the standard reminders. In the remainder using curiosity (email A) the message was reframed as "Try out our latest feature" followed by the message "check to see my doctor open appointments." In the remainder using exclusivity (email B), the message was reframed as, "You qualify for online scheduling" followed by the message, "Show me how to use this exclusive feature." It concludes that there was increased appointment completion in both the nudge-based experimental group compared to the control group. These findings are supported by the proposed explanation of Thaler and Sunstein (2008) that people hate losses and seek pleasure. In the above-discussed study, the researchers tapped into the pleasure-seeking system 1 which does not think twice about the consequences during decision-making (default mode).

Hence it can be concluded that by reframing the language using insights from behavioral economics, viz., loss aversion, curiosity and exclusivity we can nudge people to seek appointment (curiosity and exclusivity) and follow through the action (loss aversion). This finding is very important as it helps in mitigating the effect of stigma even after scheduling the appointment.

2.3. Suicide Prevention

Suicide prevention has posed to be a threat to society for time immemorial and its primary cause is low treatment-seeking rates. The study involved sending emails to university students informing them of the norm stating that based on previous research, a large percentage of people roughly around 95.6% believe that individuals having thoughts or ruminations about suicide should seek treatment and improved results can be observed based on evidence (Khazanov et al., 2022). This stems from the innate tendency to conform to social norms by not deviating from the predominant responses. The numeric evidence of other people siding with help-seeking behavior makes it the more appropriate response. The follow-up email simply contains a link to the concerned resources and interventions. Results revealed that the individuals who received the nudge stimulus were 164% more probable to open and view the second email than those who did not receive the nudge stimulus. This implies that social conformity if presented at the optimal time will bring about the desirable behavior by means of a nudge.

The EAST framework has also been useful in suicide prevention studies. EAST—meaning the desired behavior should be Easy, Attractive, Social, and Timely (*EAST: Four Simple Ways to Apply Behavioural Insights*, n.d.) all of which make the system 1 processing of cognition extremely susceptible to change. This approach attempts to reduce cognitive biases which lead to poor health outcomes for primary care patients. The characteristics of patients who are potentially at risk for suicide and absent from treatment are first identified and the reasons behind the inattentiveness are also pointed out through a qualitative interview method. Following this, the EAST frameworks are individually implemented to curb attitudinal and social barriers (that can potentially elicit suicidal ideation), which resulted in improved self-regulation, enhanced perceived social support, and prevented technical barriers to effective communication. Thus the study portrayed an effective engagement strategy to impede suicidal behavior tendencies of primary care patients.

2.4. In-Person Counseling

Gamification is a concept that involves learning from games directly or indirectly through the transfer of game elements to non-game situations. There are nine elements of a game that have been identified. These are assessment, conflict/challenge, game fiction, interpersonal interaction, action language, immersion, control, environment and rules/goals (Bedwell et al., 2012). We have identified one study that uses gamification to increase the understanding of the process of in-person counseling. This paper also hypothesized that the dynamicity of the information provides greater intensity of vividness. The concept of vividness can be explained as "Information may be described as vivid, that is, as likely to attract and hold our attention and to excite the imagination, to the extent that it is: (a) emotionally interesting, (b) concrete and imagery-provoking, and (c) proximate in a sensory, temporal or spatial way" (Nisbett, 1980). This study found that gamification increases the adoption of in-person counseling and vividness acts as a moderation. In fact, this result has been supported by other studies exploring gamification. They stated that gamification increases higher vividness (Callow et al., 2007; Müller-Stewens et al., 2017).

This study is important in the view that it demonstrates how gamification can be applied in dealing with complex problems like overcoming stigma. Adequate knowledge rids people of apprehension and negative attitudes as they actually know the process. Hence, this technique can be used by researchers in varied settings, both directly or indirectly through the use of points, badges, and leaderboards (Deterding et al., 2011).

2.5. Recruitment in Studies

Recruitment in studies is often the first and the most important stage in an experimental study like RCTs. And even more so when it comes to recruitment in an RCT in psychology. This problem was taken up by Piao and Joo (2022) who tried to circumvent it by using the EAST framework which involves BE insights and NUDGE technique. EAST is the acronym for easy, attractive, social and timely. This framework proposes that minimal effort should be required to start an activity. If the desired activity is complex it should be broken down into parts. Attractive communication or messages are required to get our attention which would otherwise be overlooked. We have already discussed in the previous sections how reframing the language using BE insights can be helpful (Aggarwal et al., 2016). The third stands for social which is more related to humans being a part of a community and striving for social approval. The last is timely which is more related to the concept of reinforcement in psychology. The timely dispensation of reinforcement is important in order to instill desirable behavior.

The primary aim of this study (Piao & Joo, 2022) was to find whether mailouts drafted in a standard format and mail outs drafted in the EAST framework resulted in differential involvement of schools. The result of this study suggests that there was a significant increase in the involvement of schools in the recruitment of participants for RCT. This finding provides a probable solution to the problem of stigma sipping into the recruitment of participants in studies.

2.6. App-Based Intervention

Digital applications can direct a person to initiate health-seeking behavior and also increase engagement with mental health apps.

Evidence of a personalized digital app called "Welbot" (Haile et al., 2020) aimed at curbing behaviors that promote sedentary behavior traits in the workplace with the help of optimally timed prompt messages. Messages like "prioritize stand up, sit less, and move more" or "reduce stress" are the primary notifications that initiate the phenomenon. The nudge stimulus for this app-based intervention provides the user with a graphic informative prompt depicting what to do, where to do, and for how long to do it. It is followed by a short instruction on how to engage in the exercise which is further preceded by a brief text highlighting the usefulness of the exercise. This framework essentially employs the idea of triggering the system 1 information processing unit which is automatic, reflexive, and relatively unconscious instead of system 2, which is reflective, conscious, and time-consuming (Saghai, 2013).

The app produces frequent detailed messages that are adequately vivid but provides scanty time for the user to decide to act differently from what the app commands. By activating the system 1 processing unit the user resorts to shallow cognitive processing owing to the lack of active attention and ends up following the prompts. The study revealed that customers using this digital app have claimed to become more aware of their unhealthy behavior patterns and more conscious while being at their workplace. This speaks to the fact that certain stimuli that instill a prominent feeling of rightness are extremely unlikely to be further attended to (Thompson, 2009). The communicative nature of this application promotes a particular "healthiness" by means of instant reinforcement texts comprising body and mind benefits after the completion of each exercise. This conditions the user further to prevent restriction to the nudge stimulus.

Studies have suggested that patient engagement was a significant indicator of the reduction of anxiety and depression (Glenn et al., 2013). User retention, daily duration of use, and the number of sessions per day determine the effectiveness of a mental health application. This leads to an equation of achieved mental points to real health progress (Cheng et al., 2019).

3. CONCLUSION

It can be concluded from the systematic review that NUDGE techniques have the potential to overcome the stigma attached to mental health problems and bring people to therapeutic settings. We can see that the EAST framework, language framing using BE insights, gamification, loss aversion, curiosity and exclusivity are the techniques that have been already explored and have received empirical evidence. Some of the techniques that have been used in the above mentioned research might seem familiar to a psychologist as these concepts have been derived from psychological literature (Kanheman, 2011). But the essence of BE is its applicability. Translating the theory into practice has made BE more effective. Hence joining hands with BE might be more effective than we can think.

The findings highlight the lack of knowledge of BE insights in the psychological literature on mental health. We were able to select only eight research papers which matched with the inclusion criteria. However, on the brighter side, the few pieces of literature that we were able to select did show that NUDGE can be a very effective tool to overcome stigma and bring people to the already existing mental health infrastructure.

3.1. Limitations

Our primary aim was to reduce any error while reviewing the literature however our study has some limitations. We had explored four databases including the Scopus database; however, we were not able to explore the databases that were not available online. Moreover, we restricted our search to articles that were published in the English language only which has the potential to overlook (although few) relevant documents in those we have excluded.

3.2. Future Implications of the Study

In this review, we had the opportunity to identify the NUDGE techniques that might be useful in overcoming stigma in mental health. But many other techniques remain unexplored. Our brain functions in default mode which has been highlighted by Thaler and Sunstein (2008), which can be exploited in getting people to seek help. For example, making stress management programs default for the employees. In that case, employees have to make an effort to opt out of the program. Future researchers are encouraged to be creative as NUDGE is not a technique per se but a label that has been used to identify techniques with similar features. This review article is a call to action to tackle stigma in mental health, which is one of the biggest problems in the world right now, using a multi-disciplinary approach.

REFERENCES

Aggarwal, A., Davies, J., & Sullivan, R. (2016). "Nudge" and the epidemic of missed appointments. *Journal of Health Organization and Management, 30*(4), 558–564. https://doi.org/10.1108/JHOM-04-2015-0061

Aghaei Chadegani, A., Salehi, H., Yunus, M., Farhadi, H., Fooladi, M., Farhadi, M., & Ale Ebrahim, N. (2013). A comparison between two main academic literature collections: Web of Science and Scopus databases. *Asian Social Science, 9*(5), 18–26. https://doi.org/10.5539/ass.v9n5p18

Auf, H., Dagman, J., Renström, S., & Chaplin, J. (2021). Gamification and nudging techniques for improving user engagement in mental health and well-being apps. *Proceedings of the Design Society, 1*, 1647–1656. https://doi.org/10.1017/pds.2021.426

Bauer, B. W., Tucker, R. P., & Capron, D. W. (2018). A nudge in a new direction: Integrating behavioral economic strategies with suicide prevention efforts. *Clinical Psychological Science, 7*(3), 612–620.

Bedwell, W. L., Pavlas, D., Heyne, K., Lazzara, E. H., & Salas, E. (2012). Toward a taxonomy linking game attributes to learning. *Simulation and Gaming, 43*(6), 729–760. https://doi.org/10.1177/1046878112439444

Boxell, O. (2020). Social context affects mental health stigma. *Open Health, 1*(1), 29–36. https://doi.org/10.1515/openhe-2020-0003

Callow, N., Roberts, R., Fawkes, J. Z. (2007). Effects of dynamic and static imagery on vividness of imagery, skiing performance, and confidence. *Journal of Imagery Research in Sport and Physical Activity, 1*, 1001. https://doi.org/10.2202/1932-0191.1001

Cheng, V. W. S., Davenport, T., Johnson, D., Vella, K., & Hickie, I. B. (2019). Gamification in apps and technologies for improving mental health and well-being: Systematic review. *JMIR Mental Health, 6*(6), e13717.

Conner, K. O., Copeland, V. C., Grote, N. K., Koeske, G., Rosen, D., Reynolds, C. F., & Brown, C. (2010). Mental Health Treatment Seeking Among Older Adults With Depression: The Impact of Stigma and Race. *The American Journal of Geriatric Psychiatry, 18*(6), 531–543. https://doi.org/10.1097/jgp.0b013e3181cc0366

Corrigan, P. W., Rafacz, J., & Rüsch, N. (2011). Examining a progressive model of self-stigma and its impact on people with serious mental illness. *Psychiatry Research, 189*, 339–343. https://doi.org/10.1016/j.psychres.2011.05.024

Crime in India 2020. (n.d.). Retrieved November 12, 2022, from https://ncrb.gov.in/en/Crime-in-India-2020

Cummings, K. M., Frisof, K. B., Long, M. J., & Hrynkiewich, G. (1982). The effects of price information on physicians' test-ordering behavior: Ordering of diagnostic tests. *Medical Care, 20*(3), 293–301.

Deterding, S., Dixon, D., Khaled, R., Nacke, L. (2011). From game design elements to gamefulness: Defining "gamification". In *Proceedings of the 15th International Academic MindTrek Conference: Envisioning Future Media Environments (MindTrek'11)*; Tampere, Finland, September 28–30, 2011; New York, NY, USA: Association for Computing Machinery; 2011

EAST: Four Simple Ways to Apply Behavioural Insights. (n.d.). Retrieved November 15, 2022, from https://www.bi.team/publications/east-four-simple-ways-to-apply-behavioural-insights

Feldman, D. B., & Crandall, C. S. (2007). Dimensions of mental illness stigma: What about mental illness causes social rejection? *Journal of Social and Clinical Psychology, 26*(2), 137–154. https://doi.org/10.1521/jscp.2007.26.2.137

Freed, G. L., Clark, S. J., Butchart, A. T., Singer, D. C., & Davis, M. M. (2011). Sources and perceived credibility of vaccine-safety information for parents. *Pediatrics, 127*(Suppl. 1), S107–S112.

Fung, K. M., & Tsang, H. W. (2010). Self-stigma, stages of change and psychosocial treatment adherence among Chinese people with schizophrenia: A path analysis. *Social Psychiatry and Psychiatric Epidemiology, 45*, 561–568.

GBD Compare. (2014, April 22). Institute for Health Metrics and Evaluation. https://www.healthdata.org/data-visualization/gbd-compare

General/keyword search. (n.d.). Harzing.com. Retrieved November 15, 2022, from https://harzing.com/resources/publish-or-perish/manual/using/use-cases/general-search

Glenn, D., Golinelli, D., Rose, R. D., Roy-Byrne, P., Stein, M. B., Sullivan, G., ... & Craske, M. G. (2013). Who gets the most out of cognitive behavioral therapy for anxiety disorders? The role of treatment dose and patient engagement. *Journal of Consulting and Clinical Psychology, 81*(4), 639.

Haile, C., Kirk, A., Cogan, N., Janssen, X., Gibson, A.-M., & MacDonald, B. (2020). Pilot testing of a nudge-based digital intervention (Welbot) to improve sedentary behaviour and wellbeing in the workplace. *International Journal of Environmental Research and Public Health, 17*(16). https://doi.org/10.3390/ijerph17165763

Held, P., & Owens, G. P. (2013). Stigmas and attitudes toward seeking mental health treatment in a sample of veterans and active duty service members. *Traumatology, 19*(2), 136–143. https://doi.org/10.1177/1534765612455227

Kahneman, D. (2011). *Thinking, fast and slow.* Farrar, Straus and Giroux.

Kahneman, D., & Tversky, A. (1979). *Prospect theory: An analysis of decision under risk.*

Khazanov, G. K., Jager-Hyman, S., Harrison, J., Candon, M., Buttenheim, A., Pieri, M. F., Oslin, D. W., & Wolk, C. B. (2022). Leveraging behavioral economics and implementation science to engage patients at risk for suicide in mental health treatment: A pilot study protocol. *Pilot and Feasibility Studies, 8*(1), 181.

Krutsinger, D. C., O'Leary, K. L., Ellenberg, S. S., Cotner, C. E., Halpern, S. D., & Courtright, K. R. (2020). A randomized controlled trial of behavioral nudges to improve enrollment in critical care trials. *Annals of the American Thoracic Society, 17*(9), 1117–1125.

Liang, S. Y., Stults, C. D., Jones, V. G., Huang, Q., Sutton, J., Tennyson, G., & Chan, A. S. (2022). Effects of behavioral economics-based messaging on appointment scheduling through patient portals and appointment completion: Observational study. *JMIR Human Factors, 9*(1), e34090. https://doi.org/10.2196/34090

Manos, R. C., Rüsch, L. C., Kanter, J. W., & Clifford, L. M. (2009). Depression self-stigma as a mediator of the relationship between depression severity and avoidance. *Journal of Social and Clinical Psychology, 28*, 1128–1143.

McFadden, P., Taylor, B. J., Campbell, A., & McQuilkin, J. (2012). Systematically identifying relevant research: Case study on child protection social workers' resilience. *Research on Social Work Practice, 22*(6), 626–636. https://doi.org/10.1177/1049731512453209

McGinn, T., Taylor, B., McColgan, M., & McQuilkan, J. (2016). Social work literature searching: Current issues with databases and online search engines. *Research on Social Work Practice, 26*(3), 266–277. https://doi.org/10.1177/1049731514549423

Müller-Stewens, J., Schlager, T., Häubl, G., & Herrmann, A. (2017). Gamified information presentation and consumer adoption of product innovations. *Journal of Marketing, 81*, 8–24. https://doi.org/10.1509/jm.15.0396

Munn, Z., Peters, M. D. J., Stern, C., Tufanaru, C., McArthur, A., & Aromataris, E. (2018). Systematic review or scoping review? Guidance for authors when choosing between a systematic or scoping review approach. *BMC Medical Research Methodology, 18*(1), Article 143. https://doi.org/10.1186/s12874-018-0611-x

Nisbett, R. E., & Ross, L. (1980). *Human inference: Strategies and shortcomings of social judgment* (1st ed., p. 62). New Jersey, NJ: Prentice-Hall.

Page, M. J., Moher, D., & McKenzie, J. E. (2022). Introduction to PRISMA 2020 and implications for research synthesis methodologists. Research Synthesis Methods, 13(2), 156-163

Pascoe, K. M., Waterhouse-Bradley, B., & McGinn, T. (2021) Systematic literature searching in social work: A practical guide with database appraisal. *Research on Social Work Practice, 31*(5), 541–551. https:// doi.org/10.1177/1049731520986857

Piao, S., & Joo, J. (2022). A behavioral strategy to nudge young adults to adopt in-person counseling: Gamification. *Behavioral Sciences, 12*(2). https://doi.org/10.3390/bs12020040

Saghai, Y. (2013). Salvaging the concept of nudge. *Journal of Medical Ethics, 39*(8), 487–493.

Sarkis-Onofre, R., Catalá-López, F., Aromataris, E., & Lockwood, C. (2021). How to properly use the PRISMA Statement. *Systematic Reviews, 10*(1), 1–3.

Sohrabi, C., Franchi, T., Mathew, G., Kerwan, A., Nicola, M., Griffin, M., Agha, M., & Agha, R. (2021). PRISMA 2020 statement: What's new and the importance of reporting guidelines. *International Journal of Surgery, 88*, Article 105918. https://doi.org/10.1016/j.ijsu.2021.105918

Thaler, R. H., & Sunstein, C. R. (2003). Libertarian paternalism. *The American Economic Review, 93*(2), 175–179. http://www.jstor.org/stable/3132220

Thaler, R. H., & Sunstein, C. R. (2008). *Nudge: Improving decisions about health, wealth, and happiness.* Yale University Press.

Thompson, V. A. (2009). *Dual-process theories: A metacognitive perspective.* Oxford University Press.

Vidourek, R., & Burbage, M. (2019). Positive mental health and mental health stigma: A qualitative study assessing student attitudes. *Mental Health Prevention, 13*, 1–6.

Wade, N. G., Post, B. C., Cornish, M. A., Vogel, D. L., & Tucker, J. R. (2011). Predictors of the change in self-stigma following a single session of group counseling. *Journal of Counseling Psychology, 58*, 170–182.

Warner, G., Osman, F., McDiarmid, S., & Sarkadi, A. (2021). Giving schools a nudge: Can behavioural insights improve recruitment of schools to randomised controlled trials?. *BMC Research Notes, 14*(1), 99. https://doi.org/10.1186/s13104-021-05509-8

3. Impaired Response Inhibition in Psychosis

Arinjoy Bhattacharjee*

Department of Psychology, University of Calcutta, Kolkata, West Bengal, India

*Corresponding author. Email: bhattacharjeearinjoy24@gmail.com

ABSTRACT

BACKGROUND: Various aspects of inhibitory control can be linked to different neurobiological, psychiatric, or behavioral implications both in the normal and pathological functioning of the brain. The range of psychotic disorders encompasses a set of conditions in which the research has unveiled significant effects on inhibitory control and overall collection of executive functions. One perspective involves the utilization of diverse approaches that can effectively investigate the different facets of inhibitory control. Simultaneously, throughout history, scientists have applied classical models to uncover the neural processes that drive this mechanism. The differences in findings also take an attempt toward explaining the pathological brain on an anatomical, physiological, as well as molecular level.

OBJECTIVE: This chapter tries to explore the existing body of literature and come up with a conceptual understanding of how the executive function of inhibitory control gets impaired in the brains of individuals suffering from disorders belonging to the psychotic spectrum and how it can be used as a parameter to separate the diagnosis of different psychotic disorders.

METHODS: The existing body of literature on response inhibition impairments in depression was reviewed. The keywords for searching included "Response Inhibition," "Psychosis," "EEG-ERP," and "Psychophysiology."

RESULTS: The P300 component can be used to differentiate various psychotic conditions.

KEYWORDS: Inhibitory Control, Psychosis, Schizophrenia, Schizoaffective Disorder, Bipolar Disorder

Chapter 3 DOI- 10.4324/9781003397175-3

1. INTRODUCTION

Psychotic disorders form a cluster of conditions that are, broadly speaking, characterized by delusions, hallucinations, disorganized thought or speech, and often accompanied by abnormal motor behavior. Most of these symptoms come under the common term of positive symptom cluster, which refers to bizarre additions to behavioral patterns. Another set of symptoms, commonly referred to as negative symptoms, refers to certain deficits in normal behavior like limited emotional expressibility or the inability to feel pleasure (anhedonia). Negative symptoms are responsible for a large amount of the morbidity associated with schizophrenia, but they are less common in other psychotic diseases (American Psychiatric Association, 2013).

Response inhibition constitutes a pivotal facet within the realm of executive functions, denoting the cognitive prowess to effectively curtail actions that have ceased to hold relevance or have strayed into inappropriateness for an individual. Succinctly put, response inhibition, alternatively referred to as inhibitory control, entails the faculty to extinguish extant cognitive processes, behaviors, or actions. This intricate phenomenon finds explication across diverse strata, encompassing motor, attentional, and behavioral tiers. Numerous experimental approaches have explored response inhibition, with prominent examples encompassing the Stop-Signal and Go/No-go Paradigms (Verbruggen and Logan, 2008). Experiments around response inhibition have been conducted and studied by various disciplines including abnormal psychology, cognitive neuroscience, and clinical neuroscience. Researchers have employed a diverse range of approaches, spanning various fields, to further explore the underlying neural and behavioral mechanisms associated with response inhibition. This exploration has encompassed not only individuals within typical and healthy populations but also those afflicted by psychiatric or neurological disorders. These insights serve as the fundamental building blocks for enhancing patient treatment strategies. For instance, Ye et al.'s study demonstrated that individuals with Parkinson's disease exhibited prolonged stop-signal reaction times and diminished stop-related engagement within the right inferior frontal gyrus (rIFG) when compared to control subjects. Additionally, these patients displayed reduced functional connectivity between the rIFG and the striatum (Ye et al., 2015). Comparable findings were observed in a study concerning individuals diagnosed with obsessive-compulsive disorder (OCD). The results indicated that when it came to selectively inhibiting motor responses, the performance of OCD patients was notably inferior compared to the control group (Bannon et al., 2002; Penades et al., 2007). Additional research has also substantiated the presence of response inhibition impairments in individuals with ADHD (Desman et al., 2008; Slaats-Willemse et al., 2003; Wodka et al., 2007), autism spectrum disorder (Kelly et al., 2021; Schmitt et al., 2019), depression (Kaiser et al., 2003), etc.

This review endeavors to center its attention on the realm of the psychotic spectrum, elucidating the implications of these disorders on the cognitive process of response inhibition. While predominant research efforts within the psychotic spectrum have gravitated toward schizophrenia and bipolar disorder, it is noteworthy that analogous outcomes have surfaced concerning various other manifestations of psychotic conditions.

2. THE BRAIN DURING RESPONSE INHIBITION

The prefrontal cortex (PFC) has been historically considered a significant origin of cognitive control and the ability to inhibit behavior triggered by stimuli, a concept dating back to early neuropsychological studies (Luria, 1966). Over the course of 70 years since Holmes' (1938) initial exploration, human neuropsychology and cognitive neuroscience have revealed a network of cortical and sub-cortical regions crucial for suppressing reactions. The research literature has pointed out the involvement of specific regions in the brain, such as the dorsolateral prefrontal cortex (DLPFC), ventrolateral prefrontal cortex (VLPFC), and anterior cingulate cortex (ACC), in the process of response inhibition. Many of these studies utilized methods like event-related BOLD-fMRI signals, with the tasks mainly revolving around variations of the stop signal (Logan, 1994; Verbruggen and Logan, 2009) or the go/no-go task (Drewe, 1975; Picton et al., 2007).

In situations of no-go trials, Liddle et al. hypothesized that the observed activity in both the bilateral PFC and the ACC displayed notable statistical significance. The results of the investigation subsequently supported this idea. Additionally, the study confirmed a similarity in the ACC activation patterns across go and no-go trials. The researchers hypothesized that the ACC had a more significant role in deciding whether to initiate or inhibit a response during task execution as a result of this commonality (Liddle et al., 2001).

Apart from these notable areas in the frontal cortex, other areas in the parietal cortex have also shown activation during no-go trials (Liddle et al., 2001; Rubia et al., 1998). The documented existence of bidirectional links between the lateral frontal cortex and the parietal cortex, as observed in studies on rhesus monkeys, aligns with the notion that the parietal association cortex might play a role alongside bilateral activation of the lateral frontal regions (Pandya and Seltzer, 1982).

3. THE PSYCHOTIC SPECTRUM

Since it was originally theorized, schizophrenia has been connected to the frontal lobes and the concept of executive functioning (Bleuler, 1958). There is evidence from three distinct categories of research that suggests the presence of an executive function deficit in individuals with schizophrenia: investigations in neuropsychology, studies involving neuroradiology, and research delving into the correlation between structural and neuropsychological discoveries among patients with schizophrenia (Suhr, 1997). A consistent collection of abnormalities have been identified by neuropsychological studies examining cognitive decline in adult and adolescent schizophrenia patients that point to frontal system dysfunction. These deficiencies include difficulties with exams like the WCST and the SCWT as well as tasks like delayed recall, conditional associative learning, word and design fluency, and others. Importantly, these investigations have not revealed any weaknesses in visual-constructive assignments, general language competency, or recognition memory (Gruzelier et al., 1988; Seidman et al., 1992; Yurgelun-Todd et al., 1988).

Studying prepotent response inhibition and interference control in individuals with schizophrenia is crucial for a variety of reasons.

- First, translational pathophysiology research has been made possible by the successful implementation of response inhibition tasks in animal models (Gilmour et al., 2013)
- Second, a number of inhibition tasks are used as biomarkers in the complex process of developing new drugs in the area of brain medicine, which calls for a thorough understanding of the pattern of impairment in schizophrenia (Carter and Barch, 2007)
- Third, endophenotypes for inhibitory impairments have been proposed by several researchers including Clementz (1998) and Hutton and Ettinger (2006). Endophenotypes, also known as intermediate phenotypes, identify a clinical disorder's hereditary susceptibility. They are investigated to learn more about how risk genes function at various research levels. The validity of the aforementioned endophenotype is crucially confirmed by the presence of an endophenotype in both patients and first-degree relatives who are clinically unaffected (Gottesman and Gould, 2003).

3.1. Impairments in Schizophrenia

Symptoms of schizophrenia include abnormalities in thinking, feeling, perceiving, and acting (American Psychiatric Association, 20013). Numerous deficits are suggested by neurocognitive research, notably in the executive functions of patients suffering from schizophrenia (Pantelis et al., 1997).

Deficits in inhibition have frequently been linked to schizophrenia. These deficiencies are frequently linked to the PFC and related networks' activity. Knowing how activities are planned and started, as well as the elements involved in stopping these actions, it is necessary in order to comprehend purposeful inhibitory control. Even though schizophrenia is not the only psychopathological disorder with inadequate reaction inhibition, its issue is very distinct from that of other psychopathological groups. Considerable consideration is given to potential brain mechanisms behind the problems in initiating inhibitory responses and voluntary activities in schizophrenia (Badcock et al., 2002).

According to a large body of research, individuals with schizophrenia-spectrum illnesses display weak reaction inhibition when doing the stop-signal task (SST) (Bellgove et al., 2006, Ross et al., 2008). The ACC, in addition to the PFC areas, is a significant area that is showing notable deterioration (Bennes et al., 1992). People with schizophrenia took part in go/no-go and stop signal exercises as part of a study where variations in cortical and subcortical activation or inactivation were highlighted. The results demonstrated that the left anterior cingulate's BOLD signal response decreased during both inhibition tasks. Additionally, there was increased activity in the thalamus and the putamen, while the left rostral dorsolateral prefrontal BOLD signal response decreased during the performance of the stop signal task. Accordingly, the findings showed that when faced with problems involving motor response inhibition, people with schizophrenia exhibited aberrant neural network patterns characterized by decreased activation in the left prefrontal region and increased subcortical activity (see Rubia et al., 1998).

Other fMRI studies have reported similar findings. Clinically high-risk participants (CHR) and schizophrenia patients (SZ) showed reduced rIFG and ACC no-go activation compared to controls in a study conducted by Fryer (2019). Based on the decelerated and inconsistent motor responses displayed, coupled with the diminished engagement of the right inferior frontal and dorsal ACC regions renowned for their role in exerting inhibitory control, as well as the protracted and variable motor reactions observed, it becomes evident that both the CHR and SZ cohorts encountered difficulties in cultivating robust and dependable prepotent responses. In stark contrast, only the control cohort exhibited a functional connectivity configuration congruent with the hypothesis that an elevated response prepotency necessitates a more pronounced segregation between regions governing inhibitory control and those constituting the default mode network during the intricate process of response inhibition.

In certain ERP investigations, it was ascertained that patients exhibited delayed peak latencies for the stop-signal N1 and P3. Additionally, the patients displayed reduced amplitudes of the go stimulus and stop-signal ERP components (N1/P3), suggesting the presence of early-stage impairment in stop-signal processing (Hughes, 2012). In other studies reduced N1 and P3 amplitudes have also been reported (Ford et al., 2001).

3.2. Impairments in Schizoaffective Disorder

Developing an accurate classification system for psychosis and mood disorders is certainly a contentious topic within the fields of psychiatry and neuroscience. The question of whether schizoaffective disorder (SA) should be distinguished from both schizophrenia (SZ) and mood disorders is a consistent subject of debate. This is due to the complex combination of pathophysiological patterns observed in individuals afflicted by this condition (Evans et al., 1999; Lake and Hurwitz,

2007). In such a scenario, a vital first step is the identification of biomarkers in mental illness. One such biomarker is response inhibition.

It is a well-known fact that patients with schizophrenia continuously show a decreased and delayed P300. This is evident from numerous event-related electrophysiology (EEG/ERP) investigations (Bestelmeyer et al., 2009; Duncan et al., 1987; Ford, 1999; Ford et al., 2010; Jeon and Polich, 2003; Groom et al., 2008; Nieman et al., 2002). Certain studies have successfully made biological and neuropsychological distinctions between these conditions with regard to response inhibition. Mathalon et al. (2010) found out that despite a considerable P300 drop in patients with schizophrenia, schizoaffective disorder patients did not show significantly reduced P300 amplitudes. This shows that whereas schizophrenia has compromised neurophysiological systems, schizoaffective disorder has intact mechanisms for allocating attentional resources to infrequent stimuli, whether they are task-relevant targets or task-irrelevant distractors.

In a study by Chun et al. (2013), the researchers assessed the potential of utilizing P300 as a biomarker to differentiate between SZ, BD, and SA in the context of response inhibition. While there are physiological differences among SZ, BD, and healthy individuals, as indicated by the outcomes of sparse logistic regression focused on particular facets of P300, the SA group remains less clearly differentiated.

3.3. Impairments in Bipolar Disorder

Impulsivity remains a prominent characteristic of bipolar disorder across all phases of the illness (Swann et al., 2007), although notably in manic and/or hypomanic episodes that interpolate periods of depression. This heightened impulsiveness could stem from atypical response restraint. Swift-response impulsiveness, which involves the incapacity to thoroughly analyze a stimulus prior to reacting, along with reward-delay impulsiveness, linked to the incapability to postpone gratification for a greater reward and connected to a hastened devaluation of postponed rewards, constitute two interrelated approaches to inhibiting responses (Evenden, 2000).

Research using neuroimaging and electrophysiological methods has revealed various structural and functional irregularities in individuals diagnosed with bipolar disorder. A study by Swann et al. (2009) implied that bipolar disorder is identified by difficulties in maintaining attention and controlling responses, which are associated with a more intense progression of the condition and could potentially serve as valuable endophenotypic markers for bipolar disorder.

In the earlier mentioned comparative research, the electrophysiological biomarker, P300, achieved success in differentiating between bipolar disorder and schizophrenia or schizoaffective disorders. The individuals with bipolar disorder displayed considerably prolonged P300 latencies in comparison to the control subjects. Notably, when the analysis involved intentionally delaying a response to no-go stimuli, individuals with bipolar disorder exhibited typical P300 augmentation but experienced delayed P300 latencies. This observation suggested that bipolar disorder is not linked to reduced cognitive resources; instead, it appears to be associated with a specific balance between response speed and accuracy. Specifically, the latency of P300 separated bipolar disorder from a control group, while the amplitude of P300 related to response inhibition in the fronto-central region differentiated bipolar disorder from schizophrenia (Chun et al, 2013).

Findings related to P300 amplitudes are consistent, in a way, that SZ and BD are not really distinguishable based on that (Menkes et al., 2022) and significantly reduced frontal N2 responses in bipolar patients, a hallmark of improper stop-signal processing, also demonstrated specific abnormalities in frontal response inhibition mechanisms. Hence, in bipolar disorder, achieving standard response inhibition might necessitate the engagement of atypical and potentially compensatory cognitive control mechanisms.

Neuroimaging results have clearly revealed the role of brain areas in BD like hypoactivation in IFG (Hajek et al., 2013; Townsend et al., 2012), frontopolar cortex on the left, and dorsal amygdala on both sides (see Kaladjian et al., 2009).

3.4. Impairments in Schizotypy and Related Personality types

Schizotypy pertains to a multidimensional configuration of personality traits that mirror the manifestations witnessed in schizophrenia. These dimensions encompass the positive or psychotic-like facet, the negative or deficit facet, and the cognitive-behavioral disorganization facet, all characteristic of schizophrenia (Grant et al., 2018). Different classifications of inhibitory control have been created due to the diverse nature of inhibition as a concept (Aron 2007, 2011). A key component of executive control is the capacity to dismiss unimportant information and prevent undesirable responses. Ettinger et al. (2018) conducted a clinical examination of non-affected first-degree family members of individuals diagnosed with schizophrenia. Assessed schizotypy traits in both these family members and individuals without any history of mental disorders for the purpose of investigating the impact of genetic factors and subtle schizophrenia-like characteristics discovered links between schizotypy traits and the capacity to perform inhibitory tasks in both the family members and the control group. These findings align with prior studies indicating cognitive impairments in individuals with schizotypy traits.

Other studies have also reported that both at the behavioral and neural levels, schizotypy patients have problems with inhibiting responses (Jia et al., 2021). The research investigated both reactive and proactive inhibitory controls, revealing distinct findings for individuals with schizotypy when compared to those without this trait. Those with schizotypy displayed notably elevated N1 amplitude across various stop signal probability levels and increased P3 amplitudes during proactive inhibition. Conversely, in instances of reactive inhibition, individuals with schizotypy demonstrated considerably prolonged stop signal reaction times and reduced N2 amplitudes in contrast to those without schizotypy. Additionally, other factors such as highly sensitive C-reactive protein (hs-CRP) and childhood trauma have been identified in studies for their potential contribution to response inhibition among individuals with schizotypy (Gong et al., 2019) These results align with previous studies that investigated the positive correlation between CRP levels and clinical manifestations of schizophrenia. Specifically, two studies indicated a connection between CRP levels and the severity of distressing symptoms (Fawzi et al., 2011, Garcia-Rizo et al., 2012).

There has not been a lot of neuroimaging research on schizotypy or schizophrenia susceptibility, however, one study undertook an fMRI investigation on prepulse inhibition (PPI) of the startle response. When a non-startling, lower-intensity stimulus with a duration of between 30 and 500 milliseconds is followed by a powerful, startling stimulus (the pulse), the response to the pulse is reduced. High psychosis propensity was associated with lower PPI and decreased activity in the IFG, insula extending to putamen and thalamus, parahippocampal gyrus, inferior parietal, and middle temporal regions (Graham, 1975).

4. DISCUSSION

This current review aimed to investigate the variations in neural mechanisms within the spectrum of psychosis concerning inhibitory control. The focus was on discerning distinctive patterns and differences among these mechanisms, aiming to use inhibitory control as a psychobiological indicator for distinguishing between different pathologies. In most of the studies, the common method for measuring response inhibition is the stop signal task, which is based on Logan's (1994) "race" model. It enables evaluation of the effectiveness of inducing an inhibitory response as well as the speed at which behavioral inhibitory (stopping) processes occur. Particularly in Psychosis, according to frontal dysfunction theories (Hill et al., 2004), the dorsal and ventral prefrontal areas are often involved in response inhibition (Blasi et al., 2006).

In the case of schizophrenia dysfunctions in the frontal and cingulate cortical areas became increasingly prominent while electrophysiology reflected delayed N1 and P3 ERP components with lower amplitudes.

Indeed, new research suggests that other areas like the rIFG have a distinct response inhibition role (Chambers et al., 2006), but the pathways implicated may vary between healthy people and those who have schizophrenia.

Our comprehension of schizoaffective disorder (SCA) in comparison to other psychiatric conditions remains limited. The conjecture suggests that it might be categorized as a variant of either bipolar disorder (BPD) or schizophrenia (SCZ). Alternatively, it could be viewed as a diverse amalgamation of both, or positioned as an intermediary point within a spectrum where affective disorders and SCZ are positioned at opposite ends (Abrams, 1984; Pope et al., 1980;

Lapensee, 1992; Taylor and Amir, 1994). In the case of schizoaffective disorders, the P300 amplitudes did not take a drop like in schizophrenia which serves as a major biomarker for distinction and to some extent tries to solve the debate of classification.

These findings align with the observation that the neurocognitive and neuroimaging deficits in schizoaffective disorder closely resemble those seen in schizophrenia rather than bipolar disorder. The inclination of schizoaffective disorder toward schizophrenia rather than bipolar disorder implies that it could be categorized as a subtype of schizophrenia or considered a constituent of the continuum spectrum model for psychosis (Madre et al., 2016).

For bipolar disorder, P300 has been successful in serving as a biomarker and also separating the anatomy and pathophysiology of the diseases.

These findings are substantiated by electrophysiological and neuroimaging research as well and are consistently carried over to other psychiatric pathophysiologies and even psychosis-propensity or schizotypy. The findings can be summarized in the following way:

Table 1. Summary of the Findings.

Psychotic Conditions	ERP Findings			
	N100		P300	
	Latency	Amplitude	Latency	Amplitude
Schizophrenia	Delayed	Lower	Delayed	Lower
Schizoaffective	N/A	N/A	Normal	Normal
Bipolar	Normal	Normal	Delayed	Higher
Schizotypy	N/A	Higher	N/A	Higher

5. CONCLUSION

By reviewing the existing body of electrophysiological and imaging findings, it can be concluded that the ERP component of P300 can be used successfully to distinguish psychotic pathologies from each other.

REFERENCES

Abrams, R. (1984). Genetic studies of the schizoaffective syndrome: A selective review. *Schizophrenia Bulletin*, 10(1), 26.

American Psychiatric Association, D. S., & American Psychiatric Association. (2013). *Diagnostic and statistical manual of mental disorders: DSM-5* (Vol. 5). Washington, DC: American Psychiatric Association.

Aron, A. R. (2007). The neural basis of inhibition in cognitive control. *The Neuroscientist*, *13*(3), 214-228.

Aron, A. R. (2011). From reactive to proactive and selective control: Developing a richer model for stopping inappropriate responses. *Biological Psychiatry*, *69*(12), e55-e68.

Badcock, J. C., Michie, P. T., Johnson, L., & Combrinck, J. (2002). Acts of control in schizophrenia: Dissociating the components of inhibition. *Psychological Medicine*, *32*(2), 287-297.

Bannon, S., Gonsalvez, C. J., Croft, R. J., & Boyce, P. M. (2002). Response inhibition deficits in obsessive–compulsive disorder. *Psychiatry Research*, *110*(2), 165-174.

Bellgrove, M. A., Chambers, C. D., Vance, A., Hall, N., Karamitsios, M., & Bradshaw, J. L. (2006). Lateralized deficit of response inhibition in early-onset schizophrenia. *Psychological Medicine*, *36*(4), 495-505.

Benes, F. M., Vincent, S. L., Alsterberg, G., Bird, E. D., & SanGiovanni, J. P. (1992). Increased GABAA receptor binding in superficial layers of cingulate cortex in schizophrenics. *Journal of Neuroscience*, *12*(3), 924-929.

Bestelmeyer, P. E., Phillips, L. H., Crombie, C., Benson, P., & Clair, D. S. (2009). The P300 as a possible endophenotype for schizophrenia and bipolar disorder: Evidence from twin and patient studies. *Psychiatry Research*, *169*(3), 212-219.

Blasi, G., Goldberg, T. E., Weickert, T., Das, S., Kohn, P., Zoltick, B., ... & Mattay, V. S. (2006). Brain regions underlying response inhibition and interference monitoring and suppression. *European Journal of Neuroscience*, *23*(6), 1658-1664.

Bleuler, E. (1958). *Dementia praecox or the group of schizophrenias*. New York: International Universities Press.

Carter, C. S., & Barch, D. M. (2007). Cognitive neuroscience-based approaches to measuring and improving treatment effects on cognition in schizophrenia: The CNTRICS initiative. *Schizophrenia Bulletin*, *33*(5), 1131-1137.

Chambers, C. D., Bellgrove, M. A., Stokes, M. G., Henderson, T. R., Garavan, H., Robertson, I. H., ... & Mattingley, J. B. (2006). Executive "brake failure" following deactivation of human frontal lobe. *Journal of Cognitive Neuroscience*, *18*(3), 444-455.

Chun, J., Karam, Z. N., Marzinzik, F., Kamali, M., O'Donnell, L., Tso, I. F., ... & Deldin, P. J. (2013). Can P300 distinguish among schizophrenia, schizoaffective and bipolar I disorders? An ERP study of response inhibition. *Schizophrenia Research*, *151*(1-3), 175-184.

Clementz, B. A. (1998). Psychophysiological measures of (dis)inhibition as liability indicators for schizophrenia. *Psychophysiology*, *35*(6), 648-668.

Desman, C., Petermann, F., & Hampel, P. (2008). Deficit in response inhibition in children with attention deficit/hyperactivity disorder (ADHD): Impact of motivation?. *Child Neuropsychology*, *14*(6), 483-503.

Drewe, E. A. (1975). An experimental investigation of Luria's theory on the effects of frontal lobe lesions in man. *Neuropsychologia*, *13*(4), 421-429.

Duncan, C. C., John, M., Morihisa, M. D., & Fawcett, R. W. (1987). P300 in Schizophrenia: State or. *Psychopharmacology Bulletin*, *23*(3-4), 497.

Ettinger, U., Aichert, D. S., Wöstmann, N., Dehning, S., Riedel, M., & Kumari, V. (2018). Response inhibition and interference control: Effects of schizophrenia, genetic risk, and schizotypy. *Journal of Neuropsychology*, *12*(3), 484-510.

Evans, J. D., Heaton, R. K., Paulsen, J. S., McAdams, L. A., & Jeste, D. V. (1999). Schizoaffective disorder: A form of schizophrenia or affective disorder?. *The Journal of clinical psychiatry*, *60*(12), 18895.

Evenden, J. L. (1999). Varieties of impulsivity. *Psychopharmacology*, *146*(4), 348-361.

Fawzi, M. H., Fawzi, M. M., Fawzi, M. M., & Said, N. S. (2011). C-reactive protein serum level in drug-free male Egyptian patients with schizophrenia. *Psychiatry Research*, *190*(1), 91-97.

Ford, J. M. (1999). Schizophrenia: The broken P300 and beyond. *Psychophysiology*, *36*(6), 667-682.

Ford, J. M., Mathalon, D. H., Kalba, S., Marsh, L., & Pfefferbaum, A. (2001). N1 and P300 abnormalities in patients with schizophrenia, epilepsy, and epilepsy with schizophrenialike features. *Biological Psychiatry*, *49*(10), 848-860.

Ford, J. M., Roach, B. J., Miller, R. M., Duncan, C. C., Hoffman, R. E., & Mathalon, D. H. (2010). When it's time for a change: Failures to track context in schizophrenia. *International Journal of Psychophysiology*, *78*(1), 3-13.

Fryer, S. L., Roach, B. J., Ford, J. M., Donaldson, K. R., Calhoun, V. D., Pearlson, G. D., ... & Mathalon, D. H. (2019). Should I stay or should I go? FMRI study of response inhibition in early illness schizophrenia and risk for psychosis. *Schizophrenia Bulletin*, *45*(1), 158-168.

Gilmour, G., Arguello, A., Bari, A., Brown, V. J., Carter, C., Floresco, S. B., ... & Robbins, T. W. (2013). Measuring the construct of executive control in schizophrenia: Defining and validating translational animal paradigms for discovery research. *Neuroscience & Biobehavioral Reviews*, *37*(9), 2125-2140.

Gong, J., Wang, Y., Liu, J., Fu, X., Cheung, E. F., & Chan, R. C. (2019). The interaction between positive schizotypy and high sensitivity C-reactive protein on response inhibition in female individuals. *Psychiatry Research*, *274*, 365-371.

Gottesman, I. I., & Gould, T. D. (2003). The endophenotype concept in psychiatry: Etymology and strategic intentions. *American Journal of Psychiatry*, *160*(4), 636-645.

Graham, F. K. (1975). The more or less startling effects of weak prestimulation. *Psychophysiology*, *12*(3), 238-248.

Grant, P., Green, M. J., & Mason, O. J. (2018). Models of schizotypy: The importance of conceptual clarity. *Schizophrenia Bulletin*, *44*(Suppl. 2), S556-S563.

Groom, M. J., Bates, A. T., Jackson, G. M., Calton, T. G., Liddle, P. F., & Hollis, C. (2008). Event-related potentials in adolescents with schizophrenia and their siblings: A comparison with attention-deficit/hyperactivity disorder. *Biological Psychiatry*, *63*(8), 784-792.

Gruzelier, J., Seymour, K., Wilson, L., Jolley, A., & Hirsch, S. (1988). Impairments on neuropsychologic tests of temporohippocampal and frontohippocampal functions and word fluency in remitting schizophrenia and affective disorders. *Archives of General Psychiatry*, *45*(7), 623-629.

Hajek, T., Alda, M., Hajek, E., & Ivanoff, J. (2013). Functional neuroanatomy of response inhibition in bipolar disorders—Combined voxel based and cognitive performance meta-analysis. *Journal of Psychiatric Research*, *47*(12), 1955-1966.

Hill, K., Mann, L., Laws, K. R., Stephenson, C. M. E., Nimmo-Smith, I., & McKenna, P. J. (2004). Hypofrontality in schizophrenia: A meta-analysis of functional imaging studies. *Acta Psychiatrica Scandinavica*, *110*(4), 243-256.

Holmes, G. (1938). Cerebral integration of ocular movements. *British Medical Journal*, *2*(4045), 107.

Hughes, M. E., Fulham, W. R., Johnston, P. J., & Michie, P. T. (2012). Stop-signal response inhibition in schizophrenia: Behavioural, event-related potential and functional neuroimaging data. *Biological Psychology*, *89*(1), 220-231.

Hutton, S. B., & Ettinger, U. (2006). The antisaccade task as a research tool in psychopathology: A critical review. *Psychophysiology*, *43*(3), 302-313.

Jeon, Y. W., & Polich, J. (2003). Meta-analysis of P300 and schizophrenia: Patients, paradigms, and practical implications. *Psychophysiology*, *40*(5), 684-701.

Jia, L. X., Qin, X. J., Cui, J. F., Zheng, Q., Yang, T. X., Wang, Y., & Chan, R. C. (2021). An ERP study on proactive and reactive response inhibition in individuals with schizotypy. *Scientific Reports*, *11*(1), 1-13.

Kaiser, S., Unger, J., Kiefer, M., Markela, J., Mundt, C., & Weisbrod, M. (2003). Executive control deficit in depression: Event-related potentials in a Go/Nogo task. *Psychiatry Research: Neuroimaging*, *122*(3), 169-184.

Kaladjian, A., Jeanningros, R., Azorin, J. M., Nazarian, B., Roth, M., & Mazzola-Pomietto, P. (2009). Reduced brain activation in euthymic bipolar patients during response inhibition: An event-related fMRI study. *Psychiatry Research: Neuroimaging*, *173*(1), 45-51.

Kelly, S. E., Schmitt, L. M., Sweeney, J. A., & Mosconi, M. W. (2021). Reduced proactive control processes associated with behavioral response inhibition deficits in autism spectrum disorder. *Autism Research*, *14*(2), 389-399.

Lake, C. R., & Hurwitz, N. (2007). Schizoaffective disorder merges schizophrenia and bipolar disorders as one disease—There is no schizoaffective disorder. *Current Opinion in Psychiatry*, *20*(4), 365-379.

Lapensèe, M. A. (1992). A review of schizoaffective disorder: I. Current concepts. *The Canadian Journal of Psychiatry*, *37*(5), 335-346.

Liddle, P. F., Kiehl, K. A., & Smith, A. M. (2001). Event-related fMRI study of response inhibition. *Human Brain Mapping*, *12*(2), 100-109.

Logan, G. D. (1994). On the ability to inhibit thought and action: A users' guide to the stop signal paradigm.

Luria, A. R. (1966). Higher cortical functions in man, London. *Tavistock Publications. Yudovich, F*, *1*, 1959.

Madre, M., Canales-Rodríguez, E. J., Ortiz-Gil, J., Murru, A., Torrent, C., Bramon, E., ... & Amann, B. L. (2016). Neuropsychological and neuroimaging underpinnings of schizoaffective disorder: A systematic review. *Acta Psychiatrica Scandinavica*, *134*(1), 16-30.

Mathalon, D. H., Hoffman, R. E., Watson, T. D., Miller, R. M., Roach, B. J., & Ford, J. M. (2010). Neurophysiological distinction between schizophrenia and schizoaffective disorder. *Frontiers in Human Neuroscience*, 70.

Menkes, M. W., Andrews, C. M., Suzuki, T., Chun, J., O'Donnell, L., Grove, T., ... & Tso, I. F. (2022). Event-related potential correlates of affective response inhibition in bipolar I disorder: Comparison with schizophrenia. *Journal of Affective Disorders*, *309*, 131-140.

Nieman, D. H., Koelman, J. H. T. M., Linszen, D. H., Bour, L. J., Dingemans, P. M., & De Visser, B. O. (2002). Clinical and neuropsychological correlates of the P300 in schizophrenia. *Schizophrenia Research*, *55*(1-2), 105-113.

Pandya, D. N., & Seltzer, B. (1982). Intrinsic connections and architectonics of posterior parietal cortex in the rhesus monkey. *Journal of Comparative Neurology, 204*(2), 196-210.

Pantelis, C., Barnes, T. R., Nelson, H. E., Tanner, S., Weatherley, L., Owen, A. M., & Robbins, T. W. (1997). Frontal-striatal cognitive deficits in patients with chronic schizophrenia. *Brain, 120*(10), 1823-1843.

Penades, R., Catalan, R., Rubia, K., Andres, S., Salamero, M., & Gasto, C. (2007). Impaired response inhibition in obsessive compulsive disorder. *European Psychiatry, 22*(6), 404-410.

Picton, T. W., Stuss, D. T., Alexander, M. P., Shallice, T., Binns, M. A., & Gillingham, S. (2007). Effects of focal frontal lesions on response inhibition. *Cerebral Cortex, 17*(4), 826-838.

Pope, H. G., Lipinski, J. F., Cohen, B. M., & Axelrod, D. T. (1980). Schizoaffective disorder: An invalid diagnosis? A comparison of schizoaffective disorder, schizophrenia, and affective disorder. *American Journal of Psychiatry*.

Ross, R. G., Wagner, B., Heinlein, S., & Zerbe, G. O. (2008). The stability of inhibitory and working memory deficits in children and adolescents who are children of parents with schizophrenia. *Schizophrenia Bulletin, 34*(1), 47-51.

Rubia, K., Russel, T., Taylor, E., Bullmore, E., Brammer, M., Williams, S., ... & Sharma, T. (1998). Brain activation in schizophrenia during performing a go-no-go task in functional magnetic resonance imaging (fMRI). *Schizophrenia Research, 29*(1-2), 112-113.

Schmitt, L. M., Bojanek, E., White, S. P., Ragozzino, M. E., Cook, E. H., Sweeney, J. A., & Mosconi, M. W. (2019). Familiality of behavioral flexibility and response inhibition deficits in autism spectrum disorder (ASD). *Molecular Autism, 10*(1), 1-11.

Seidman, L. J., Talbot, N. L., Kalinowski, A. G., McCarley, R. W., Faraone, S. V., Kremen, W. S., ... & Tsuang, M. T. (1991). Neuropsychological probes of fronto-limbic system dysfunction in schizophrenia: Olfactory identification and Wisconsin Card Sorting performance. *Schizophrenia Research, 6*(1), 55-65.

Slaats-Willemse, D., Swaab-Barneveld, H., De Sonneville, L., Van Der Meulen, E., & Buitelaar, J. A. N. (2003). Deficient response inhibition as a cognitive endophenotype of ADHD. *Journal of the American Academy of Child & Adolescent Psychiatry, 42*(10), 1242-1248.

Suhr, J. A. (1997). Executive functioning deficits in hypothetically psychosis-prone. *Schizophrenia Research, 27*(1), 29-35.

Swann, A. C., Steinberg, J. L., Lijffijt, M., & Moeller, F. G. (2008). Impulsivity: Differential relationship to depression and mania in bipolar disorder. *Journal of Affective Disorders, 106*(3), 241-248.

Swann, A. C., Lijffijt, M., Lane, S. D., Steinberg, J. L., & Moeller, F. G. (2009). Severity of bipolar disorder is associated with impairment of response inhibition. *Journal of Affective Disorders, 116*(1-2), 30-36.

Taylor, M. A., & Amir, N. (1994). Are schizophrenia and affective disorder related?: The problem of schizoaffective disorder and the discrimination of the psychoses by signs and symptoms. *Comprehensive Psychiatry, 35*(6), 420-429.

Townsend, J. D., Bookheimer, S. Y., Foland-Ross, L. C., Moody, T. D., Eisenberger, N. I., Fischer, J. S., ... & Altshuler, L. L. (2012). Deficits in inferior frontal cortex activation in euthymic bipolar disorder patients during a response inhibition task. *Bipolar Disorders*, *14*(4), 442-450.

Verbruggen, F., & Logan, G. D. (2008). Response inhibition in the stop-signal paradigm. *Trends in Cognitive Sciences*, *12*(11), 418-424.

Verbruggen, F., & Logan, G. D. (2009). Models of response inhibition in the stop-signal and stop-change paradigms. *Neuroscience & Biobehavioral Reviews*, *33*(5), 647-661.

Wodka, E. L., Mark Mahone, E., Blankner, J. G., Gidley Larson, J. C., Fotedar, S., Denckla, M. B., & Mostofsky, S. H. (2007). Evidence that response inhibition is a primary deficit in ADHD. *Journal of Clinical and Experimental Neuropsychology*, *29*(4), 345-356.

Ye, Z., Altena, E., Nombela, C., Housden, C. R., Maxwell, H., Rittman, T., ... & Rowe, J. B. (2015). Improving response inhibition in Parkinson's disease with atomoxetine. *Biological Psychiatry*, *77*(8), 740-748.

Yurgelun-Todd, D., Craft, S., O-Brian, C., Kaplan, E., & Levin, S. (1988). Wisconsin Card Sort in schizophrenic and manic depressive illness. *Journal of Clinical and Experimental Neuropsychology, 71*.

4. The COVID-19 Pandemic, Anomie, and Social Media Use: Trigger Effects on Violent Behavior in Young Adults

Sai Mounish Reddy, J.V. Sri Roshan Anand and Soujanya Devarakonda
Administrative Staff College of India, Hyderabad 500034, India

ABSTRACT: Due to COVID-19 pandemic, there were significant changes in the lifestyle of individuals, which were completely unheard of, new social norms such as social distancing and restriction of individual freedom resulted along with consequences of isolation has significantly affected behavioral trends like increased aggression in children and increased domestic violence. During the COVID-19 pandemic, when computers became not just a source of pleasure or recreation but also a required and regular method of education and communication with others, social media played a significant part in the lives of children and teenagers. Social media networks have led a path of changes from physical to online violence where technology has played a vital role facilitating digital abuse impacting a series of dangerous behaviors. As a result, they experienced increased levels of psychological stress, anxiety, and depression as well as health issues like headaches, a sense of helplessness, a lack of satisfaction with their activities, overwhelming nervousness, depressed mood, reduced activity, and sleep issues. The erroneous information spread via the internet sparked alarm and worry, and this misinformation was linked to frantic purchasing of needless medical supplies and medications, which also led to stressful states and unfavorable feelings and thoughts, putting the general public's mental health at risk. Adolescents with depressed traits including low self-esteem performed poorly in school, had trouble learning the material, acted in an antisocial manner, and were more likely to consume drugs. For those who may be at risk for perpetrating interpersonal harm, including gang and clique-related violence, social media became a place to share frustrations and threats about future acts of violence.

Chapter 4 DOI- 10.4324/9781003397175-4

1. INTRODUCTION

The social life of youth has been disrupted and the developmental tasks have been halted in profound ways due to COVID-19 pandemic. During the time of physical distancing and isolation, people of all ages are experiencing loneliness and the youth especially are being most affected (Laursen and Hartl, 2013). Social media can do many things starting from interaction with peers to killing boredom, but it will be different from traditional, in person interactions (Nesi et al., 2018).

COVID-19 pandemic has significantly changed the usage of social media. It is used by everyone and mostly by Gen Z and the Millennials (Calhoun and Gold, 2020). Throughout the coronavirus pandemic, children and young adults had to be on silent mode restricting themselves to many of the recreational activities they used to do. Due to these restrictions of the pandemic, the usage of all electronic gadgets and gaming consoles were their only refuge especially to the youth while staying back at home (Blakemore, 2018). The majority of students had to study online, which significantly increased their network usage time and increased their chance of developing Internet addiction. On the positive side, platforms such as Twitter, Instagram, Snapchat, and TikTok became lifesavers for those who felt isolated and marginalized (Colzato et al., 2019). Social media has offered a distinctive opportunity to have information regarding health and safety during the pandemic and many of the youth took advantage of this (Anderson and Jiang, 2018). Connecting people especially the youth to their local communities during the pandemic through social media so that people can experience a sense of global community by taking on various difficulties was a huge benefit (Taylor, 2020).

Furthermore, overuse of apps exposes youth to many mental and psychological problems such as cyber bullying, tech addiction, health problems, increased aggression, and many more. Sleep is one of the critical aspects to regulate emotions and stress and to maintain healthy immune system during the COVID-19 pandemic (Palmer and Alfano, 2017; Park et al., 2016). Due to the pandemic, schools were closed which resulted in the late waking of children, even though there were virtual classes. However, absence of the routine class work and the scheduled timing has affected their slumber time which will have the impact on both physical and mental well-being (Becker et al., 2017).

Social media allows people of any age to curate their images whatever they post through which cyber bullying may increase. For example, many people posted themselves in various attires so that they will not miss seeing their peers during the pandemic (Choukas-Bradley et al., 2020). There is a major negative impact on the mood and anxiety of youth which has been well grounded in many of the before literatures (Twenge et al., 2018). Negative social comparisons will be crafted among the peers as youth will view images and videos of the peers regarding their lifestyles and start to compare. Due to this, they start to feel pressure for likes and views (Nesi and Prinstein, 2019; Yau and Reich, 2019).

Various sites have shown mixed responses on the social media effects on youth's mental health based on the relationship between their screen time and wellbeing (Odgers and Jensen, 2020).

Many studies have found notable and a minor amount of correlation among social media usage, melancholy, and discontent (Holland and Tiggemann, 2016; Twenge and Campbell, 2018; Vannucci et al., 2020) and few others did not find any such associations among these factors (Cohen et al., 2017; Jensen et al., 2019; Przybylski, 2019).

It results in less time to do healthy, real-world activities. One more major effect of social media is the communication among youth (Kilford et al., 2016; Schriber and Guyer, 2016). Creating relationships needs direct communication, which is lacking due to the overuse of social media.

But on the other side, there were adverse effects of the pandemic on child health and wellbeing including risk of experiencing violence. Mental health is also impacted and, social media and depression are closely linked (Griffiths and Meredith, 2009).

When friendship is done online and through texts, youth are being intimidated (Nesi and Prinstein, 2019). They have no idea of what the other person is feeling as they are not directly communicating.

The overuse of social networking sites may cause the nervous system to go into fight-or-flight mode, aggravating conditions including attention-deficit/ hyperactivity disorder (ADHD), depression, and anxiety (Leigh-Hunt et al., 2017). In addition, there is a range of negative health outcomes due to social media isolation including depression, suicide and poor physical health (King and Merchant, 2008; Leigh-Hunt et al., 2017). Recent findings suggests that most of the youth keep thinking about audience of social media even when they are offline which will have serious implications on mental well-being (Choukas-Bradley et al., 2020).

According to the research, it is shown that when youth are depressed due to any issue, they tend to look at social media more often (Griffiths, 2003). Children, parents, teachers and every individual needs to understand the impact of social media usage by adolescents and youth, especially the risks on their mental health (Anderson and Jiang, 2018; Hamilton et al., in press).

This pandemic has created new normal using social media which has both positive and negative effects. Therefore, in this paper we would like to discuss the following questions.

1. For what extent did social media have accurate results or info about the COVID-19 period and how it impacted the youth?
2. What are the factors which made youth attract social media on a larger scale during COVID-19?
3. What are the aggressive behaviors inhibited by the youth especially due to gaming?

2. LITERATURE REVIEW

Social media and the impact on the youth's change in behavior have a relation and a significant impact on their lives (Rettie, 2002). Messages that reach them and the target groups in real time will generate changes in them. There is a great contact with different kinds of social media in the youth of today's generation and they grow with that every day. Most of them live in the digital world; they tend to use the internet before or after doing anything. They have very less engagement with their peers, but their social connection will be more.

Due to the overuse of social media, there has been an increased generation gap between parents and children according to Livingstone and Bober (2005).

Lusk said that social media could be used by students for academic purposes where they can learn and enhance their communication skills. New web tools are also available in social media through which they can advance their learning skills (Lusk, 2018).

Brady, Holcomb, and Smith also stated that social media has provided good platform ways for everyone for education and e-learning even through social media. Lusk and these three have the same opinion on social media regarding learning (Brady, 2008; Holcomb, 2010; Smith, 2016).

According to Tappscott, the youth of today are referred to as the .net generation. Independence, free expression of views, sensitivity to issues, emotional openness has been the main characteristics of this generation (Tappscott, 1998).

Kalpidou, Costin, and Morris stated that there is a relationship between social media and grades and Jacobsen and Forest stated that there are negative effects on grades due to the distraction of social media (Costin, 2010; Morris, 2017).

According to Berson and Berson, most of the active agents are young people or the youth who create, adapt, manipulate, and help to spread different ideas through communication technologies (Berson and Berson, 2005).

3. THEORETICAL FRAMEWORK

This study uses three communication theories:
1. The Magic Bullet Theory
2. The Gratification Theory
3. Technological Determinism Theory

3.1. The Magic Bullet Theory

According to this hypothesis, each social media message is a bullet sent straight from the media gun to the viewer's brain. During COVID-19 pandemic, youth used a lot of social media where the messages and feed whether it is important or not had a great impact on the viewers. This theory is also called as "Hypodermic

Needle Theory" (Nwabueze, 2014). Messages are directed toward the passive audience and the audience respond immediately (Communication Studies, 2017).

3.2. The Gratification Theory

This idea enlightens us as to why and how people select particular media for particular objectives. According to this, audiences are responsible for integrating social media feed into their lives (Katz et al., 1974). During COVID-19 pandemic, youth are responsible to meet their desires and achieve gratification by using social media. Gratification received from the use of Facebook groups is socializing, information, self-seeking and entertainment. Gratification received from twitter groups includes both mass media functions and interpersonal communications.

3.3. Technological Determinism Theory

This theory explains that learning, feeling, and thinking of individuals happen the other way we do because they receive the messages through current technology that is available (Toffler, 1980). The emerging technology the social media has requires individuals to listen and engage more often. People should then use the information supplied to them on social media in their daily lives to analyze it (Cocking, 1999).

4. DISCUSSION

1. What are the factors which made youth attract social media on a larger scale during COVID-19?

    Social media has attracted youth on a larger scale due to various factors such as information regarding the pandemic, movie industry, and for the skill development.

    Due to the unprecedented events of pandemic, there has been a lot of uncertainty among the public. They must stay informed about what is happening around them to ensure safety. This can be done only through news where social media has played a vital role (Steinberg, 2019). The main problem arises because youth mindset is dangerous and will lead to the perception that the infection is not that harmful to them and their families (Buis and Thompson, 1989). This kind of misinterpretation of data and information made the transmission of the virus much easier (Hamilton et al., 2019).

    The movie industry related information also made the youth addicts to the social media. Reels were first introduced on Instagram in 2019 due to the excess usage of social media by youth and adolescents (Taylor, 2020). It provides unique ways to experiment with different interests and creative expressions so that they can show themselves and explore opportunities on their own terms (Davis and Weinstein, 2017).

Social media helped to bridge the gap between the relations during the pandemic. The public has communicated through every social media application to sustain subjective wellbeing and relations (Carpenter, 2018). But on the other side, browsing behaviors and the people young adults communicate with has been linked to decreased positive effect, depressive behaviors (Escobar-Viera et al., 2018; Lup et al., 2015). During the pandemic, this has become even worse as the social media usage has increased tremendously and the fear of missing out increased among the youngsters (Verduyn et al., 2019).

Mental health problems have increased in young people majorly because of gaming. Gaming was introduced in social media also and the youth during the pandemic has become addicted to online gaming which has created a huge problem (Duan et al., 2020; Wang et al., 2020). Research has not only shown mental health problems but also said that the perceptions and attitudes of a child are being transformed (Fitzpatrick et al., 2020). Anxiety levels have also been increased among the youth, which is a part of mental health.

In the modern era, social networks are influencing the student community, and this is a part of everyday life for everyone. For example, sites such as LinkedIn and Facebook not only give entertainment but also educational stuff (Lee and Horsley, 2017, 2019). There has been a lot of skill development among the youth during the pandemic. People who have the interest have spent the time wisely by learning new things every day. But some of the young adults have become addicted to social media not intentionally (Brand et al., 2019). Today, young adults are relying on information which is easily available through social media (McEvoy and Gilbertz, 2017). They are not verifying the authenticity of the data (Gyamfi, 2017). There are more adverse effects of this rather than positive effects. One of the most undesirable effects of social media is that people are becoming dependent on it (Jasso and Shelly, 2018). Not only dependency, but also privacy has become one of the concerns. Youth has become more interested in invading the privacy of the other person (Adolfsson, 2019). All these behaviors affect the future of the child as well as the society (Abbas and Aqeel, 2019). According to research there is a tremendous decrease in the ability to focus of a social media addicted individual when compared to a non-addicted individual.

2. For what extent did social media have given accurate results or info about the COVID-19 period and how it impacted the youth?

There are lot of effects on the youth due to the misinformation from social media which increases their fear, anxiety, stress, depression, and many more.

Direct access of unknown content was given by many of the social media platforms such as Instagram, YouTube, and Twitter which gave rise to question the information by the users. Hate speech also has been increasing at a constant rate by which youth has been affected (Schild and Velásquez, 2020). The traditional news which the public will get through the news channels and newspapers is shifted and this impacts the social perceptions and the narratives which influences in decision making, policy making, public debates and many more which will make the issues more controversial (Schmidt, 2019).

Algorithms were made which facilitated content promotion and then this information will spread across various profiles (Kulshrestha, 2017). Users across social media will try to acquire the information and form groups to share their own narratives which will become more toxic (Cinelli, 2020). Misinformation will be easily proliferated across public (Vicario, 2019). Some research also shows that inaccurate information will be spread more easily than fact based real information (Vosoughi, 2018). This increased during the pandemic at a high rate and the public started to believe every information without bothering about the facts, which became a huge problem.

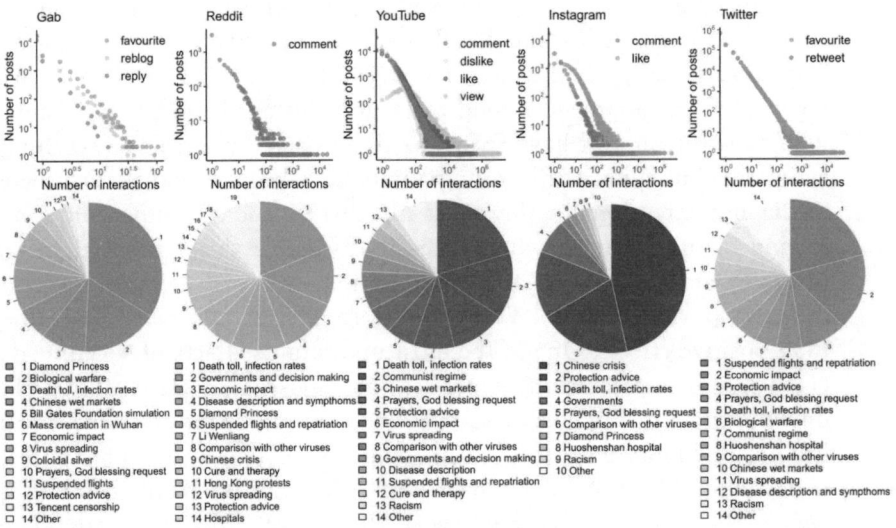

Source: Cinelli et al. (2020).

Who shares the information is the most important question and the most important aspect to look upon. An expert source such as WHO if shares the correct information and addresses misinformation, there would be a great relaxation to the public (Vraga, 2017). That would be more effective to reduce misperceptions which will be spread across the social media accounts without any valid source (Balog-Way, 2020).

Latest updates were provided about the virus across different countries but overloading the public through the data and inputs resulted in the increase of fear and anxiety among them. Especially during the pandemic, when news and information were not valid enough the public has turned to social media to get the correct information (Oh et al., 2015). The impact of mass media was pointed out to a great extent during many of such prior infections such as Avian flu and H1N1 flu (Fung et al., 2011; Oh et al., 2015).

There are different ways to counteract misinformation such as, engagement of scientists so that they can share knowledge which is not only natural but also desirable for the society during the pandemic stage. Social media can increase visibility through this as evidence will be shown for every statement they release (Luc, 2020).

Expert fact checking can be done in which scientist will be the weapon who can counteract the fake news. It can be crowdsourcing or a professional fact checker; they can share their knowledge and scrutinize the news (Pennycook, 2019).

Interaction with the public by sharing correct data, detecting misinformation will raise an alarm to the public and they will get to know about the accurate news (Schünemann, Clark, 2020, 2019).

3. What are the aggressive behaviors inhibited by the youth especially due to gaming?

Outdoor activities and direct human interactions have dramatically decreased as a result of the coronavirus epidemic. As a result, playing video games more frequently may become a coping mechanism for stress and fear of getting sick.

When compared to reality, virtual reality might be more alluring and can serve as an escape from the complexity of daily life. Health authorities recently ordered people to stay at home to stop the COVID-19 pandemic, which led to an increase in video game exposure among kids and teenagers. (Smirni et al., 2021)

Scientists through experiments proved a direct relationship between gaming exposure and differences in brain activity (Murray et al., 2006; Mathiak and Weber 2006).

Overuse of gaming or computer or mobile stations implies excessive time consumption and addiction to it (Gentile, 2009; Rehbein, et al., 2010; Tejeiro Salguero and Bersabe Moran, 2002). Correlations between individuals addicted to games and the presence of psychiatric symptoms have been identified (Cao and Su, 2006; Chan and Rabinowitz, 2006; Jang et al., 2008; Kim et al., 2006).

Some of the strongest damages resulting from the addiction of electronic games are isolation and introversion. The personality of the individual and their relationship with others will be at stake due to their exhibition of negative human behavior which will indicate the inability and their disinterest in engaging with people surrounding them (Chiang et al., 2019; Gentile et al., 2011).

Children many become apathetic sitting for longer periods in front of the electronic screen, which results in various problems such as stress, poor eyesight and also obesity. Mental brain disorders can also happen especially the thinking capability and the concentration of the child will be at risk (Mundy et al., 2020; Chen et al., 2020)

Video game addiction also destroys the child's psychological development exposing him or her to negative feelings such as distress, fear, tension which will lead to insomnia. The exposure to withdrawal anxiety also increases in a child due to the terrifying contents of electronic games. A few examples of withdrawal anxiety disorders are excessive anxiety, severe sadness, impaired concentration and many more. They can also develop the desire for cruelty and violence by continuously involving themselves in these games (Wan and Chiou, 2006; Lissak, 2018).

This addiction not only has an impact on an individual but also their family. For instance, a South Korean man in August,2005 died by playing the online game Starcraft for 50hrs straight in an internet café (BBC, 2005). This has happened due to lack of sleep and no food intake. There will be tremendous consequences to the gamer if he or she is addicted to it. To create time for gaming, young adults skip sleep, socializing, exercise and many other activities (Young, 2004).

Conversely, many cross-sectional and intervention studies have shown that the intensive use of some types of video games leads to significant improvements in many cognitive domains and behaviors. Video games are even considered as "virtual teachers" and effective and "exemplary teachers" (Gentile and Gentile, 2008; Greitemeyer and Osswald, 2010).

Violent video games would increase aggressive thoughts, affect and behavior, physiological persistent alertness, and desensitize players to violence and to the pain and suffering of others. This would support a perceptual and cognitive bias to attribute hostile intentions to others as a result of repeated exposure to justified and enjoyable violence (Anderson et al., 2010).

4. CONCLUSION

There are many trigger effects on violent behavior of youth due to social media. Social media has become an easy access to share problems, frustration and any sort of issue. Youth are finding social relations here where there is no authenticity.

Youngsters are going through many psychological problems such as depression, anger, frustration and many more due to the addiction of social media. Initially, during COVID-19 the public used social media to know the information but later on its usage increased as an entertainment and gaming platform. There are ways to curb social media addiction, but the individual must be interested in the first place.

REFERENCES

Anderson, C. A., & Bushman, B. J. (2001). Effects of violent video games on aggressive behavior, aggressive cognition, aggressive affect, physiological arousal, and prosocial behavior: A meta-analytic review of the scientific literature. Psychological science, 12(5), 353-359.

Bavel, J. J. V., Baicker, K., Boggio, P. S., Capraro, V., Cichocka, A., Cikara, M., ... & Willer, R. (2020). Using social and behavioural science to support COVID-19 pandemic response. Nature human behaviour, 4(5), 460-471.

Birim, B. (2016). Evaluation of corporate social responsibility and social media as key source of strategic communication. Procedia-Social and Behavioral Sciences, 235, 70-75.

Chen, L., Lu, X., Yuan, J., Luo, J., Luo, J., Xie, Z., & Li, D. (2020). A social media study on the associations of flavored electronic cigarettes with health symptoms: observational study. Journal of medical Internet research, 22(6), e17496.

Chiang, J. T., Chang, F. C., Lee, K. W., & Hsu, S. Y. (2019). Transitions in smartphone addiction proneness among children: The effect of gender and use patterns. PloS one, 14(5), e0217235.

Cinelli, M., Quattrociocchi, W., Galeazzi, A., Valensise, C. M., Brugnoli, E., Schmidt, A. L., ... & Scala, A. (2020). The COVID-19 social media infodemic. Scientific reports, 10(1), 1-10.

Gentile, D. A., Choo, H., Liau, A., Sim, T., Li, D., Fung, D., & Khoo, A. (2011). Pathological video game use among youths: A two-year longitudinal study. Pediatrics, 127(2), e319-e329.

Gentile, D. A., & Gentile, J. R. (2008). Violent video games as exemplary teachers: A conceptual analysis. Journal of Youth and Adolescence, 37(2), 127-141.

Greitemeyer, T., & Osswald, S. (2010). Effects of prosocial video games on prosocial behavior. Journal of personality and social psychology, 98(2), 211.

Hamilton, J. L., Nesi, J., & Choukas-Bradley, S. (2020). Teens and social media during the COVID-19 pandemic: Staying socially connected while physically distant.

Ifinedo, P. (2016). Applying uses and gratifications theory and social influence processes to understand students' pervasive adoption of social networking sites: Perspectives from the Americas. International Journal of Information Management, 36(2), 192-206.

Kim, K., Kim, J., & Reid, L. N. (2017). Experiencing motivational conflict on social media in a crisis situation: The case of the Chick-fil-A same-sex marriage controversy. Computers in Human Behavior, 71, 32-41.

Lee, A. R., & Horsley, J. S. (2017). The role of social media on positive youth development: An analysis of 4-H Facebook page and 4-H'ers' positive development. Children and Youth Services Review, 77, 127-138.

Lissak, G. (2018). Adverse physiological and psychological effects of screen time on children and adolescents: Literature review and case study. Environmental Research, 164, 149-157.

Ma, Z. (2021). The role of narrative pictorial warning labels in communicating alcohol-related cancer risks. Health Communication, 1-9.

McEwan, B. (2013). Sharing, caring, and surveilling: An actor–partner interdependence model examination of Facebook relational maintenance strategies. Cyberpsychology, Behavior, and Social Networking, 16(12), 863-869.

Mundy, L. K., Canterford, L., Hoq, M., Olds, T., Moreno-Betancur, M., Sawyer, S., ... & Patton, G. C. (2020). Electronic media use and academic performance in late childhood: A longitudinal study. PLoS One, 15(9), e0237908.

Murthy, D., & Gross, A. J. (2017). Social media processes in disasters: Implications of emergent technology use. Social science research, 63, 356-370.

Neubaum, G., Rösner, L., Rosenthal-von der Pütten, A. M., & Krämer, N. C. (2014). Psychosocial functions of social media usage in a disaster situation: A multi-methodological approach. Computers in Human Behavior, 34, 28-38.

Patchin, J. W., & Hinduja, S. (Eds.). (2012). Cyberbullying prevention and response: Expert perspectives. Routledge.

Pennington, N. (2021). Communication outside of the home through social media during COVID-19. Computers in human behavior reports, 4, 100118.

Ramirez Jr, A., Sumner, E. M., & Spinda, J. (2017). The relational reconnection function of social network sites. New Media & Society, 19(6), 807-825.

Reddick, C. G., & Norris, D. F. (2013). Social media adoption at the American grass roots: Web 2.0 or 1.5?. Government Information Quarterly, 30(4), 498-507.

Riehm, K. E., Holingue, C., Kalb, L. G., Bennett, D., Kapteyn, A., Jiang, Q., ... & Thrul, J. (2020). Associations between media exposure and mental distress among US adults at the beginning of the COVID-19 pandemic. American Journal of Preventive Medicine, 59(5), 630-638.

Scheff, S., & Schorr, M. (2017). Shame nation: The global epidemic of online hate. Sourcebooks, Inc..

Smirni, D., Garufo, E., Di Falco, L., & Lavanco, G. (2021). The playing brain. the impact of video games on cognition and behavior in pediatric age at the time of lockdown: a systematic review. Pediatric Reports, 13(3), 401-415.

Snippe, E., Simons, C. J., Hartmann, J. A., Menne-Lothmann, C., Kramer, I., Booij, S. H., ... & Wichers, M. (2016). Change in daily life behaviors and depression: Within-person and between-person associations. Health Psychology, 35(5), 433.

Teng, Z., Pontes, H. M., Nie, Q., Griffiths, M. D., & Guo, C. (2021). Depression and anxiety symptoms associated with internet gaming disorder before and during the COVID-19 pandemic: A longitudinal study. Journal of Behavioral Addictions, 10(1), 169-180.

Van Zoonen, W., Verhoeven, J. W., & Vliegenthart, R. (2017). Understanding the consequences of public social media use for work. European Management Journal, 35(5), 595-605.

Wan, C. S., & Chiou, W. B. (2006). Psychological motives and online games addiction: Atest of flow theory and humanistic needs theory for Taiwanese adolescents. CyberPsychology & Behavior, 9(3), 317-324.

Zhong, B., Huang, Y., & Liu, Q. (2021). Mental health toll from the coronavirus: Social media usage reveals Wuhan residents' depression and secondary trauma in the COVID-19 outbreak. Computers in human behavior, 114, 106524.

5. Evolution of India's E-commerce Fashion House

Shaik Nouman[1], Pulivendula Preethi Reddy[2], Ranga Harshitha Gupta[3] and Dr. Subhadarshini Khatua[4]

1. INTRODUCTION

July 2020, the startup founder Prabhas Raju has a big question to deal. Government of India has bought a new rule for the E-commerce in the country. Prabhas Raju started a startup named "MY ZONE" in the year 2017 which provides good quality customised clothing's and accessories. Initially the company focused on customising T-shirts and over the time they expanded their market into customising of different kinds of clothing outfits and accessories too. Ram Charan and Pavan Kalyan are the two directors of the company. Initially in the year 2017, company sales were less too moderate as it was a startup. Later, when the company started expansion of their market, they find gradual increase in the growth of their sales. And company never compromised in the quality of their products and services they provided to the customers. MY ZONE has its own website, mobile application and social media marketplaces which is easily accessible by the customers and company had collaborated with different logistics companies for the fast and safe delivery in both the rural and urban areas.

Prabhas Raju is an entrepreneur who has five years of fashion industry expertise. He was born and raised in Hyderabad where he developed interest towards the fashion. He earned his bachelor's degree from National institute of fashion technology, Chennai. And his Master of Business Administration from Woxsen School of Business. He worked in different fashion brands like Puma and Nike as product stylist followed by the graduation. However, he has a strong ambition to create his own work and make a positive impact of the world he decided to pursue MBA. After graduating from the Woxsen university, his knowledge of fashion and interest towards the business leads to startup idea where MY ZONE is founded.

Ram charan and Pavan Kalyan are the two directors of the company. MBA graduates from Woxsen university. Each one of them are experts in different fields like logistics, finance, marketing, sales and strategy making. Prabhas,

Pavan Kalyan and Ram charan were the dean's list students who worked together in different real-time projects during their masters. Ram and Pavan were more passionate to work in a startup where they will more scope to learning which can be helpful to upgrade their skills and be successful in the future. Before joining in MY ZONE, Ram and Pavan worked as senior managers in area of marketing and logistics.

On 23 July 2020, Prabhas came across a news regarding "change in the Consumer Protection (E-Commerce) Rules".

1.1. New Government Policies

The Ministry of Consumer Affairs, Food and Public Distribution, Government of India, introduced the Consumer Protection (E-Commerce) Rules, 2020, to regulate the e-commerce sector and protect online customers. These regulations went into effect on July 23, 2020 and apply to all e-commerce firms operating in India, including international entities providing goods or services to Indian consumers (Department of Consumer Affairs, n.d.).

> Definition of E-commerce entity: The guidelines describe an e-commerce entity as anyone who owns, maintains or manages a digital or electronic facility or platform for online transactions.

> Mandatory registrations: E-commerce firms must register with the Department for Promotion of Industry and Internal Trade (DPIIT) within a certain time frame. Failing to register might result in fines or potentially registration termination.

> Disclosures: E-commerce firms must make certain disclosures on their platform, such as the seller's identity, the country of origin of products or services, the total price of the product, the seller's return, refund, and exchange policies and the projected delivery time.

> Fair-trade practices: E-commerce businesses must avoid engaging in unfair trade practices such as pricing manipulation or providing incorrect or misleading information.

> Grievance redressal: E-commerce businesses must have a grievance redressal procedure to settle consumer complaints within a certain time frame.

> Marketplace entities' liability: Marketplace entities are not responsible for the quality of goods or services supplied by vendors on their platform. They are accountable, however, if they fail to make required disclosures, engage in unfair trade practices or fail to handle consumer complaints.

There are few issues facing by e-commerce companies in India due to change in the rules by the government. For example:

Higher compliance costs: E-commerce businesses must register with the DPIIT, provide disclosures on their platform and set up a grievance redressal procedure, which may necessitate additional resources and investments.

Noncompliance liability: If e-commerce enterprises do not follow the rules, they may face penalties or have their registration cancelled, which may harm their reputation and commercial operations.

Effect on pricing plans: To achieve compliance with the guidelines, e-commerce enterprises may need to change their pricing methods, which may impair their competitiveness and profitability.

International e-commerce enterprises offering goods or services to Indian consumers are also subject to the restrictions, which may impose additional compliance obligations.

Effect on marketplace entities: E-commerce organisations that operate as marketplaces may be held liable if they fail to make required disclosures or participate in unfair trade practices, which may harm their relationships with vendors and customers.

Prabhas decided to have a board meeting after getting to know about the new policies. Same day he had meeting with Ram and Pavan. Where he conveyed the changes and then they discussed the future where they decided to expand their business gradually and set a benchmark in the Indian fashion E-commerce market in past. Myntra one of the main competitors in Fashion e-commerce market in India came up with the merger offer to MY ZONE company, but Prabhas was in dilemma whether to accept the Myntra's merger offer or to continue as independent business.

2. DIGITAL ECO-SYSTEM

Myntra has had a significant impact on India's digital ecosystem. By offering clients a convenient, secure and dependable online purchasing experience, the company has helped to change the country's e-commerce market. Myntra has also contributed to the expansion of the Indian e-commerce business, which is expected to reach $84 billion by 2021. In addition, Myntra has inspired other firms to enter the e-commerce market. It has offered a forum for entrepreneurs to promote their products and services, hence creating a vibrant and competitive market. Myntra has experienced phenomenal growth in recent years. The company has over 4 million active consumers and is one of India's most visited e-commerce websites. It also boasts over 1 million daily app downloads and is currently the country's third most popular e-commerce app.

By providing a frictionless shopping experience, Myntra has been able to establish a devoted consumer base. The corporation has also been successful in establishing a significant brand presence in the country, with millions of individuals reached by its campaigns and commercials. Myntra, the founders believed, should be more than just an online store. They desired to develop an integrated shopping experience that would allow customers to quickly look for, compare and purchase products from the comfort of their own homes. To reach out to its customers, Myntra has implemented a complete digital marketing approach. The company has been extensively investing in advertising and marketing, as well as actively utilising social media channels such as Facebook, Instagram and Twitter to create interesting content and drive sales. Myntra has also prioritised personalisation and targeting to ensure that its advertising reach the intended demographic. Data has been used by the organisation to better understand client behaviour and preferences and to personalise its ads accordingly.

Myntra, which was formed in 2007, is one of India's digital economy pioneers. Since then, Myntra has evolved to become one of India's major e-commerce platforms, offering a diverse range of products and services to meet a wide spectrum of client needs. The digital economy in India has seen substantial growth and development over the last decade, with the establishment of a diverse range of e-commerce platforms delivering products and services to clients across the country. These platforms have enhanced access to goods and services, allowing customers to make more informed purchasing decisions. Myntra has been at the forefront of this expansion and is a key part of India's digital ecosystem. The digital ecosystem has become an essential component of the Indian economy, experiencing tremendous expansion over the last decade. The rise of cellphones, the internet and digital platforms has resulted in an infusion of new enterprises and opportunities in the country. Myntra is one such company that has benefited from India's digital transformation. Myntra's platform provides clients with clothes, footwear and lifestyle items as one of the largest and most popular e-commerce sites. In recent years, the company has attracted millions of subscribers and experienced significant growth. Myntra, an online fashion shop based in Bangalore, was started in 2007. It has expanded over the years to become one of India's most successful e-commerce companies, with a presence in over 1,000 cities and towns across the country. Myntra has been at the vanguard of the Indian retail revolution, offering customers an unrivalled shopping experience.

Myntra's growth is undeniably tied to the expansion of the Indian digital ecosystem. India has the world's second-highest internet penetration rate, with over 600 million internet users. Nevertheless, internet availability and speeds have significantly improved, with the number of mobile connections expected to increase by 24% by 2020. Myntra has been able to reach out to customers in remote, rural areas and offer them with access to a wide choice of trendy products because of this. Nonetheless, India's digital environment still confronts certain hurdles. Many rural and tiny villages still lack access to internet and digital services. This has an impact on Myntra's growth because it is unable to contact clients in these areas. Furthermore, the lack of dependable payment and delivery methods means that there are still barriers to internet shopping in India.

Despite these obstacles, Myntra has been able to capitalize on India's digital economy. The company has concentrated its efforts on expanding its presence in major cities and urban areas, where internet connectivity is more widespread. Myntra has also collaborated with physical stores to improve the shopping experience for customers. This has allowed the organisation to reach a bigger consumer base while also improving service. Furthermore, Myntra has concentrated on providing a seamless purchasing experience. The company has introduced features such as "Try Before You Buy", which allows customers to try on clothing before making a purchase. In addition, "Express Delivery" guarantees speedier delivery times and more efficient returns. These characteristics have helped Myntra become one of India's most popular e-commerce sites.

Myntra has also been able to benefit on the Indian digital ecosystem's growing preference for online payments. According to the Reserve Bank of India, digital payments would increase by 40% by 2020, reaching $1.1 trillion. This has enabled Myntra to provide customers with a convenient and secure online payment platform, allowing them to easily pay for merchandise. Myntra has also benefited from the increasing popularity of social media in India. Several effective marketing strategies on social media sites such as Facebook and Instagram have reached millions of potential customers. Myntra has also used social media sites to communicate with existing consumers, offering personalised recommendations and discounts. India is experiencing a rapid move towards digitalisation, with an increasing number of enterprises moving their operations online. This transformation is particularly visible in the retail industry, with e-commerce behemoths like Myntra emerging as market leaders. Myntra has an extraordinary chance to extend its reach and consumer base thanks to the Indian digital ecosystem.

According to The Economic Times, the Indian e-commerce business would be worth $200 billion by 2026. This provides Myntra with a once-in-a-lifetime opportunity to extend its client base and reach across the country. Overall, India's digital economy provides tremendous possibility for expansion, and Myntra has been quick to capitalise on this. The company has tried to improve and grow its presence in rural areas in order to provide clients with a better and more convenient shopping experience. Despite significant hurdles, Myntra has been able to make the most of India's digital ecosystem.

3. INTRODUCTION TO COMPETITORS: MYNTRA

Myntra is an Indian e-commerce platform focused on fashion and lifestyle items. Mukesh Bansal, Ashutosh Lawania and Vineet Saxena launched it in 2007 in Bengaluru, India as an online platform for personalised gifts and items. At the beginning, the company experienced various hurdles, including limited capital and a lack of industry knowledge. However, the founders were successful in obtaining investment from several investors and establishing Myntra as a significant player in the Indian e-commerce business, with a concentration on fashion and lifestyle products. The company has evolved and increased its product offerings over the years, as well as implemented new features to improve the consumer experience. The company serves as an online marketplace for buying and selling apparel, footwear and accessories for men, women and children. It also sells a variety of home decor and furnishing items.

Myntra's primary concentration is on fashion and leisure products, and it offers a diverse range of brands and products. Customers may easily browse and purchase products thanks to the company's user-friendly website and mobile app. It also has features like tailored suggestions, search filters and a wishlist to assist shoppers locate the things they want. One of Myntra's distinguishing aspects is its diverse product offering, which comprises both international and Indian brands. In addition, the company places a significant emphasis on customer service, with features such as quick returns and exchanges, cash on delivery and a 30-day return policy. Myntra maintains multiple retail locations across India in addition to its online business. These businesses allow shoppers to try on and purchase things in person. This allows buyers to have a more immersive shopping experience and get a better feeling of the fit, quality and feel of the products they are purchasing (Kushwaha, 2023)

Myntra has expanded tremendously over the years to become one of India's biggest e-commerce enterprises. Its product offerings have also extended to include a wide range of areas such as electronics, home appliances and personal care. In addition, the organisation is always developing and releasing new features to enhance the client experience. To summarise, Myntra is a leading Indian e-commerce company that focuses on fashion and leisure products. Myntra has been a popular destination for online shoppers in India because of its diverse product offering, user-friendly infrastructure and strong emphasis on customer care. Myntra's offline store presence allows customers to try and buy things in person, which is an extra benefit. In addition, the organisation is always developing and releasing new features to enhance the client experience.

3.1. National Scenario

India is a country full of potential. The country's economic status has greatly improved since its independence, and it is now emerging as one of the world's top nations. It can also be compared to a marketing behemoth with a population of more than 100 crores and a growth rate of more than 6%. As a result, it is easy to see why internet purchasing has grown quickly in India in recent years.

Myntra has come a long way and is considered India's pioneer in the online fashion industry. As technology expands to the most remote communities and countless work opportunities are made available to unemployed youth, an increasing number of individuals are becoming aware of, and have the means to purchase, expensive and luxurious products on the internet. Vineet Saxena, Ashutosh Lawania and Mukesh Bansal, three IIT-Kanpur alumni, founded the company in 2007. Myntra ascended to the top of the B2B marketplace for personalised gifts in three years. It evolved into a model for a fashion retail platform in 2011, which Flipkart purchased in 2014 (TechWelkin, n.d.). Ananth Narayan took over as CEO of Myntra in 2015. Myntra shut down its desktop website based on the traffic and orders received through its mobile app—more than 90% of traffic and 70% of orders. Myntra reopened in 2016, exactly one year after it was shut down. While mobile applications do not provide the same visual and interactive experience as computers, the website generates 15% of the company's revenue. The next year, it acquired a competitor, Jabong, and expanded its GVM run rate to $1 trillion. It expanded at an average rate of 80% in 2018.

3.2. Global Scenario and International Expansion

Myntra's worldwide expansion strategy involves launching Myntra Fashion brands throughout the Middle East. The company has collaborated with the region's leading e-commerce sites, noon.com and namshi.com, to provide Indian-made products to the region's millions of fashion-conscious customers. Myntra Fashion Brands' entry into the market reflects a big gamble on the casual wear categories, which account for a sizable portion of the company's domestic market activities. This aligns with current trends, which indicate a recent shift towards T-shirts, comfy bottoms and activewear, resulting in a focus on related sectors. There has been a huge lifestyle change as a result of the ongoing epidemic, which is driving this trend shift. More than 75% of the styles Myntra exports to these regions are designed in India (Editor, 2020).

"We are thrilled to announce our international expansion. This is Myntra's next stage of development and a significant turning point in our journey to date. According to our research, there are numerous potential routes around the world, which give significant opportunities for the online fashion sector. The Middle East stood out among them due to audience demographics, particularly high cellphone penetration, fashion parallels and a substantial population of Indian heritage. We foresee a 5X growth in the connection over the next two years. Myntra will continue to explore new markets in order to build brands with a strong international reputation and expand their international client base. (GANDHI, n.d.)

4. GOVERNMENT POLICIES TOWARDS E-COMMERCE

The Indian Government's legislation and regulations have had a significant impact on the development of the country's e-commerce company. These policies aimed to level the playing field for all market participants, protect the interests of consumers and support the growth of the e-commerce sector. One of the primary policies is the foreign direct investment (FDI) policy, which regulates the flow of foreign capital into the e-commerce sector. This legislation allows for 100% FDI in the e-commerce marketplace model but not in the inventory-based approach. Foreign corporations can invest in e-commerce platforms that act as intermediaries between buyers and sellers, but they cannot own the goods. This legislation has benefited in drawing foreign investment into the e-commerce sector, resulting in the expansion of the country's numerous e-commerce platforms (Department for promotion of industry and international trade, n.d.). Another key policy is the RBI's e-commerce guidelines, which control the operation of the country's e-commerce activities. Client identity, transaction limits and data security are all covered by these standards. They also require e-commerce businesses to implement strict security protocols to protect user data. The guidelines have increased customer trust in e-commerce by contributing to the safety and security of online transactions (Reserve Bank of India, n.d.).

The Indian Government has also introduced the Consumer Protection (E-Commerce) Rules, 2020, with the purpose of protecting online shoppers' rights. These rules require e-commerce companies to provide detailed information on the products and services they offer, as well as a clear refund and return policy. They also prohibit online businesses from engaging in fraudulent or misleading practises, such as misrepresenting a product or service. The limits have aided in increasing the openness and responsibility of e-commerce businesses, which has enhanced consumer trust in the sector (Department of Consumer Affairs, n.d.).

Furthermore, the government has issued the National Policy on Electronics 2019 (NPE 2019) to encourage the growth of India's electronics sector, notably the e-commerce sector. The policy intends to promote the growth of the electronics industry by providing incentives for firms to build manufacturing facilities in the country and supporting the development of a skilled workforce. The policy has contributed to the increased availability of electronic products in the country, culminating in the growth of the e-commerce sector. The government has also taken initiatives in recent years to foster the growth of small- and medium-sized businesses (SMEs) in the e-commerce industry. This includes the establishment of the "Digital India" project, which aims to empower SMEs with access to digital technology and platforms in order to expand their reach and competitiveness. The initiative has contributed to an increase in the number of SMEs active in the e-commerce industry, resulting in increased market competitiveness and innovation (Digital India, n.d.).

4.1. Mergers

Myntra, an Indian e-commerce startup, has gone through a number of mergers and acquisitions in recent years. These transactions have assisted the company in expanding its operations and strengthening its position in the Indian e-commerce market.

One of Myntra's most major acquisitions was its 2014 acquisition by Flipkart, another Indian e-commerce behemoth. Myntra was able to profit from Flipkart's massive consumer base and substantial logistics network as a result of this agreement. Myntra was also able to extend its product offerings and reach more customers as a result of the acquisition.

Myntra purchased Jabong, another large Indian apparel e-commerce company, in 2016. This acquisition aided Myntra in broadening its client base and product choices, as well as gaining a stronger presence in the Indian e-commerce sector. Myntra was also able to exploit Jabong's expertise in sectors such as fashion, as well as unite the two companies' logistics and supply chain operations, as a result of the acquisition (Nair, n.d.)

In 2017, Myntra collaborated with Roadster, a leading Indian fashion brand, to introduce a new product line. This collaboration enabled Myntra to provide customers with a diverse choice of trendy and high-quality products, as well as to improve its position in the Indian fashion sector.

Wit works, a Bengaluru-based technology business that specialises in developing smart wearables, was bought by Myntra in 2018. Myntra was able to expand its product offerings and join the smart wearables sector as a result of the acquisition.

Myntra announced a strategic alliance with the fashion e-commerce site Roposo in 2019. This collaboration enabled Myntra to broaden its consumer base and reach more customers in Tier II and Tier III cities that are currently underserved by e-commerce firms.

Myntra announced the acquisition of HRX, a fitness and wellness start-up founded by Hrithik Roshan, in 2020. This acquisition enabled Myntra to broaden its product line to include fitness and wellness products.

These acquisitions and mergers have enabled Myntra to expand its company and enhance its position in the Indian e-commerce market. Myntra has increased its user base and product offers by purchasing other companies, giving it a stronger foundation in the Indian e-commerce market. Furthermore, these acquisitions have enabled Myntra to capitalise on the qualities of the companies it has acquired, such as Jabong's fashion expertise and Roadster's brand reputation.

Finally, Myntra has experienced several mergers and acquisitions in recent years, which have aided the company in expanding its business and strengthening its position in the Indian e-commerce sector. Several companies, such as Flipkart, Jabong, Roadster and Wit works, have been bought by the corporation to extend its client base and product offerings and build a stronger footing in the Indian e-commerce sector. Furthermore, these acquisitions have enabled Myntra to capitalise on the qualities of the companies it has acquired, such as Jabong's fashion expertise and Roadster's brand reputation. (Das, n.d.)

4.2. Partnership

The company has a number of agreements and collaborations with different fashion and retail brands and enterprises. Jabong, another big Indian fashion e-commerce portal, is one of Myntra's important partners. Myntra purchased Jabong in 2016, and the two companies have subsequently merged to establish one of India's leading fashion e-commerce companies. Myntra also collaborates with a number of well-known fashion and lifestyle companies, including Puma, Nike, Adidas, Levis, Tommy Hilfiger and many others. These collaborations enable Myntra to offer a diverse choice of products from these businesses to its customers (https://startuptalky.com/myntra-subsidiaries-companies/, n.d.). Myntra collaborates with a number of technological companies in addition to fashion labels. Myntra, for example, has teamed with Microsoft to use Microsoft Azure as its cloud computing platform. This collaboration enables Myntra to expand its operations and enhance the functionality of its e-commerce platform (Microsoft, n.d.).

Myntra also collaborates with a number of payment providers, including Paytm, Google Pay, PhonePe and others, to give customers with a variety of payment alternatives. These collaborations make it simple for users to make purchases on Myntra. Myntra also features exclusive collaborations with various designers, including Rohit Bal, Manish Malhotra and Ritu Kumar. These collaborations enable Myntra to provide clients with special collections and designs. Myntra has a number of partnerships and collaborations with prominent fashion and lifestyle brands, technology firms, payment providers and designers in order to provide a diverse selection of products and services to customers and to develop its e-commerce platform.

5. KEY CHALLENGES

Since its inception, Myntra has experienced numerous obstacles. Competition from larger, more established e-commerce companies such as Amazon and Flipkart has been a major challenge. Myntra has had to put in a lot of effort to establish itself as a competitive alternative in a market dominated by these behemoths (Ghosh, 2018). Building and sustaining a good supply chain has been another problem. Myntra has had to ensure that it can source and distribute products swiftly and effectively with a vast choice of products and a constantly rising consumer base. This has necessitated substantial expenditures in logistics and infrastructure. Myntra likewise encountered difficulties in establishing a devoted consumer base. With so much competition in the e-commerce business, it can be difficult to stand out and build a strong brand. To attract and retain customers, Myntra has had to focus on providing a high-quality customer experience and developing a great reputation (Jain, 2019).

Myntra also had to face technological and scaling obstacles. As Myntra expanded, it needed to invest in more complex technology to support its operations and handle a high number of clients and transactions. Significant expenditures in areas such as artificial intelligence, machine learning and data analytics have been necessary (Sharma, 2021). Furthermore, Myntra has had to navigate India's complex legal and regulatory framework. The Indian Government has implemented various e-commerce-related rules and regulations, and Myntra has had to guarantee that it complies with all of them. Finally,

Myntra has faced hurdles in the offline industry, as the majority of Indians still prefer to buy from physical stores. Myntra is attempting to address this by creating physical locations in collaboration with other retailers. Overall, Myntra has experienced various hurdles since its inception, but the firm has been able to establish itself as a major competitor in the Indian e-commerce market through strategic investments and a focus on customer experience.

5.1. AI Algorithm

Personalisation is a crucial area in which Myntra use AI. To generate individualised product recommendations to specific clients, the company employs machine learning algorithms to examine user data such as browsing and purchase history. This improves customer engagement and revenues. Myntra's adoption of a unique AI Algorithm benefits brands and their acquisition branding. This AI algorithm is created in such a way that it refers to the top acquisition brands. When a customer goes on the home page and begins searching for required products such as shoes, shirts or any other, Myntra's Ai algorithm prioritises acquisition brands in order to improve the branding and sales of in-house brands. Unless the customer searches for the goods using the precise brand name, the results will be inaccurate.

Myntra also employs AI for picture identification. The startup has created a system that automatically tags and categorises products based on their photographs using deep learning algorithms. This enables for more efficient and accurate product search and browsing. Myntra also use artificial intelligence to optimise its supply chain and logistics. To make better use of resources and enhance delivery times, the organisation use algorithms to analyse data on client demand, inventory levels and shipment schedules. In addition, Myntra uses AI to detect and prevent fraud. Machine learning algorithms are used by the organisation to evaluate client data and find patterns that may suggest fraudulent behaviour. This helps to safeguard the company's and its customers' financial interests. Finally, Myntra employs AI to improve customer service. The organisation has built a chatbot that understands consumer enquiries and provides relevant responses using natural language processing (NLP). Overall, Myntra employs a variety of AI algorithms to assist its operations and enhance the consumer experience, ranging from personalisation to logistics, fraud protection to customer care (Desai).

5.2. Private Label Brands and Their Ambassadors

Myntra, an Indian e-commerce company, has used several brand ambassadors to promote its products and brands. Some of the notable brand ambassadors of Myntra include:

Hrithik Roshan: Bollywood star, Hrithik Roshan, has served as a brand ambassador for Myntra's Roadster and HRX brands. He has appeared in various advertisements and advertising campaigns.

Jacqueline Fernandez: Jacqueline Fernandez, a Bollywood actress, has served as Myntra's Mast & Harbour brand ambassador. She has been in various advertisements and advertising campaigns.

Kriti Sanon: Kriti Sanon, a Bollywood actress, is the brand ambassador for Myntra's dress berry collection. She has been in various advertisements and advertising campaigns.

Kiara Advani: Kiara Advani, a Bollywood actress, has served as Myntra's All About You brand ambassador. She has been in various advertisements and advertising campaigns.

Shraddha Kapoor: Bollywood actress, Shraddha Kapoor, is the brand ambassador for Myntra's Anouk line. She has been in various advertisements and advertising campaigns.

Varun Dhawan: Varun Dhawan, a Bollywood actor, has served as Myntra's Roadster brand ambassador. He has appeared in various advertisements and advertising campaigns.

Several well-known celebrities and influencers have also been employed by Myntra to promote its products and companies. These brand ambassadors assisted Myntra in raising brand recognition, attracting new consumers and increasing customer loyalty. Myntra has promoted its products and brands with celebrities such as Hrithik Roshan, Jacqueline Fernandez, Kriti Sanon, Kiara Advani, Shraddha Kapoor, Disha Patani and Varun Dhawan. They have assisted the company in raising brand awareness, attracting new customers and increasing client loyalty. (myntra, n.d.)

5.2.1. Business Model

Customers can shop for a wide range of fashion and lifestyle products from numerous national and international brands, as well as Myntra's own private-label products, on Myntra's platform. The company acts as a middleman between clients and sellers, linking them to enable product buying and selling. Myntra does not maintain any inventory under this approach, instead acting as a middleman for clients to purchase things from various sellers. Clothing, footwear] and accessories for men, women and children, as well as home decor and beauty products, are available from Myntra. The company provides a diverse range of items from national and international brands, as well as its own private-label products. Private-label products from Myntra are designed and manufactured in-house and are priced lower than other products on the platform. Myntra's platform is intended to make it simple for customers to find and buy the things they desire. The website and mobile app are user-friendly and include a variety of filters and sorting options to assist clients in finding the products they need. Myntra also accepts a variety of payment methods, such as credit and debit cards, net banking and cash on delivery.

Customers will also enjoy a convenient and hassle-free shopping experience thanks to the company's logistics and shipping services. Myntra provides speedy delivery services and real-time tracking information to clients so they can keep track of their orders. Myntra's revenue is mostly derived from commissions on sales made through its platform. When a customer buys a product from a vendor on Myntra's marketplace, the company earns a commission. The commission % varies depending on the product category and the seller. Myntra also earns money from advertising and sponsorships. The organisation collaborates with a variety of brands and businesses to market their goods and services on its platform. Revenue is also generated by the company's loyalty program which rewards customers for making purchases on the platform. Myntra has diversified its revenue streams through strategic acquisitions and collaborations, in addition to its marketplace business model. To increase its product choices and attract a larger client base, the company has purchased various fashion and lifestyle brands and partnered with other e-commerce companies. Refer to the image (Figure 1)

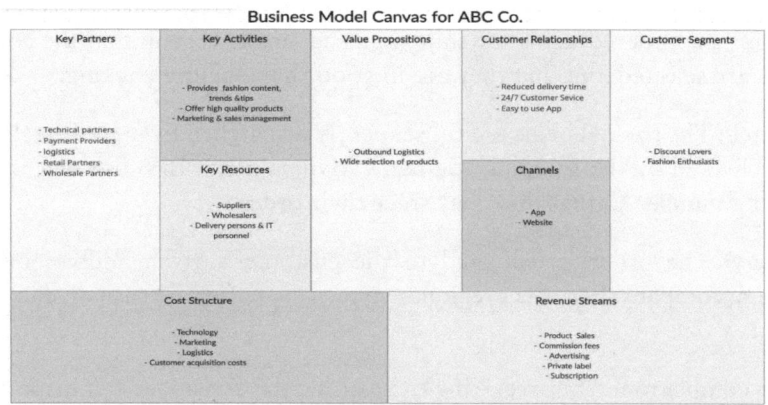

Figure 1: Business model.

5.2.2. Pricing Strategy

The company's loyalty program, which rewards customers for making purchases on the platform, also generates revenue. In addition to its marketplace business model, Myntra has diversified its revenue streams through smart acquisitions and collaborations. The company has purchased several fashion and lifestyle brands and worked with other e-commerce companies to expand its product offerings and attract a broader client base. Furthermore, when a new product is produced and the product is in high demand during the offer season, Myntra generates fake scarcity. In this case, Myntra displays out of stock for that specific product due to strong demand and offer pricing; nevertheless, after a few hours, customers can view the availability of the product again at a different price. It is important to note that Myntra may employ various price strategies concurrently and may adjust its pricing strategy in response to market conditions.

5.2.3 Order Fulfillment Process

Myntra's app features over 15,000 products for men and women, including apparel, shoes and accessories. Myntra employs a defined approach for order fulfilment to provide an overview of the process.

Order placement: Consumers can order using the Myntra website or mobile app. Users can go through a variety of things and add them to their shopping basket. When they are ready to buy, they can proceed to the checkout and pay.

Order confirmation: Myntra sends an order confirmation email or SMS to the customer after the payment is processed. This comprises information such as the order number, the items ordered and the estimated delivery date.

Picking and packing: Myntra's warehouse team then selects, packs and prepares the ordered items for shipping. They make certain that the correct things are selected and that they are in good shape before packing.

Dispatch: The packed items are subsequently dispatched from the warehouse and delivered by the shipping company. Myntra offers the customer with a tracking number so that they may trace their order.

Delivery: The order is delivered to the customer's given address by the shipping company. For select regions, Myntra also offers a "cash on delivery" option.

Order completion: After receiving their order, the consumer can inspect the things and contact Myntra if there are any difficulties or anomalies. Once everything is in order, the order is deemed complete.

This is on the front end. Let's have a brief understanding of the backend.

Inventory management: Myntra keeps a precise inventory of all the products in stock. This includes details about the product, stock levels and supplier information. This assists them in ensuring that they have enough inventory to fulfil consumer orders.

Order processing: When you place an order, it is received by the Myntra order processing team. They examine the availability of the products, verify the order details and update the inventory accordingly. They also assign the order a unique order number and notify the consumer of the projected delivery date.

Picking and packing: The Myntra warehouse crew then selects products from inventory in accordance with the order and packs them for dispatch. They make certain that the correct products are selected and that they are in good shape before packing.

Dispatch: The order is dispatched from the warehouse to the shipping carrier for delivery after it has been packed. The team also provides the customer with order tracking information.

Payment processing: To process client payments, Myntra employs a secure payment gateway. They also feature a "cash on delivery" option for select regions.

Order tracking: Myntra allows customers to track their orders using the website/app. The team also checks order status and keeps the consumer informed of any delays or concerns.

Return and refund: In the event that a customer is dissatisfied with their order or receives a damaged goods, Myntra provides a simple return and refund policy. To ensure client satisfaction, the crew manages the return and refund process effectively.

The most significant aspect is the brand's utilisation of technology and other applications. Because it is a private corporation, it does not publicly reveal information regarding its primary competencies. The latest collaboration with Flipkart, as well as previously acquired logistics companies and software, makes the process considerably smoother.

Myntra may utilise various types of software to aid in the order fulfilment process, such as warehouse management systems, shipping and logistics software or enterprise resource planning (ERP) software, in addition to inventory management software. These systems aid in the automation and simplification of different stages of the order fulfilment process, including selecting, packing and shipping.

Myntra also follows four key models: Myntra JIT, Myntra PPMP, Myntra FBM and Myntra Pretr/Omni. These models now assist Myntra speed its growth and sales by integrating with other technologies for each part of the business, such as consumers, order management, sellers and so on. These procedures are typically associated with the sale's backend. (Lopienski, n.d.) Refer to the image (Figure 2)

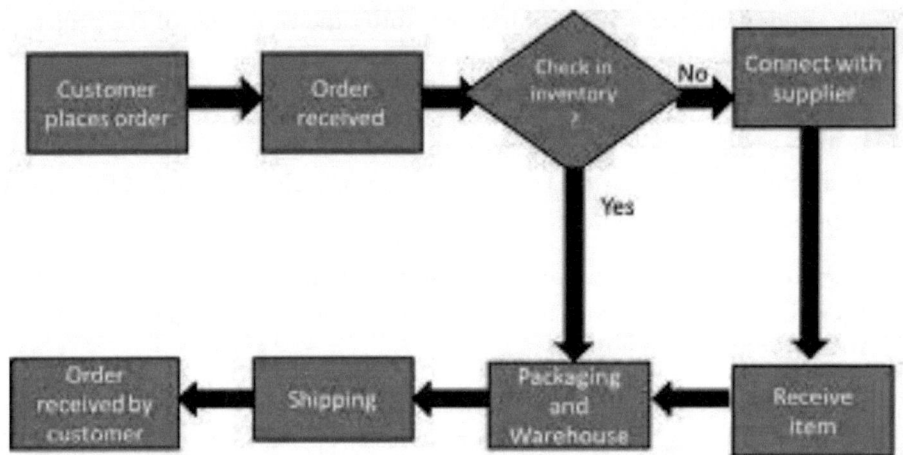

Figure 2: Order fulfillment process.

Forward and reverse logistics:

Forward logistics involves the process starting from the warehouse to the end customers and this is briefly as follows:

Order processing: Myntra's order processing team verifies the order data and issues the order a unique order number. They also give the buyer an expected delivery date.

Picking and packing: Myntra's warehouse team selects products from inventory in accordance with the order and packs them for shipping. They make certain that the correct products are selected and that they are in good shape before packing.

Dispatch: The order is dispatched from the warehouse to the shipping carrier for delivery after it has been packed. The team also provides the customer with order tracking information.

Transportation: The order is subsequently delivered to the customer's specified address by the shipping carrier. Depending on the region and service availability, Myntra uses a variety of shipping companies such as India Post, Bluedart, DHL and FedEx.

Delivery: The order is delivered to the customer's given address by the shipping company. For select regions, Myntra also offers a "cash on delivery" option.

Order completion: After receiving their order, the consumer can inspect the things and contact Myntra if there are any difficulties or anomalies. The order is complete if everything is in order.

Myntra also employs a variety of tools and technologies to track packages and provide real-time information to customers (Rashi, n.d.).

Reverse logistics refers to the process of returning, repairing or disposing of purchased things. Myntra most likely has a reverse logistics procedure in place to manage client returns and exchanges.

The process of reverse logistics in Myntra may include the following steps:

Return initiation: Consumers can request a return by contacting Myntra's customer care team or by using the return option in their account on the website or mobile app.

Return authorisation: Myntra's customer care staff verifies the return request and allows the return if it fits the conditions of the company's return policy.

Pickup and transportation: Myntra will arrange for the return of the items from the customer's location. After that, the product is returned to the warehouse or the approved service centre.

Inspection and processing: Myntra's team inspects returned items to ensure they are in good condition and match the return policy conditions. If the product is qualified for a refund, exchange or repair, the team will handle the request.

Refund or exchange: If a refund is requested by the consumer, Myntra executes the refund and notifies the customer. If the client requests an exchange, the team handles the exchange and notifies the customer of the revised delivery date.

Disposal or Reprocessing: If the product is not eligible for a refund or exchange, the team will dispose of or reprocess it.

But, more importantly, the partners and teams with whom they've worked and partnered in the past make them one of the most trustworthy companies in India. These partners can also be known as:

Suppliers: Myntra obtains products from a wide range of vendors, including manufacturers, wholesalers and distributors. They collaborate with a wide number of vendors to provide a diverse selection of items to their clients.

Warehouses and logistics partners: Myntra stores merchandise in warehouses and employs logistics partners to transport products from the warehouse to customers. They also have a network of logistical partners to ensure timely and effective product delivery.

Payment and financial partners: Myntra collaborates with financial partners to provide clients with secure payment choices. They also collaborate with banks and financial institutions to facilitate financial transactions and cash-on-delivery options.

Technology partners: Myntra's e-commerce platform is most likely supported by several technology partners, including inventory management software, warehouse management systems, and shipping and logistics software.

Marketing and advertising partners: To advertise its products and services to customers, Myntra collaborates with marketing and advertising partners. They also sell their items through numerous digital platforms including social media.

Customers: Myntra's mission is to give clients with a diverse selection of fashion and lifestyle products at reasonable pricing. They are also aiming to make purchasing more frictionless and user-friendly (Rashi, n.d.).

6. CONSUMER BEHAVIOUR TOWARDS E-COMMERCE

Numerous significant modifications and alterations in customer behaviour in India have occurred as a result of e-commerce. Customers in India were initially driven to online shopping because of its convenience and the possibility of finding items at lower rates than in traditional brick-and-mortar stores. Nonetheless, customer behaviour in India has shifted over time to place a higher value on the online purchasing experience, with a greater emphasis on personalised recommendations, quick delivery and simple returns. This trend has been fuelled by several causes, including the rise of online marketplaces like Flipkart and Amazon, as well as the broad availability of mobile internet connections in India. One of the most significant developments in Indian consumer behaviour has been the rise of mobile shopping. According to a Google and KPMG report (KPMG, n.d.), by 2021, India's mobile internet users will number 628 million, with mobile commerce accounting for more than 70% of all e-commerce transactions in the country. The increased availability of low-cost cell phones, as well as the growing public popularity of mobile apps, has fuelled this trend.

Another development that has changed consumer behaviour in India is the rise of social commerce. This refers to the use of social media platforms such as Facebook and Instagram to facilitate online purchases. Social commerce has gained in popularity in India, particularly among younger consumers. Aside from these tendencies, there have been changes in customer behaviour in terms of the types of items purchased online. Users in India, for example, primarily purchased books and electronics online in the early days of e-commerce. But, in recent years, there has been a significant growth in the number of consumers purchasing fashion and beauty products, as well as home goods and furnishings, online. According to a PwC report, India has one of the world's fastest-growing e-commerce marketplaces, with an estimated $64 billion in 2019 and a projected $200 billion by 2026.

Online markets such as Flipkart, Amazon and Myntra have fuelled this growth. These marketplaces offer a wide range of products, from electronics and books to fashion and household goods, making it easy for buyers to get what they need. To recapitulate, client behaviour towards e-commerce in India has shifted considerably throughout the years. Customer behaviour has shifted away from convenience and inexpensive prices and towards the shopping experience, personalisation, quick delivery and simple returns. The growth of mobile and social commerce, as well as the increased use of the internet and cellphones, has all played a role in this transformation. With the Indian e-commerce market expected to grow significantly in the coming years, it will be intriguing to see how customer behaviour changes in response to new technology and changing market conditions.

7. TECHNOLOGY AND MYNTRA'S OPPORTUNITY AND GROWTH IN INDIA

Myntra, an Indian e-commerce startup, has benefited greatly from technological advancements. Myntra has achieved great growth in India over the last several years, which can be credited to its successful use of technology to better reach and service its customers.

Myntra was formed in 2007 by a group of entrepreneurs and was primarily focused on product customisation before shifting its attention to online shopping. Myntra had extraordinary growth as India's internet infrastructure and access improved. Myntra had developed itself as one of the country's top e-commerce startups in less than two years.

Myntra's success has been credited largely to its use of technology. The organisation was able to customise its products and tailor its offers to the demands of its clients by utilising AI and machine learning techniques. Furthermore, by employing predictive analytics to deliver relevant recommendations and discounts, Myntra was able to develop personalised shopping experiences for its customers. Myntra also used technology to improve its delivery services, allowing it to provide consumers with fast and effective deliveries.

Myntra was able to personalise the shopping experience for each user by utilising machine learning algorithms. This allowed them to provide their clients with a more intuitive shopping experience and boost their chances of making a deal. Technology has also enabled Myntra to grow its footprint in rural India by providing products targeted to their requirements and preferences. Furthermore, Myntra has used the power of the internet to establish a virtual retail network via which retailers can connect to the site. This has created a fresh route for Myntra to grow and broaden its reach. Additionally, Myntra has made its items more accessible to Indian clients by utilising technology. Myntra has enabled clients to shop online from the comfort of their own homes by using the power of the internet. This has allowed them to reach a wider audience.

Myntra's success is also due to its ability to remain ahead of the curve and anticipate the requirements of its clients. The organisation has been able to spot developing market trends and promptly adjust to them. For example, Myntra was fast to capitalise on the growing popularity of online payment methods like UPI and developed a simple and secure payment platform for its clients. Similarly, the corporation recognised the power of digital marketing early on and devised a successful digital marketing plan to reach its target demographic.

Myntra's success is proof of the power of technology. Myntra has been able to successfully reach and serve its customers, provide personalised shopping experiences and remain ahead of the competition by using the power of technology. As India continues to develop and expand its digital infrastructure, Myntra's success is expected to be copied by other Indian e-commerce startups.

8. CONCLUSION

Prabhas Raju started a startup named "MY ZONE" which provides good quality customised clothing's and accessories. Initially, the company focused on customising T-shirts, but over time they expanded their market into customising of different kinds of clothing outfits and accessories. Ram Charan and Pavan Kalyan are the two directors of the company, who are MBA graduates from Woxsen University and have five years of fashion industry expertise. MY ZONE has its own website, mobile application and social media marketplaces, and has collaborated with different logistics companies for the fast and safe delivery in both the rural and urban areas. The Ministry of Consumer Affairs, Food and Public Distribution, Government of India, introduced the Consumer Protection (E-Commerce) Rules, 2020, to regulate the e-commerce sector and protect online customers. These regulations apply to all e-commerce firms operating in India, including international entities providing goods or services to Indian consumers. E-commerce firms must register with the Department for Promotion of Industry and Internal Trade (DPIIT) within a certain time frame, make disclosures on their platform, avoid unfair trade practices and have a grievance redressal procedure to settle complaints. Marketplace entities are responsible for the quality of goods or services supplied by vendors on their platform but are accountable if they fail to make required disclosures, engage in unfair trade practices or fail to handle consumer complaints. There are several issues facing e-commerce companies in India due to the new rules by the government, such as higher compliance costs, noncompliance liability and pricing plans. International e-commerce enterprises offering goods or services to Indian consumers are also subject to the restrictions, which may impose additional compliance obligations. Prabhas decided to have a board meeting after getting to know about the new policies and discussed the future where they decided to expand their business gradually and set a benchmark in the Indian fashion e-commerce market. Myntra came up with a merger offer to MY ZONE company, but Prabhas was in dilemma whether to accept or continue as independent business?

REFERENCES

Department for promotion of industry and international trade . (n.d.). Retrieved from https://dpiit.gov.in/foreign-direct-investment/foreign-direct-investment-policy

Department of Consumer Affairs. (n.d.). Retrieved from https://consumeraffairs.nic.in/acts-and-rules/consumer-protection

Digital India. (n.d.). Retrieved from https://digitalindia.gov.in/

Editor, I. R. (2020). *Indian Retailer Bureau*. Retrieved from https://www.indianretailer.com/news/myntra-set-to-expand-its-international-footprint-in-middle-east-market.n10128

Das, M. (2021). *Startup talky*. Retrieved from https://startuptalky.com/myntra-subsidiaries-companies/

Desai, P., *et al.* (2022). Artificial intelligence in strengthening the operations of ecommerce based business. *International Journal of Advanced Research in Management and Social Sciences. 2022 Interdisciplinary Research in Technology and Management (IRTM)*.

Gandhi, F. (2020). *The Hindu Business Line*. Retrieved from https://www.thehindubusinessline.com/companies/myntra-looks-to-expand-overseas-presence/article32219459.ece

Ghosh, P. (2018). Competitive Strategy of Myntra. *Journal of Commerce and Accounting Research*.

https://startuptalky.com/myntra-subsidiaries-companies/. (n.d.). *startup talky*. Retrieved from https://startuptalky.com/myntra-online-fashion-store/#:~:text=Myntra%20has%20partnered%20with%20Aditya,store%20on%20November%2016%2C%202021.

Jain, S. (2019). Online fashion retailing in India: Trends, opportunities, and challenges. *Journal of Fashion Marketing and Management: An International Journal*.

Kushwaha, V. P. (2023). *Startup Talky*. Retrieved from https://startuptalky.com/myntra-online-fashion-store/

Microsoft. (2020). *Myntra accelerates its digital transformation journey with Microsoft Cloud*. Retrieved from Microsoft: https://news.microsoft.com/en-in/myntra-digital-transformation-microsoft-cloud/

Myntra. (n.d.). *Myntra blogs*. Retrieved from https://blog.myntra.com/

Nair, A. (2016). *your story*. Retrieved from https://yourstory.com/2016/07/myntra-acquires-jabong

Sharma, N. (2021). Understanding the dynamics of counterfeit products in the online retail industry: a study of Myntra.

Rashi. (2022). Reverse logistics VS forward logistics: understanding the difference. Retrieved from shiprocket fulfillment: https://fulfillment.shiprocket.in/blog/reverse-logistics-vs-forwardlogistics/#:~:text=Reverse%20Logistics,What%20is%20Forward%20Logistics%3F,and%20shipping%20to%20the%20consumer.

6. Effect of CRM on retail POS (point of sale) system

Mantosh Mahapatra, Parag Raut, and Aksh Dharaskar
Woxsen University, Hyderabad, Telangana, India

ABSTRACT: One of the fastest expanding industries in India is retail. To keep customers, retailers are implementing a variety of CRM activities. Most of the organized retailers in India use effective loyalty programs to promote repeat business. CRM programs offer a wealth of customer expectations, attitudes, and behavior data. This data may be utilized to improve and plan client retention efforts. In this field, state-of-the-art technologies are still not particularly common in India compared to other developed economies. The purpose of this study is to look into the influence of Customer Relationship Management (CRM) on retail point-of-sale (POS) systems. A Systematic Literature Review (SLR) technique was used to collect and analyze relevant papers in order to achieve this aim. The study examines the theoretical basis of CRM and POS systems, their interdependence, and the probable benefits of merging CRM and POS systems in retail contexts. According to the SLR results, CRM integration with POS systems may boost customer happiness, loyalty, and continuation, as well as improve productivity and profitability. The report finishes by discussing the SLR findings' implications for retailers and additional study prospects. Store management, transaction management, customer retention, and operations management are the four categories addressed.

KEYWORDS: Geospatial analytics, mPOS terminal, RFID technology, customer data, dataanalysis

1. INTRODUCTION

The retail sector is responsible for most of a country's economic growth and success. The retail sector is one of the most lively in the country, and it is noted for its quick development as a result of a significant number of new entrants. The retail industry is primed to emerge as the future's promise dueto its growth potential. Retailing is all about a customer's direct interaction with a shop. The

Delhi High Court defined "retail" as a sale for ultimate consumption or sales to the final client (ssss, 2019).

When conducting more business with current customers, remember that "the customer is a leader." A management concept that asserts that the best approach to achieve a company's goals is to recognize and satisfy the requirements and desires of its customers, both directly and implicitly.

The contemporary customer has grown substantially, and as a result, so have their options; the client is now more concerned with a great buying experience than with cost. Pricing is no longer the primary drive and differentiator for customers. In today's competitive environment, if a consumer isn't adequately happy with a business, the customer is unlikely to return. The importance of client retention and loyalty is demonstrated here. Consumer preferences are moving as a result of a continually changing environment of economic, marketing, and demographic forces. Big multinational firms compete in today's retail markets, and customer relationship management systems are their primary weapon. CRM operations are a complete strategy of recruiting, retaining, and developing a client base.

CRM may be a cross-functional, customer-driven, and technologically integrated management of business operations strategy that strengthens relationships (Chen and Popovich, 2003). Customer relationship management (CRM) is the firm's effort to establish a long-term, financially advantageous relationship with its customers for the benefit of both the customer and the company. CRM is defined as the infrastructure that allows the identification of and growth in customer value, as well as the appropriate tools to drive valuable consumers to remain loyal – indeed, to buy again (Berkowitz, 2006; Dyché, 2002). This is divided into two categories: analytical and operational. The distinction is that operational CRM focuses on acquiring data, planning for it, and implementing strategies at customer touchpoints, or locations where the customer and business interact. Analytical CRM is largely concerned with data processing and interpretation, as well as strategy development in the back office. CRM programs aid in the acquisition and retention of lucrative clients, the expansion of the customer base, and the use of pricing signals to entice less profitable customers to stay and become profitable (Milakovich, 1995).

2. APPLICATION OF CUSTOMER RELATIONSHIP MANAGEMENT (CRM) INITIATIVES IN RETAIL

CRM initiatives in particular necessitated taking into account the complete client base. By utilizing CRM methods, a firm may employ its core competencies to provide high-value services to a broader client base (Association of Traders of Maharashtra v. Union of India, 2005). CRM helps firms to better understand customer wants, manage connections more effectively, and plan for the future (Chen and Popovich, 2003). CRM is built on the premise that once a client is secured, keeping in touch with them will benefit both the firm and the customer, resulting in a win-win situation. Companies may be able to enhance recurring business, cross-sell possibilities, and obtain extensive customer information for future CRM projects by implementing loyalty programs (Berkowitz, 2006).

Customer relationship management is the establishment, expansion, maintenance, and optimization of long-term, mutually beneficial interactions between customers and businesses (Dyché, 2002). Based on their defensive character and long-term emphasis, loyalty cards differ from other sales promotional methods used as CRM tools (Milakovich, 1995). Customers benefit from participating in such programs since they may be rewarded for their loyalty. Gmarket is the firstgrocery chain to employ loyalty cards (in 2001). According to previous studies, financial incentives and emotional attachment have a good influenceon client retention (Geib et al., 2006).

To strengthen CRM activities, a well-organized business must first collect consumer data. They accomplish this by utilizing the membership card system, which provides all critical customer information. To appreciate client expectations and aspirations, the organization should take advantage of the information processes and capabilities at its disposal (Dominici and Guzzo, 2010). Personal informationfrom customers is used to notify them of any special offers or limited-time discounts given by the shop. Researchers were drawn to this sector to do a study on the creation of one-of-a-kind, inventive, and effective loyalty programs (Liu, 2007), as well as program service quality criteria (Fox and Stead, 2001).

Businesses may now collect massive amounts of data on individual customers. It is feasible to measure client attitudes, preferences, and expectations by tracking client purchases and responses to marketing initiatives. Tesco intends to use its Loyalty Card as a significant instrument in its marketing strategy (Sharp and Sharp, 1997). Tesco is one of the prosperous retailers that makes extensive use of customer information. Tesco is frequently cited as a successful case study in textbooks and business journals (Verhoef, 2003). A loyalty program, according to the work, is a perk that allows consumers to earn points for recurrent business with a firm (Vorhies and Morgan, 2005).

Businesses should utilize the information processes and capabilities at their disposal to understand their customers' needs and aspirations (Bridson et al., 2008). This will only enhance their efficacy and efficiency in establishing and maintaining solid client connections. A CRM workflow can easily be stated as follows: It iscritical to capture basic client information such as name, address, gender, and age. Yet, transactional information such as the date, time, item, and value, among other things, is critical at every point of interaction with the client. Information is usually necessary to complete this data and establish a meaningful link to the applications. Marketers use this data and information to follow customer interests and preferences, which must be thoroughly investigated. Furthermore, they make an effort to link purchasing trends to transaction data. CRM initiatives help firms to collect relevant customer-related data. The emphasis then switches to learning how to delight them and determining how and why consumers engage with the firm. CRM efforts improve contact between organizations and customers by establishing technological and non-technical communication networks.

To analyze the pattern of buying, modern cutting-edge technologies such as RFID and PSA are used. Throughout the purchase process, the customer's journey is scanned with this technology. This makes it easy to display items along the intended path and is likely to attract more attention. The success of modern organized retail enterprises is primarily linked to the convergence of technology, personalization, and, as a result, customization.

The application of complicated modelling and data analysis tools, as well as behavior prediction, to estimate future consumer behavior using prior customer behavior data. Marketing strategy is primarily reliant on the findings of product propensity-to-buy analytics and product affiliation research, which demonstrate that certain items are commonly purchased by one type of consumer while other specific products are frequently purchased by another type of client. CRM data mining activities provide merchants with insights and information about their most important customers, resulting in increased revenue (Omar et al., 2009).

If the merchant receives up-to-date information from the store on loyalty programs and special offers, the merchant will be able to create a strong relationship with the consumer and, as a result, influence the customer's purchasing decisions (Humby and Hunt, 2003). CRM initiatives enable the organization to accelerate this process by identifying smaller, more homogenous groups of clients, sometimes known as consumer segments or sub-segments. By identifying the client sub-segments and understanding their desires, the merchant may make tailored interactions with customized offers that more closely fit the expectations of each subgroup. While developing such interactions, the necessity that they adhere to the overarching brand concept may occasionally provide a dilemma. This is because a store that does not do so will mislead customers about the brand's

Unrivalled personnel service is one of the characteristics that contribute to a service's quality (Ibid.). Furthermore, scientific research demonstrates the importance of staff service quality in a range of service scenarios, including deal-related quality (Supra 8).

Several consumers say that there is a little gap between providers due to a lack of customized care and attention (Supra 12). Finally, emerging notions of relevance through branded relationships and targeted interactions with members make the new portrayal of loyalty programs more unsettling (Chevalier and Mayzlin, 2006). It is critical to build an online relationship with customers since their profit tends to improve with time and they have a history of making significant purchases as their confidence develops (Bellizzi and Bristol, 2004).

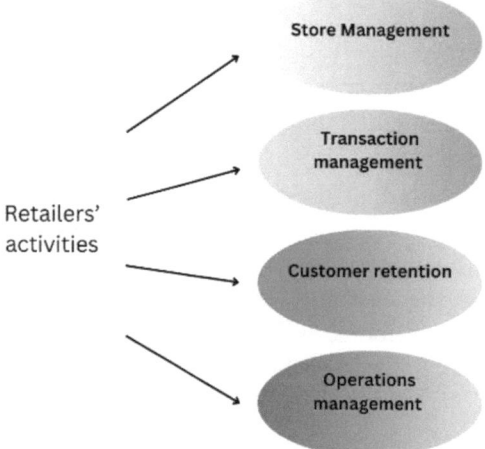

Retailers' main BI&A applications and their value creation mechanisms

3. STRUCTURED TRENDS OF CRM IMPLEMENTATION

3.1. Store Management

3.1.1. Data capture

For *text-based data capture*, store managers can use point-of-sale (POS) systems, inventory management software, and customer relationship management (CRM) software. These tools allow the store to track sales, inventory levels, and customer behavior, which can be used to make informed decisions about product offerings, pricing, and promotions. Text-based data capture can also include manual data input,such as recording customer feedback or noting product defects.

Video-based data capture can also be used in-store management to monitor customer behavior, identify potential security issues, and improve store layout and merchandising. Video cameras can be strategically placed throughout the store to capture footage of customers as they browse, helping store managers identify popular areas and products and areas that may need improvement. Video data can also be used to analyze customer behavior, such as how long they spend in certainstore sections or how they interact with specific products.

4. DATA ANALYSIS

Deep learning machine learning uses artificial neural networks to model and solve complex problems. In-store management, deep understanding can be used to analyze and identify patterns and data sets, and insights that can be used to optimizestore operations, improve inventory management, and personalize the customer experience.

Store optimization refers to the use of technology and data to improve the overall performance of a retail store. This can include everything from optimizing store layout and merchandising to improving supply chain management and reducing waste.

Geospatial analytics involves the use of location-based data to analyze patterns and trends. In-store management and geospatial analytics can be used to analyze foot traffic, identify popular areas of the store, and optimize store layout and productplacement.

RFID is a technology that uses radio waves to track and identify objects. In-store management, RFID can be used to track inventory in real time, reducing the need formanual inventory checks and improving inventory accuracy. RFID can also personalize the customer experience by allowing retailers to track customer behavior and preferences and offer personalized recommendations and promotions.

5. ACTION

Checkout lines: By capturing and analyzing data on checkout lines, retailers can identify and mitigate bottlenecks that cause customer delays and frustration. For example, retailers can use data on checkout line length and wait times to optimize staffing levels and improve customer service. They can also use data on the types ofproducts commonly purchased to optimize the checkout process and increase sales.Retailers can increase sales and revenue by optimizing checkout lines and creating personalized product recommendations. Customers are more likely to make additional purchases and return to the store if they have a positive checkout experience and receive relevant product recommendations.

Product recommendations: By capturing and analyzing customer behavior and preferences data, retailers can create personalized product recommendations that improve the shopping experience and increase sales. For example, retailers can usedata on customer purchase history and browsing behavior to recommend products to customers. They can also use customer demographics and location data to customize recommendations based on local trends and preferences. By optimizing checkout lines and creating personalized mobile sites, retailers can improve the customer experience and build loyalty. Customers are likely to return to the store if they have a seamless and personalized shopping experience.

Personalized mobile site: By capturing and analyzing data on mobile browsing behavior, retailers can create personalized mobile sites that improve the shopping experience and increase sales. For example, retailers can use customer browsing history and location data to recommend products and promotions tailored to the customer's interests and location. They can also use data on mobile browsing behavior to optimize the mobile site layout and improve the site's usability. By capturing and analyzing data on checkout lines, product recommendations, and personalized mobile sites, retailers can identify areas where they can better allocate resources to create more value. For example, they can optimize staffing levels based on checkout line data and allocate marketing resources found on customer browsing behavior.

6. MECHANISM TO CREATE VALUES

Leveraging existing assets: By capturing and analyzing customer behavior and store operations data, retailers can identify areas where they can leverage existing assets to create more value. For example, retailers can use data on customer preferences to optimize product placement and improve sales. They can also use data on foot traffic and customer behavior to optimize store layout and enhance the shopping experience.

Better allocation of resources: By capturing and analyzing data on store operations, retailers can identify areas where they can better allocate resources to create more value. For example, retailers can use sales trends and customer behavior data to optimize staffing levels and improve customer service. They can also use data on inventory levels and product demand to optimize restocking and prevent stockouts.

6.1. Transaction Management

6.1.1. Data capture

Transaction management in CRM (customer relationship management) retail-based POS (point-of-sale) systems based on *text data* can help retailers better understand their customers and optimize their business operations.

When a customer makes a purchase at a retail store, the POS system captures data on the transaction, including the items purchased, the time and date of the transaction, and the payment method used. This data can be used to build a customer profile, which can include information such as purchase history, preferred payment methods, and demographic information.

By analyzing this data, retailers can gain insights into their customer's buying habits, preferences, and behavior. This can help them to target their marketing efforts better, improve their product offerings, and optimize their business operations.

Transaction management in CRM retail-based POS systems can also help retailers to manage their inventory more effectively. By tracking sales data in real-time, retailers can identify trends and patterns in customer buying behavior, allowing them to make more informed decisions based on facts about inventory management and product ordering.

In addition, transaction management in CRM retail-based POS systems can help retailers to provide a more personalized shopping experience. Using the data captured through the POS system, retailers can offer targeted promotions and product recommendations to individual customers based on their purchase historyand preferences.

Overall, transaction management in CRM retail-based POS systems based on text data capture can help retailers improve their business operations, better understand their customers, and provide a more personalized shopping experience.

7. DATA ANALYSIS

Transaction management on CRM using machine learning, data optimization, cloudcomputing, predictive technology, and RFID can help retailers to optimize their business operations and improve the customer experience.

Machine learning algorithms are used to analyze transaction data captured by the CRM system, identifying patterns and trends in customer behavior. This can help retailers to better understand their customers, personalize their marketing efforts, andoptimize their inventory management and supply chain operations.

Data optimization techniques such as data cleansing and normalization can be used to ensure that the transaction data captured by the CRM system is accurate and consistent. This can help retailers to make more business decisions based onreliable data.

Cloud computing can be used to store and process large amounts of transaction data captured by the CRM system. This can provide retailers with real-time insights into customer behavior and business operations, allowing them to make more informed decisions and react quickly to changing market conditions.

Predictive technology can be used to anticipate future trends and customer behavior based on historical transaction data. This can help retailers to optimize their product offerings, marketing campaigns, and business operations.

RFID technology can be used to track inventory in real time, allowing retailers to manage their supply chain operations more efficiently and reduce the risk of stockouts or overstocks. RFID can also provide customers with a more personalizedshopping experience by allowing retailers to track customer behavior and offer targeted promotions and product recommendations.

Overall, transaction management on CRM using machine learning, data optimization, cloud computing, predictive technology, and RFID can help retailers to optimize their business operations, improve the customer experience, and stay competitive in an increasingly digital retail environment.

8. ACTION

Cash registers: Many retailers, particularly smaller businesses, still use traditional cash registers. They are standalone devices allowing cashiers to input transaction data and process payments.

Barcode scanners: Barcode scanners scan the barcode on a product, allowing thePOS system to retrieve product information and pricing automatically.

Card readers: Card readers are used to read credit and debit cards, allowing customers to make electronic payments.

Touchscreen displays: Touchscreen displays are increasingly common in modernPOS systems, providing a more intuitive and user-friendly interface for cashiers.

Mobile devices: Many retailers now use mobile devices such as smartphones and tablets as part of their POS systems, allowing employees to process transactions andmanage inventory from anywhere in the store.

POS Software: POS software is used to process transactions and manage inventory. It typically includes features such as sales tracking, inventory management, and reporting. E-commerce integration software: E-commerce integration software integrates POS systems with online stores, allowing retailers tomanage inventory and process orders across multiple channels.

9. MECHANISM TO CREATE VALUE

Increased brand recognition: A CRM system can help create value by improving brand recognition through targeted marketing campaigns, personalized communications, and social media engagement. By analyzing customer data, companies can identify trends, preferences, and behaviors to create more effective marketing strategies and build stronger brand loyalty. Additionally, CRM systems canhelp companies track and respond to customer feedback, which can enhance brand reputation and perception.

Accelerating sales and optimizing business performance: Another way to createvalue in CRM is by streamlining sales processes and optimizing business performance. A CRM system can help businesses track customer interactions and sales cycles, identify opportunities for cross-selling and upselling, and automate key sales tasks such as lead generation and follow-up. This can lead to increased efficiency, productivity, and revenue generation.

Centralizing customer data: CRM systems can provide value by centralizing customer data, making it easily accessible and actionable for sales and customer service teams. This can help companies better understand their customers, includingtheir preferences, behaviors, and purchase history. By leveraging this information, businesses can improve customer service, tailor their offerings to customer needs, and develop more effective marketing strategies.

Customizable payment solutions: Some CRM systems offer customizable payment solutions, which can create value by simplifying customer payment processes and reducing payment processing costs for businesses. These solutions can include features such as online payment portals, mobile payments, and recurring billing options. Companies can improve customer satisfaction and retention by offering convenient and flexible payment options while reducing administrative overhead.

mPOS terminal market growth: Mobile Point of Sale (mPOS) systems are becoming increasingly popular in retail and service industries, providing value by allowing businesses to accept payments on the go, reducing checkout wait times, and improving customer service. CRM systems can integrate with mPOS solutions, providing additional value by centralizing payment and customer data, automating inventory management, and simplifying sales reporting and analysis. As the mPOSmarket continues to grow, businesses that invest in CRM and mPOS solutions can gain a competitive edge by offering a more seamless and convenient customer experience.

9.1. Customer Retention

9.1.1. Data capture

When utilized in conjunction with a Customer Relationship Management (CRM) system, datacollection through video, audio, and text may be a potent tool for client retention. Here are a few techniques for doing this:

Video: Utilize video to record customer comments, endorsements, and product demos. Youmay utilize these movies to inform marketing campaigns, sales pitches, and customer support initiatives by integrating them into your CRM.

Audio: When recording customer service contacts, such as phone conversations and voicemails, use audio. You may see patterns and areas for development in your customer service division by examining this data. The audio may also be used to customize consumerinteractions, for example, by bringing up prior talks.

Text: Text may be used to record consumer information such as contact details, pastpurchases, and preferences. Product suggestions, customer support encounters, andtargeted marketing efforts may all benefit from this data.

You may get a thorough grasp of your clients and their demands by combining these three different sorts of data-collecting techniques. You may use this to find potential customers forupselling, cross-selling, and retention campaigns. Also, it can assist in the early detection ofpossible concerns, which can enhance client loyalty and satisfaction.

Each organization must prioritize customer retention since it is more economical to keep existing clients than to get new ones. By strengthening customer experience and loyalty, data analysis employing process optimization, AI, and machine learning may significantlyimprove customer retention.

To employ data analysis, process optimization, AI, and machine learning for client retention when utilizing CRM follow these steps

Data organization and collection: The first stage is to compile any customer information that is readily available in the CRM system, including purchase history, transaction information, demographic information, and customer feedback. This information has to be set up and arranged so that it can be examined quickly.

Determine key performance indicators (KPIs): KPIs are crucial measurements that gauge how well a company is performing in achieving its objectives. Customer happiness, churn rate, lifetime value, and customer engagement are some typical KPIs for customer retention.

Evaluate client behavior: By utilizing AI and machine learning, you may examine consumer trends and behavior to pinpoint the elements that have an impact on repeat business. Predictive analytics, for instance, may be used to spot clients who are likely to leave and take preemptive steps to keep them.

Enhance customer experience: By streamlining and improving the customer journey through process optimization. This might involve enhancing customer service, lowering wait times, and improving website navigation.

Personalized communication: Use AI to make consumer communications more relevant to their preferences, past purchases, and behavior. This might involve creating tailored email campaigns, targeted advertising, and personalized suggestions.

Introduce a loyalty program: Introduce a program that rewards patrons for their continued business. Employ machine learning to determine which incentives and rewards are most successful for certain client categories. Results should be measured and tracked over time. KPIs may be used to track and measure the success of your customer retention initiatives. Use the knowledge you have obtained from this study to gradually hone and improve your strategy.

In conclusion, by boosting customer experience and loyalty, data analysis employing process optimization, AI, and machine learning may assist enhance customer retention. You can develop a data-driven strategy for customer retention that will help your company in the long run by defining key performance indicators, reviewing customer behavior, improving the customer journey, customizing communication, putting a loyalty program in place, and tracking outcomes.

10. MECHANISMS TO CREATE VALUES USING THE ACTIONS

In retail management, location-based technologies may significantly enhance client retention. Retailers may track consumer movements and offer location-based messaging or promotions by utilizing GPS or Bluetooth beacons. This can assist shops in developing a customized and pertinent purchasing experience that entices customers to come back.

These are some examples of how location-based technology might be utilized in retail management to increase client retention:

Employ location-based technologies to deliver clients targeted discounts and promotions based on their present location. For instance, a consumer may get a push notice offering adiscount on their subsequent purchase if they are close to a business.

Employ geofencing: To give clients customized messages when they enter or leave acertain place. This can entail extending a warm welcome to them or giving them driving instructions to the closest store.

Making tailored product suggestions depending on the customer's present location by using location data. For instance, a consumer in the electronics department of a store may get suggestions for similar goods.

Personalized help: may be given to customers by tracking their movements inside the business using location data. For instance, a sales associate may be informed to offer more assistance if a customer is spending a lot of time in a certain area of the business.

Loyalty programs: Track customer visits using location data, then reward or discountthem based on how frequently they visit. This can boost client retention and encourage repeat business.

Use mobile applications: to give clients location-based information, such as the whereabouts of items in-store or the standing of their online purchases. This can enhancethe general shopping experience and motivate people to purchase again.

In conclusion, location-based technology has the potential to significantly increase client retention in the retail industry. Retailers may provide customers with a more customized and relevant purchasing experience that motivates them to come back by implementing promotional campaigns, geofencing, tailored suggestions, enhanced customer service, loyalty programs, and mobile applications. Retailers may develop a data-driven strategy for customer retention that can lead to long-term success by utilizing CRM data and location-based technologies.

10.1. Operation Management

10.1.1. Data capture using audio, video and text

When utilizing CRM for retail management, data capturing through video, audio, and text may be a potent tool. Here are some strategies for utilizing these technologies to collect dataand enhance the client experience:

Video: Businesses may monitor consumer behavior in-store using video and gather information about how they engage with displays and merchandise. For instance, video cameras may be positioned in key areas to watch foot traffic and observe how visitors navigate the business. The layout of stores may be optimized using this data, which wouldalso enhance the entire shopping experience.

Audio: Businesses may utilize audio to record consumer thoughts and comments. Voice assistants, chatbots, and other communication channels can all be used for this. Retailersmay enhance their product choices and customer service by studying these encounters tolearn more about client preferences and pain areas.

Text: Text-based platforms including social media, email, and messaging applications allow retailers to collect data. Retailers may monitor these platforms for customer feedback in real-time and promptly address questions and concerns by doing so. As a result, consumer loyalty and satisfaction may increase.

All things considered, data capturing utilizing text, audio, and video might be a useful tool forCRM operation management in retail management. Retailers may improve the customer experience and spur corporate growth by utilizing these technologies to get important information about consumer behavior and preferences.

11. DATA ANALYSIS USING PROJECT MANAGEMENT CONNECTED DEVICE DATA

Using project management, analyze data Using connected device data from CRM in retail management entails leveraging such data to enhance the efficiency of a retail company. To implement this strategy, data must be gathered from a variety of linked retail-related devices,including point-of-sale (POS) systems, inventory management systems, customer loyalty programs, and other sensors.

To put this strategy into practice, a project management framework may be used to plan andcarry out the project. The structure ought to incorporate the following crucial actions:

Plan the project's goals, parameters, and schedule. Determine the measurements, data sources, and key performance indicators (KPIs) that will be employed to gauge the project'ssuccess.

Data gathering and analysis: Gather and examine data from connected devices to obtainan understanding of the functioning of the retail industry. To find patterns and trends, this may include employing technologies like data mining, machine learning, and statistical analysis.

CRM implementation: To keep track of customer contacts and transactions, adopt a customer relationship management (CRM) system. To offer a holistic view of the customerjourney, the CRM system should be linked with the data gathered from the connected devices.

Setting up a system to measure KPIs and provide reports will give stakeholders visibility intothe project's development and outcomes.

Use the learnings from the data analysis to continually enhance the performance of yourretail business. This may entail altering the design of the store, the products it sells, the prices it charges, the promotions it runs, and other elements of the company.

Retailers may leverage CRM and linked device data to better understand consumerbehavior and run their businesses by using this project management approach.

11.1. Cloud technology and predictive technology

Using project management, analyze data Businesses aiming to enhance their operations and customer experience can greatly benefit from employing cloud technology and predictive technologies in retail management.

Businesses may manage their data analysis projects more effectively with the use of cloud project management solutions. Teams may work together on projects, monitor development in real-time, and make sure that deadlines are reached. Also, this technology may aid in data management and storage, making it simpler to access and analyze the data.

CRM in retail management may use predictive technologies to help organizations make more educated decisions about their clients. Businesses may forecast client requirements and preferences by evaluating customer data, such as purchase history and surfing habits. They can target their marketing efforts and offer experiences that are more customized as a result.

Overall, firms may enhance their operations, boost customer happiness, and spur revenue development by combining project management cloud technology with predictive technology employing CRM.

11.2. Project management machine learning, and process optimization

The critical elements of retail management include data analysis, project management, machine learning, process optimization, and CRM. When used together, they may assist retail firms in streamlining their processes, understanding their clients better, and boosting profitability. The following are some retail management applications for each of these elements:

Analyzing data: Retail companies produce a ton of information, such as sales numbers, consumer demographics, inventory levels, and more. Businesses may acquire insights into their operations and make data-driven choices by examining this data. For instance, they can assess which goods are doing well and which are not, which marketing initiatives are most successful, and how to best price their products.

Project management: Retail organizations sometimes have several ongoing projects, such as introducing a new product line, building a new location, or upgrading their website. These projects may be completed on schedule, within budget, and to the necessary quality standards with the help of good project management.

Machine learning: In retail management, machine learning may be used to automate procedures, increase the accuracy of forecasts, and improve customer satisfaction. For instance, companies may customize marketing efforts, forecast customer behavior, and improve pricing using machine learning algorithms.

Process optimization: By streamlining their operations, retail companies may save costs, boost productivity, and enhance customer happiness. They may, for instance, automate theircustomer service procedures and enhance their inventory management.

CRM: Systems for managing customer relationships may help firms get a deeper understanding of their clients, boost client happiness, and boost revenue. Businesses may tailor their marketing efforts, deliver better customer service, and find new sales possibilities by tracking client interactions and preferences.

Businesses may improve their competitiveness and long-term performance by utilizing analysis of data, project management, machine learning, process optimization, and CRM inretail management.

12. ACTIONS

There are various ways that CRM systems might gain from combining with companymanagement software and cameras in operation management:

Better customer service: Business management software may provide the CRM system with real-time information on inventory levels, shipping and delivery schedules, and other crucial parameters. Customer satisfaction may be raised by providing correct and timely information to customers via customer care professionals.

More sales opportunities: The CRM system may uncover consumer buying trends and behaviors by evaluating data from cameras used in operation management. Sales teams may use this information to customize their presentations and find new business prospects.

A rise in operational efficiency: Supply chains, industrial processes, and other operationsmay all be recorded on camera in operation management. This data may be utilized to enhance customer service and spot sales opportunities by interacting with the CRM system.

Improved customer insights: The CRM system may record a more thorough view of consumer behavior and preferences by connecting with business management programs and cameras in operation management. This information may be utilized to tailor marketing efforts, increase product selections, and improve customer support.

In general, firms may gain useful insights and possibilities for development by combining CRM systems with business management software and cameras in operation management.Businesses may improve operations, boost customer happiness, and boost revenue by utilizing these technologies.

13. MECHANISMS TO CREATE VALUES

13.1. Omni-channel sales network

Customer relationship management (CRM) is used in retail management to develop an omnichannel sales network. This process necessitates developing values for customers, staff, and the firm. Here are various methods for producing value in this situation:

Use CRM to collect client information and customize the purchasing experience across channels. Consider using a customer's purchase history and browsing habits to suggest goods that are likely to be of interest to them. Customers are more likely to return for more purchases when they feel appreciated and understood.

Integration: To provide a seamless customer experience, integrate all sales channels. Make sure clients can simply switch between channels, for as from online to in-store, without losing any information or functionality. This boosts client loyalty and pleasure.

Transparency: Provide customers access to real-time data on order status, shipment estimates, and product availability. This lessens uncertainty and worries while fostering client trust.

Collaboration: is encouraged throughout the firm to develop a consistent, customer-centric strategy among various departments and teams. Empower staff members to share consumer ideas and comments to promote innovation and ongoing development.

Employee empowerment: is the key to finding a fast and efficient solution to client complaints. Provide them with the instruction, equipment, and resources they require to provide first-rate customer service across all channels.

Data analytics: Make use of data analytics to understand client behavior and preferences and to pinpoint places where the sales network needs to be improved. Make data-driven decisions using this information to improve customer experience and increase revenue.

In general, a customer-centric strategy that combines many channels promotes cooperation and equips staff to provide great customer care is necessary to create value in an omnichannel sales network. Retailers can increase customer happiness, loyalty, and income by personalizing the shopping experience and learning about consumer behavior using CRM and data analytics.

13.2. Self-sustaining working environment

Customer Relationship Management (CRM) in retail management involves value creation for staff, clients, and the company to create a self-sustaining work environment. Here are various methods for producing values in this situation:

Training and development: Provide staff members with enough opportunity for training and development to advance their abilities. They can do their tasks more efficiently as a result,and it also gives them the chance to develop their careers.

Recognize employees: Employees that do exceptionally well and reach their objectives by implementing recognition and incentive programs. Employee engagement, motivation, and morale are increased as a result, which boosts output and increases work satisfaction.

Encourage employees: to share their experiences and suggestions for improvement in theform of feedback. This promotes an environment where staff members feel respected andthat their ideas are taken seriously.

Encourage staff cooperation and teamwork to accomplish shared objectives. Employee collaboration and a feeling of community are fostered as a result, which boosts productivityand job happiness.

Encourage people to take responsibility for their job and make choices that benefit the company and its clients. Employees become more trustworthy and confident as a result,which boosts productivity and job happiness.

Data analytics: Make use of big data to understand employee performance and engagementand to spot workplace improvement opportunities. Use this data to guide data-driven decisions that improve employee happiness and provide profitable outcomes.

Overall, focusing on staff training and development, incentives and recognition, feedback, cooperation, empowerment, and data analytics is necessary to create value in a self-sustaining working environment. Retailers may increase employee happiness, productivity,and retention, which will lead to a more enduring and prosperous firm. This is done by providing an atmosphere that empowers and inspires workers to achieve their objectives.

13.3. Maintaining benchmark for quality control

Utilizing CRM in retail management requires producing value for consumers and seeing to itthat their demands are addressed in order to maintain the standard for quality control in operations management. In this context, the following techniques can be utilized to producevalue:

Customer segmentation: By grouping clients according to their requirements, preferences,and purchasing patterns, businesses may offer tailored services and more focused marketing initiatives, which in turn can boost client happiness and loyalty.

Customized communication: Retailers may use CRM system data to offer consumers tailored messages via email, SMS, or social media, which can strengthen bonds and boostcustomer engagement.

Feedback from customers: Gathering consumer feedback may assist merchants in identifying areas for improvement and making the required adjustments to fulfil customer expectations. CRM systems may be used to track customer satisfaction ratings and collectfeedback. Retailers may discover areas for improvement and make adjustments to boost efficiency and quality by routinely examining and analyzing operational procedures and performance indicators.

Training and development: Offering employees the chance for training and development can assist them in gaining the abilities and information necessary to deliver high-quality customer service and enhance overall operational performance.

Retail stores can make sure that they are providing their customers with high-quality services by utilizing these mechanisms to maintain the benchmark for quality control in operations management using CRM. This can help to increase customer satisfaction and loyalty, which in turn may serve to promote company expansion.

13.4. Reducing operating costs

By utilizing CRM in retail management, operational expenses may be decreased by discovering methods to streamline procedures and boost productivity while maintaining a high standard of customer care. In this context, the following techniques can be utilized to produce value:

Process streamlining: Reducing waste and inefficiencies can assist in minimizing operating expenses by analyzing and simplifying operational processes. CRM systems may automate specific tasks like shipping and delivery and resource allocation, which can assist save the cost of manual labor.

Data analysis: Retailers may detect trends and patterns in consumer behavior by analyzing customer data from CRM systems. These patterns can then be utilized to inform data-driven choices and improve operational procedures so that they can better satisfy customer demands.

Predictive insights: Forecasts may be used to estimate demand and optimize inventory levels, which can assist save storage and carrying expenses while preserving product availability for consumers.

Supplier management: Reducing procurement costs and increasing profit margins can be accomplished by negotiating better terms and pricing with suppliers. CRM systems may be used to monitor supplier performance and spot potential cost-saving possibilities.

Employee engagement: inspired and efficient staff members can assist save labor expenses and enhance overall operational effectiveness. CRM systems may be used to monitor employee performance, offer suggestions and rewards, and boost morale.

By providing excellent client service, retailers may save operating expenses in operations management by utilizing CRM in retail management. This can assist increase profitability and assure the long-term viability of the company.

13.5. Better utilization of assets

Improved use of resources in supply chain management Employing CRM in retail management entails figuring out how to use resources – such as inventory, equipment, and personnel – most effectively and profitably possible. In this context, the following techniquescan be utilized to produce value:

Inventory management: may assist companies in preventing stockouts and overstocking, which can result in a loss of sales and an increase in storage expenses. Real-time inventorytracking and replenishment strategy optimization are both possible with CRM systems.

Maintenance of the equipment: Keeping the equipment in good condition will help it last longer and cost less to fix or replace. To reduce downtime and assure peak performance,CRM systems may be used to plan preventative maintenance and track equipment utilization.

Workforce management: Retailers may prevent staffing issues and staff shortages, which could also result in squandered labor expenses or missed sales, by effectively scheduling and managing workers. Systems for customer relationship management (CRM) may be used to monitor employee availability and skills and improve scheduling to guaranteeappropriate staffing levels.

Retailers may better match their offers and resources with consumer demands by understanding the behavior and preferences of their customers. CRM systems may be used to evaluate consumer information and find ways to employ resources more effectively,for as by enhancing product selections or shop designs.

Operational analytics: By analyzing operational data from CRM systems, merchants may find inefficiencies and areas where they can improve, including wait times or order fulfilmentdelays.

Retailers may optimize asset utilization in operations management by utilizing CRM in retailmanagement, which can assist boost productivity, saving expenses, and boosting profitability. In the end, this can aid merchants in maintaining their competitiveness in the market and giving their clients better value.

13.6. Improving user experience

Finding methods to make it simpler and more comfortable for customers to connect with thestore and make purchases is a key part of enhancing the experience for users in operationsmanagement utilizing CRM in retail management. In this context, the following techniques can be utilized to produce value:

Personalization: By utilizing CRM data, merchants may provide customers with a more tailored purchasing experience by proposing goods and services which are pertinent to their preferences and previous purchases. As a result, consumer loyalty and satisfaction may increase.

Support across several channels: Channels such as chat, email, and the phone, can assist clients in receiving the assistance they require in the manner that suits them best. Systems for customer relationship management (CRM) may be used to monitor customer experiences across channels and make sure that support requests are addressed effectively.

Checkout process simplification: By streamlining and streamlining the checkout process, you may decrease cart abandonment and boost conversion rates. The behavior of customers throughout the checkout process may be tracked using CRM systems to find areas for improvement.

Product availability: Giving customers actual details about item availability will assist them in making knowledgeable buying decisions and help them avoid being frustrated by out-of-stock products. Inventory levels may be monitored and recent data on product availability can be provided via CRM systems.

Establishing a loyalty program: This can inspire customers to make higher purchases and encourage repeat business. CRM systems may be used to monitor customer loyalty and offer incentives and rewards based on previous purchasing patterns.

Utilizing CRM in retail may help businesses boost customer happiness, brand loyalty, and profitability by enhancing the user experience for operations management. In the end, this can aid merchants in maintaining their competitiveness in the market and giving their clients better value.

13.7. Prioritizing work, survey progress and identifying problem areas

Applying CRM in store management involves discovering strategies to optimize resource allocation and pinpoint areas where improvements may be made. These tasks include prioritizing work, evaluating progress, and detecting problem areas. In this context, the following techniques can be utilized to produce value:

Task management: Retailers may make sure that work is performed on time and as effectively as possible by using CRM tools to monitor and prioritize activities. By doing so, you may prevent delays and make sure that assets are distributed properly.

Progress monitoring: By utilizing CRM systems to track task and project progress, merchants may identify areas for improvement and ensure that work is progressing as intended. This can help shops take corrective action and ensure that problems are rectified before they be more serious.

Communication: Ensuring that team members are communicating effectively and striving toward the same objectives may be achieved by using CRM tools. This can make things clearer and guarantee that work is done quickly.

Retailers may find patterns and chances for improvement by using CRM systems to evaluate data on consumer behavior and operational performance.

This may support data-driven decision-making on the part of merchants and guarantee that resources are spent in areas that will have the biggest impact.

Utilizing CRM in retail, merchants may prioritize tasks, monitor progress, and pinpoint trouble spots by employing these methods, which can boost productivity, cut expenses, and boost profits. In the end, this can aid merchants in maintaining their competitiveness in the market and giving their clients better value.

14. CONCLUSION

It is feasible to grasp the ambitions of consumers through customer relationship management in a way similar to how the product or service was first given to them. Another measure of customer satisfaction is work ethic pride, which often encourages employees to meet customers' requests and expectations. Yet, personnel errors, delays in the delivery of products or services, or other difficulties may harm a customer's future encounters with the organization. CRM issue identification and resolution help enhance customer efficiency ratios. This is because pleased customers often make more purchases than disappointed customers. As a result, the relative cost per client falls, resulting in positive economics and profits once more. Customer relationship management practices that increase business retention may greatly benefit a firm. According to actual evidence, it can help an organization's financial health and, as a result, its success. CRM strategies must have a customer-centric management approach that identifies and categorizes profitable customerswhile promoting satisfaction and adherence. This benefits both the organization and the customers. Because of its efficacy, it permits the efficient use of human and technological resources to fulfil even more customer desires, thereby serving a larger and better market. Retailers must prioritize delivering a cohesive, omnichannel customer experience, harnessing data-driven insights to personalize products, and continually enhancing customerengagement to create value through CRM and POS integration. Retailers must also spend on staff learning and instruction to ensure that these tools are used successfully.

Overall, integrating CRM and POS systems may assist merchants in gaining an edge overtheir competitors by delivering value for both consumers and the business. Retailers can improve the consumer experience, establish loyalty, and drive company success by using the power of technology and information.

REFERENCES

Association of Traders of Maharashtra v. Union of India. (2005). (79) DRJ 426.

Bellizzi, J., & Bristol, T. (2004). A review of grocery store loyalty programs in one important US market. *Consumer Marketing Journal*, 21(2), 144–154.

Berkowitz. (2006). *Customer relationship management. 8 Common goals for a CRM programme. What are key drivers of customer satisfaction?* Retrieved April 11, 2016, from: https://www.successcenter.com/.

Brady, M., & Cronin, J. (2001). A hierarchical approach to some new ideas on conceptualizing perceived service quality. *Journal of Marketing, 65*(3), 34–49.

Bridson, K., Evans, J., & Hickman, M. (2008). Assessing the relationship between loyalty programme attributes, store satisfaction and store loyalty. *Journal of Retailing and ConsumerServices, 15*, 364–374.

Chen, J. I., & Popovich, K. (2003). Understanding customer relationship management (CRM): People, process, and technology. *Business Process Management Journal, 9*(5), 672–688.

Chevalier, J., & Mayzlin, D. (2006). Online book reviews and the impact of word-of-mouth marketing on sales. *Marketing Research Journal, 43*(3), 345.

Dominici, G., & Guzzo, R. (2010). Customer satisfaction in the hotel industry: A case study from Sicily. *IJMS International Journal of Marketing Studies, 2*(2), 3–12.

Dyché, J. (2002). *The CRM handbook: A business guide to customer relationship management*. Reading, MA: Addison-Wesley.

Fox, T., & Stead, S. (2001). *Customer relationship management: Delivering the benefits, CRM (UK) and SECOR consulting, New Malden, White Paper*. Retrieved March 20, 2013, from: http://www.iseing.org/emcis/EMCISWebsite/EMCIS2011%20Proceedings/SCI10.pdf.

Geib, M., Kolbe, L., & Brenner, W. (2006). CRM collaboration in financial services networks a multi-case analysis. *Journal of Enterprise Information Management, 19*(5–6), 591–607.

Humby, C., & Hunt, T. (2003). *Scoring points: How Tesco is winning customer loyalty*. London: Kogan Page.

Ibid.

Javalgi, R. G., & Moberg, C. R. (1997). Service providers' consequences for customer loyalty. *Journal of Services Marketing, 11*(2–3), 165–180.

Liu, Y. (2007). The long-term impact of loyalty programmes on consumer purchase behaviour and loyalty. *Journal of Marketing, 71*, 19–35.

Milakovich, M. E. (1995). *Improving service quality: Achieving high performance in the public and private sectors*. Delray Beach, FL: St. Lucie Press.

Murphy, E. C., & Murphy M. A. (2002). *Leading on the edge of chaos: The 10 critical elements for success in volatile times*. Prentice Hall Press.

Omar, N. A., Nazri, M. A., & Saad, H. S. (2009). What customers really want: Exploring service quality dimensions in a retail loyalty programme. *UNITAR e-Journal, 5*(1), 68–81.

Rowley, J. (2005). The four Cs of a devoted consumer. *Marketing Planning & Intelligence Market Intelligence & Strategy, 23*(6), 574–581.

Sharp, B., & Sharp, A. (1997). Loyalty programs and their impact on repeat-purchase loyalty patterns. *International Journal of Research in Marketing, 14*(5), 473–486.

Sirohi, N., Mclaughlin, E., & Wittink, D. (1998). For a grocery retailer, a model of consumer perceptions and store loyalty intents. *Journal of Retailing, 74*(2), 223–245.

Supra 12.

Supra 8.

Verhoef, P. (2003). Understanding the effect of customer relationship management efforts on customer retention and customer share development. *Journal of Marketing, 67*, 30–45.

Vorhies, D. W., & Morgan, N. A. (2005). Benchmarking marketing capabilities for sustainable competitive advantage. *Journal of Marketing, 69*, 80–94.

7. A Case Study on Failed Startups: How & Why They Failed?

Pooja Chauhan, Naidu Pvs
Woxsen University

ABSTRACT:

INTRODUCTION: According to CB Insights, within 20 months, more than 70% of newly funded startups fail. When asked why their businesses failed, many business owners cited a lack of capital, entering the wrong market, insufficient research, poor partnering choices, inadequate marketing, and a lack of expertise as contributing factors.

OBJECTIVE: The study's goals are to (1) learn more about why so many new firms fail and (2) examine some instances of companies that never got off the ground.

METHODOLOGY: The researcher examines case studies of failed startups to determine what went wrong in their business failures.

RESULTS: Business founders cite a lack of capital, entering the incorrect market, an absence of research, poor relationships, inefficient marketing, and a lack of expertise in the field as causes of failure.

KEYWORDS: startups, business ideas

1. INTRODUCTION

Most people have a very wrong idea about how hard it is to start a new business. It's not often that a company is so well suited to its market that it can coast along with very little effort. So, business owners say that some of the reasons their companies fail are that they run out of money, go into the wrong market, do not do enough research, make bad partnerships, do not markettheir products or services well enough, or do not know enough about the industry (Bajwa et al., 2017). New enterprises are always more precarious than those that have been around for a while. CB Insights says that after 20 months, more than 70% of new businesses that have received funding have failed. According to another piece of research, among the new businesses in the United States: 20% ofcompanies fail during their initial year, 50% during the first five years, and 70% within the first 10 years of operation (Nair and Blomquist, 2019).

Chapter 7 DOI- 10.4324/9781003397175-7

But why do new businesses often fail? This topic, yet few provide responses that are supported by statistics. Our mission is to pinpoint and discuss the most important takeawaysfrom this experience.

1.1. Background

According to the results of a poll that was conducted with more than 150 company owners, around 70% of startup companies will end up failing. Within the first 25 months after beginning their firm, over 66% will face the possibility of their business failing. Nearly 77% of respondents who were in a position where they might fail claimed that it was at least partially due to COVID-19 (Bryan and Hovenkamp, 2020). The unfortunate reality of new businesses and new goods is that at least 3–4% of all new items fail. This has the potential to ruin people's savings, the cash flows and profit and loss statements of companies, and the confidenceof the teams involved (Yin *et al.*, 2019). So, business owners say that some of the reasons their companies fail are that they run out of money, go into the wrong market, do not do enough research, make bad partnerships, do not market their products or services well enough, or do not know enough about the industry (McDonald and Eisenhardt, 2020). Setting objectives, doing correct research, liking the job you do, and being committed are all effective ways to reduce the likelihood of failing.

2. LITERATURE REVIEWS

Kalyanasundaram (2018): In entrepreneurial ecosystems, 90% of new businesses fail in their first year, which makes each failed business an orphan. Due to this tendency, businesses in the ecosystem face greater consequences for failure. extensively studied, but the lessons that can be learned from failed businesses and the direction they may provide to the ecosystem are less well known. Successful business research has been extensively studied, but the lessons that can be learned from failed businesses and the direction they may provide to the ecosystem are less well known. Using a case study method, we can tell the difference between successful and unsuccessful technology startups in Bangalore by looking into the personalities of the founders and the inner workings of the companies to figure out what went wrong. We look at one type of exit strategy that founders of failed startups used so that we can learn important things about why their businesses failed. Researchers found that the most important differentiators between successful and unsuccessful startups are the following: the time to a minimum acceptable product cycle, the time for revenue realization, the complementary skillsets of the founders, the age of the founders with their domain expertise, the personality type of the founders, the attitudetowards financial independence, and the eagerness to avail support and guidance at critical stages. So, lessons have been learned that could help people who want to start their own businesses make less money from their mistakes.

Giardino *et al.,* (2014): It is common knowledge and widely appreciated that a select group of high-tech startup companies has been instrumental in fostering technological advancement and economic expansion in recent decades. But it is also common knowledge that there's a high failure rate among startups because they take on too much risk for too little profit. Thus, it is intriguing to note that the literature mostly ignores the multiple lessons that may be gained from investigating the experiences of failed businesses in favor of focusing on successful firms and quantitative studies searching for drivers of success. To address this gap and make a meaningful contribution to the field, this study proposes a scalable and repeatable approach for extracting startup failure patterns from unstructured post-mortem document collections. The application of this approach to a massive database of 214 post-mortem reports from startups is a further potentially related contribution. Statistics describing the causes of startup failure reveal the need of a well-defined business development plan.

Cantamessa *et al.,* (2018): Startups in the software industry are young businesses with a focus on developing innovative products. Startups need efficient procedures to deal with their specific issues, since resources and time are highly limited and a single unsuccessful project might put the company out of business. Yet, very few scientific research have even attempted to examine failure's early-stage features. Our hope is that this research will help us better grasp the reasons behind the demise of so many software startups in their formative stages. This inquiry of the current state of the art included first doing a literature review, and then employing a multiple- case study methodology. The findings, presented in a behavioral framework, show how misalignment between management plans and their implementation may lead to failure.

Strategies highlight the importance of learning about the issue and how to best solve it, but in practice, product development is often prioritized for a rapid market launch to test product/market fit.

2.1. Research Gap

The researcher went on recent business research topics for a business research paper and discovered that the most important topic of today is running startups and their failure. Startups are businesses that launch a new product or service in response to market demand. In today's time, there are a number of startups on the market, but they tend to fail in the first year or fourth year. The topic has very little research and a lack of information regarding why startups fail so vigorously. Therefore, the researcher takes up this research topic to meet the gaps and provide an in-depth explorative method with a few case studies of some failed startups.

2.2. Research Questions

1. What factors cause certain startup businesses to cease operations?
2. Why do so many startups fail during their initial year in operation?

The researcher takes up the above research questions to find out/to take case studies of some start up and look at their reasons of closure and the researcher plans to study some of the failed startups to high light few common points that entrepreneurs make in their startups.

2.3. Importance of the Study

The chapter has its own importance in the present time, as people are coming up with new startups every day, but they are failing simultaneously. Therefore, this chapter uses case studies of some failed startups to study the reasons for their failure and understand some of the major reasons why the startups are failing constantly. The researcher provides an in -depth explorative analysis of the reasons for failure of startup and discusses some recommendations. This chapter can be used by business personnel as a reference paper and for enthusiastic readers or entrepreneurs who want to start their own startup in the future. However the main and foremost reason of the startups being failed is entering into wrong market or not knowing their target audience or assuming that their products or services would be liked by the consumers without having any market research, opening a business is easy but sticking to and making a profit out of it needs experience, expertise and research.

2.4. Research Objectives

The following research objectives are undertaken for the study-

- To get an understanding of the reasons why so many businesses just getting started fail.
- To study case studies of companies who were unable to launch their operations successfully.
- The purpose of this article is to shed light on some of the most common reasons of business failure in startups.

2.5. Scope and Limitation

The purpose of this chapter is to figure out why a new business fails. In today's time, most people want to start their own business, but they do not know how to start, how to survive, or what to present to the market. This chapter provides a scope for learning about the failure of startups, and the researcher provides an in-depth analysis of some case studies to underline the failure reasons. However, this research has limitations as it uses the study methods and qualitative approach and therefore, the results are oriented on the basis of the information gathered by the researcher. The results are not accurate and should only be used for knowledge enhancement. In addition, this research study is limited to understanding the reasons for failure of startups only.

3. RESEARCH METHODOLOGY

The research methodology is the most important section of a research paper; it provides the ideas of the methods to be used and the process to be followed for collecting data and analyzing the data for further results and interpretation. Here, the researcher uses a secondary method, and collects data through a case study method (Pandey and Pandey, 2021). The researcher examines case studies of failed startups to determine what went wrong in their business failures. The researcher gets the information from secondary sources because it is not possible to get it from first-hand sources. The researcher gathers information from authentic business sites, new publications, and journals and articles. The research methodology helps the researcher to plan how to investigate the data, here, the researcher collects the information of different startups andstudies the reasons for their failure. It would assist readers in comprehending the researcher's approach and strategy in reaching the final conclusion.

3.1. Research Method and Design

The research methods employ the techniques through which the data is collected by the researcher. According to the research questions and objectives, the research employs qualitative research methods and conducts case studies of various failed startups to understand their failure in the market. The case studies helped the researchers understand the common reasons for startup failure in the business world. The research design shows the nature of the analysis process, here the researcher takes explorative research to provide an in-depth analysis of the reasons (Mishra and Alok, 2022) The explorative design is the simplest and best way to describe and analyze data in a research; in this case, the researcher is trying to figure out why some startups fail in their early years. The explorative design would conduct analysis in its original form and provide real findings.

3.2. Research Approach

The researcher here opts for a case study research approach to provide the real life context of business startup failure. The researcher studies some failed startups to understand thecommon reasons for their business failure (Zhang, 2022). This provides critical analysis of different startups cases and can be used for future business entrepreneurs as a lesson for their business decisions. The case study approach provides a different perspective to the chapter, it includes the statistical observation as well as conclusions from the information gathered from the case studies.

4. ANALYSIS OF STUDY

4.1. What factors cause certain startup businesses to cease operations

The case studies of some failed startups and reasons for their failure:

1. **IProf**—With its seven-inch touch screen, iProf makes available a variety of instructional materials such as films, digital notes, and quizzes (Singh, 2022). The program, which costs Rs.14,990, is marketed as a supplemental learning module that provides pupils with a means of reviewing the material covered in class.
   What went wrong with iProf?

   - Schools' unwillingness to adapt their curriculum.
   - The internet is a treasure trove of free resources.
   - A competitor's rising popularity and market dominance, exemplified by Byjus, ultimatelyproved fatal.

2. **BabyBerry**—The startup offered services centred on children's health and wellbeing, such as a digital vaccine, health records and nutrition management, and doctor discovery. The service provided individualized information from professionals in fields including pediatrics, nutrition, and psychology.
   To what end did BabyBerry's efforts fail?

   - While costs went up just twice as much, revenue dropped by a factor of three for the firm.

   Unfortunately, there was no viable business model supporting it, and people just acceptedthe free services as is.

Ran out of money	37%
No financing/investor interest	31%
No business plan or model	25.5%
Lost focus	23.9%
Not the right team	22.8%
Was outcompeted	21.7%
No market need	20.1%
Pricing/Cost issues	19.8%
Poor marketing	19.8%

Failure to pivot	19.3%
Poor UX	18.8%
Ignored customers	17.9%
Failed geographical expansion	17.7%
Legal challenges	16%
Disharmony among team/investors	16%
Unsuccessful pivot	13.9%

(Reasons for startup failure)

3. **Flashdoor**—In a single button press, Flashdoor delivers clean clothes right to your door. It creates a centralized hub for service providers, clients, and delivery personnel to interact. What went wrong with Flashdoor?

   • Difficulty in turning a profit due to high operating expenses and shipping fees.
   • Until the firm achieves a significant size, poor unit economics will prevent it from turning a profit.
   • As resources were limited, expansion was not possible.

4. **Jabong**—The goal of Jabong, a defunct online shopping destination, was to provide customers with convenient access to a wide variety of quality, brand name, and reasonably priced footwear and clothing options supplied by its partners. At a fire sale price of $70 million in 2016, Flipkart purchased Jabong from Global Fashion Group. Flipkart, on the other hand, opted to end Jabong's operations in June 2020 so that it could focus on developing its premium fashion marketplace, Myntra (Stephan, 2021).
   What went wrong with Jabong?

   • The number of people using Jabong every day decreased by 10.61% in January 2020, while the number of people downloading the app went down by 12.71%. As a result, their sales are negatively impacted.
   • Another factor was that Myntra, backed by its parent company Flipkart's deep pockets, was eating away at Jabong's market share.
   • Because Myntra has the financial resources to absorb losses, its prices are lower than those of Jabong.

5. **Wooplr**—People were able to use Wooplr to learn about and identify current fashion, cuisine, and interior design trends according to their own

interests, geographic area, and social networks. Wooplr allows its users to set up social media-based shops from which they may sell Wooplr-catalogued products (CB Insights, 2022).

To what extent did design flaws cause Wooplr to fail?

- Due to its inability to get sufficient backing from the appropriate funds, the project ultimately collapsed. No one was willing to invest in the firm again, not even the most important backers.
- The company has also tried to find an exit with the help of the e-commerce sitesLimeroad, Flipkart, and Club Factory. These transactions did not go through.

6. **Shopo**—Shopo is a marketplace where you can purchase both modern and antique Indian designer and handcrafted goods. Thiagarajan started the company to provide a commission-freemarketplace where individuals from all across India could meet to trade goods and services with one another. After purchasing Shopo in May 2013, Snapdeal reintroduced the platform in 2015.Snapdeal had stated the previous year that they would be investing $100 million in Shopo over the course of two years (Failory, 2022).

   Exactly what went wrong with Shopo?

   - Many users of Shopo felt their concerns were not being addressed.
   - There were issues with refunds, the refund process, the shipping method, and the quality of the products themselves.
   - Spending on advertising and running the business has outpaced earnings.
   - Snapdeal, which was experiencing substantial losses, decided to shut down the Shopo business as a means of cost reduction and cash conservation.

4.2. Why do so many startups fail during their initial year in operation?

The reasons startup failure are as follows-

- Solo Efforts—Although starting a business on your own might be an exciting prospect, it is not always worth the difficulties and risks involved. Collaboration and cooperation, rather than the efforts of an individual, are what make a firm successful. Starting a business is difficult in general, but doing it alone in a highly competitive market may be disastrous. Setting up a firm is a difficult task that requires a lot of time, energy, and thecollaboration of many smart people (Griva et al., 2021). Solo entrepreneurs typically have a treacherous road ahead of them. It is hard to do it all by yourself when you've had some setbacks and blows.

1 in 5 businesses fold in year one	The health care and social assistance industry have 50% chance of surviving after 5 years	The pandemic forced 25% of the startups to close their business between March 2020 and February 2021.
50% fail by the fourth year mark	The construction industry after 5 years reduces its chance of survival to 40.5%.	
80% of the startup last fewer than 20 years	In 2006, more than 7 lakh startups were launched. However, 2 lakh business failed by the end of 2008 due to recession.	

(Startups Failure Statistics)

- Difficult Business Strategy—The company model is typically overcomplicated by inexperienced entrepreneurs due to poor presentation, unclear goals, and unreasonable expectations. Business models are the foundation of every company, no matter how large or little. As the most basic component, it must be carefully considered at all times. There are too many moving parts in the business concept, making it difficult to understand. The improper objectives, the wrong benchmarks, and the uncertainty that result from such a strategy are all that are needed to kill a firm in its infancy (Kriz *et al.*, 2022).
- Making Too Quick of a Return on Your Investment-Spending too much money at the start of manufacturing may be disastrous for a company, even if they succeed in getting their product to market. Numerous options exist for reducing the initial investment required to start a business. When first starting out, many firms make the naive error of spending all of their capital. These business owners accept funding without having a clear plan for how to put the money to use. As soon as they obtain a substantial sum of money, they make the mistake of trying to expand the business, which leads to dire consequences and their eventual demise.
- Being Unfocused and directionless – How come the same resources and opportunity may lead to such diverse results for businesses in the same sector? In spite of the fact that they both have a whole day to get their jobs done, one of them is unsuccessful while the other is successful. There is a very easy explanation for the company's failure: it has no clear goals or objectives. This is a typical reason for a business to fail in its early stages. A lack of focus may be disastrous for a company's success (Audretsch *et al.*,

2021). Whenever a company's board of directors makes a choice without first considering the potential consequences, the company makes a misstep. Whatever the choice may be about, it has the potential to harm the company. An example of going in the wrong way is investing in public relations and social media before having a marketable product ready for sale.

- Miscalculation of Scale-One of the leading reasons of a company failing is attempting to expand too quickly. Based on an analysis of more than 3,000 failed firms, startup Genome found that insufficient and premature growth was the primary reason for failure for budding enterprises. Prematurely growing organizations are doomed to fail. They waste cash in excess of what is necessary because they believe money can buy happiness. Despite having the means to do the right thing, these companies' management goes about it in the wrong sequence (Dee *et al.*, 2019). Expensive advertising, renting office space, refining a product with advanced technology, and other typical activities of huge corporations need substantial capital outlay. Without adequate capital, a company cannot meet these demands.

7. RESULTS

After conducting a comprehensive analysis of unsuccessful startup initiatives, we identified the following as the primary causes of project failure:

- Difficulties with Marketing (56%)—Insufficient product-market fit is by far the most serious issue, and it was marketing missteps that caused the most deaths. Do not put in a lot of effort or money until you know for sure that customers will buy from you.
- Issues inside the team itself (18%)—The greatest killers are things like a lack of subject understanding, marketing knowledge (and a strategy), technological knowledge, and business knowledge. Internal strife, a lack of drive, and an absence of commitment are other types, if less lethal, causes of failure to deliver (Singh, 2022).
- Money Issues (16%)—Although financial difficulties were cited by 16% of founders as the cause of their project's demise, more than half of those questioned had no budget and 75% had to finance the venture themselves. This is because ideas may be tested and validated without spending a lot of money (you need effort). Growing a proven idea costs money, which is why most firms struggle in their latter stages.
- Technology Issues (4%)—Despite the fact that the majority of tech-based firms we spoke with relied on some kind of innovation to get off the ground, we seldom saw a game-changer. Overspending on costly technology (development time) before validating marketing assumptions is the most common blunder (Kalyanasundaram, 2018).

- Issues with Operations (2%)—In our experience, software firms like the ones we spoke about seldom have any significant operational issues. However, companies developingreal-world goods may find this to be different.
- Issues with the Law (6%)—The importance of this factor is often exaggerated, and it is seldom the deciding factor in a negative outcome. However, legal hurdles persist in tightly regulated sectors like the food and financial industries (Bajwa *et al.*, 2017).

8. CONCLUSIONS

Numerous startups will crop up, and just as many will fail. Businesses that prioritize theircustomers over rapid expansion, collaboration over competition, and long-term success over short-term gains are the ones most likely to endure. If you can keep your firm from making the same mistakes that other startups do, you will have a far better chance of seeing it through to long-term success. Even though you have already made these mistakes, it's never too late to take precautions and find good solutions to the problems you are facing. Startups in their infancy (idea stage) are particularly vulnerable and often fail. For this reason, accurate figures on the percentage of initiatives that fail are difficult to come by. Most startups get their first funding from the company's creators and their personal networks rather than through venture capital firms or other organizations that track data. In order of prevalence, the following are the primary causes of startup failure: marketing, team, finances, technology, operations, and legal.

Undoubtedly, startups carry a high degree of danger, but with high probability also comes high reward. Not only might there be monetary gain, but also growth and new ideas may be developedto better the lives of people everywhere.

REFERENCES

Audretsch, D. B., Belitski, M., and Caiazza, R. (2021). Start-ups, innovation andknowledge spillovers. *The Journal of Technology Transfer*, 46(6), 1995-2016.

Bajwa, S. S., Wang, X., Nguyen Duc, A., and Abrahamsson, P. (2017). "Failures" to be celebrated: an analysis of major pivots of software startups. *Empirical Software Engineering*,22(5), 2373-2408.

Bryan, K. A., and Hovenkamp, E. (2020). Startup acquisitions, error costs, and antitrust policy. *The University of Chicago Law Review*, 87(2), 331-356.

Cantamessa, M., Gatteschi, V., Perboli, G., and Rosano, M. (2018). Startups' roads to failure. *Sustainability*, 10(7), 2346.

CB Insights Research. Retrieved December 19, 2022, from https://www.cbinsights.com/research/biggest-startup-failures/

CB Insights. (2022, December 1). *239 of the biggest, costliest startup failures of all time.* Center.

Dee, N., Gill, D., Lacher, R., Livesey, F., and Minshall, T. (2019). A review of research on the role and effectiveness of business incubation for high-growth start-ups.

Failory (2022, March 26). *17 failed Indian Startups & Analyses on why they failed*. Failory. Retrieved December 19, 2022, from https://www.failory.com/startups/india-failures

Giardino, C., Wang, X., and Abrahamsson, P. (2014, June). Why early-stage software startups fail: a behavioral framework. In *International conference of software business* (pp. 27-41). Springer, Cham.

Griva, A., Kotsopoulos, D., Karagiannaki, A., and Zamani, E. D. (2021). What do growing early-stage digital start-ups look like? A mixed-methods approach. *International Journal of Information Management*, 102427.

Kalyanasundaram, G. (2018). Why do startups fail? A case study based empiricalanalysis in Bangalore. *Asian Journal of Innovation and Policy*, 7(1), 79-102.

Kriz, A., Rumyantseva, M., and Welch, C. (2022). How science-based start-ups and their entrepreneurial ecosystems co-evolve: A process study. Industrial Marketing Management, 105,439-452.

McDonald, R. M., and Eisenhardt, K. M. (2020). Parallel play: Startups, nascent markets, and effective business-model design. *Administrative Science Quarterly*, 65(2), 483-523.

Mishra, S. B., and Alok, S. (2022). Handbook of research methodology.

Nair, S., and Blomquist, T. (2019). Failure prevention and management in business incubation: practices towards a scalable business model. *Technology Analysis & Strategic Management*, 31(3), 266-278.

Pandey, P., and Pandey, M. M. (2021). *Research methodology tools and techniques*. Bridge Retrieved December 19, 2022, from https://about.crunchbase.com/blog/failed-startups-and-lessons-learned/

Singh, H. (2022, August 18). *50 famous failed startups in India: Mini case studies [2020-21]*. HelloMeets Blog. Retrieved December 19, 2022, from https://hellomeets.com/blog/failed- startups-of-india/

Stephan, D. (2021, October 12). *7 failed startups and the lessons learned*. Crunchbase.
Yin, Y., Wang, Y., Evans, J. A., and Wang, D. (2019). Quantifying the dynamics of failureacross science, startups and security. *Nature*, 575(7781), 190-194.

Zhang, Y. (2022). Research methodology. In *Assessing Literacy in a Digital World* (pp. 51-71). Springer, Singapore.

8. An Empirical Study on Disadvantages and Dissatisfaction of Online Shopping of Apparels

K. Dileep Kumar, P. Manish, E. Surya and Subhadarshini Khatua
Woxsen University, Hyderabad, Telangana, India

ABSTRACT: With the increasing technology, improved marketing strategies of the companies increased creativity in all means of trade there is a tremendous change in shopping methodologies. Online shopping of clothes is one of the outcomes that has come out with the increased paraments in the increasing trend mentioned above. Though the online shopping trend has increased, it is of prime importance to know what customer is feeling with it, what are the expectations of the customer from the E-commerce industries, to have a look on this, in this research paper, the researcher has conducted a survey in which fixed set of questions are being asked to the participants in the survey to dig out the above said objectives regarding online shopping. Critical analysis and evaluation are done for the data recorded in the survey and reached to appropriate conclusion.

KEYWORDS: Technology, E-commerce, Online Shopping, Critical Analysis

1. INTRODUCTION

Technology and innovativeness are the two aspects that move hand in hand and are responsible for the ease in human life. Positive growth in technology and innovativeness are directly proportional to the human life ease and vice versa. With the advanced technology, the conventional way of shopping has been replaced by the new trend of online shopping which has been established to greater extent from the past couple of years. Various E-commerce websites have increased a lot like flipchart, amazon, Myntra. The spread and reach of these websites to the common population has elevated over the span of time. Various reasons responsible for this are the marketing strategies by them like summer sale, winter sale, which offers tremendous discount on the shopping. This makes the customer attracted toward the website's foe. Second reason being the reach of these websites to the target customer by various means like advertisements on YouTube, google, newspaper, etc. One other reason being the availability of

Chapter 8 DOI- 10.4324/9781003397175-8

android mobiles to most of the population who possesses the potential to purchase online caused the trend of online shopping to grow to greater extent. In the year 2020, the COVID-19 spread fear has resulted the reduction in offline shopping and people have preferred to do shopping online instead going in the market and increasing the risk of Covid (Vasić et al., 2019)

Increase in technology: As the increasing technology and its handiness to the common population have made this very much possible to do shopping on a single touch. This is one of the reason responsible to act as a catalyst for online shopping.

Improved marketing strategies: There are constant sales going on one or the other online shopping websites which continually offer various profitable deals that keep the customers attracted toward online shopping. Marketing strategies like advertisements on YouTube, newspapers, social networking websites like Facebook, and Instagram, continually engrave the mindset to buy online in the target customers' mind. This ultimately results in the rise in sales of these products online and ultimately the profitability of the E-commerce websites.

Timesaving: Shopping online saves a great proportion of time in going and returning to the market. That can be the added advantage of online shopping over conventional offline shopping in the markets.

Wide variety of one class of item: There exists a wide variety of items of the various companies possessing various unique characteristics and prices are available on the websites that make the search easier and time-saving. Along with timesaving, a wide range of variety of clothes or items to be purchased can be explored without visiting multiple stores.

COVID-19 fear: In the year 2020, online shopping has got some special importance because of the COVID-19 crisis all over the world. Partial lockdowns and restricted conditions to market shopping are something that made online shopping a first preference by the customers. This was done to prevent social gatherings and thus the spread of the COVID-19.

Ease of shopping due to the availability of mobile: Nowadays mobile has become the need of the hour. Many people possess smartphones which makes online shopping easy at a single touch which in turn raises the trend of online shopping.

2. DISADVANTAGES/DISSATISFACTION OF ONLINE SHOPPING

Delivery delay: Many times, it has been observed that there exists a greater delay in the delivery than the ideal said period of delivery. This doesn't happen in conventional shopping in the markets. The customers can take up the product/service immediately once they procure it.

No bargaining possible: As the prices are fixed on the online shopping platforms, no bargaining is possible on this platform which causes the customer to suffer. Though the discounts are given, the magnitude of the discount is fixed, and no reduction is possible on the online platforms which can be done in the market depending on the volume of shopping from the single shop. Even many times, a discount is given only if the customers pay through a credit card or pay through a particular medium of pay app that restricts the customer to take that advantage many times as due to the non-availability of that method of pay.

No real and actual feel of the product: When shopping online, the actual feel of the product cannot be taken. Especially when the thing comes to clothing. In the actual market, we can feel and see the products. In case of clothing, we can try the various items and take the actual appearance feel of the clothes being chosen which helps in choosing the items that really suit the personality. This may be not possible in online shopping.

Lack of relaxation during conventional shopping: Most people enjoy shopping by going to the is market and exploring different places of shopping. This is the hobby of many people and they enjoy doing so and get relaxed by doing the same. In online shopping, this experience is missing.

Frauds: Many times, it has been observed, the online shown product is something different, and actual delivery looks a bit different. This makes the customer totally disappointed and dissatisfied with the customers. This is not possible in market shopping (Shopping et al., 2021).

3. LITERATURE REVIEWS

It is always a fear that exists in the minds of the common population while adopting a new thing, an approach for which a conventional approach is being used currently (Nebojša Vasić, 2018). The views and expectations about the views about the particular stuff can be known by studying the existing process beingused for the said stuff. This will help the news process establish in place of the conventional one to know the expectation of the target population about the new process. In this research paper, the researcher has studiedsome areas of Serbia country where there is no that much developed online shopping. According to researchers, a particular model needs to be developed, which will let the expectation about the online shopping of the population be known in order to spread it with ease in the long term. In this chapter, the researcher developed a model which includes 26 products under study. These 26 products are categorized under the 7 classes. These 7 classes are security, information availability, shipping, quality, pricing, time, and customer satisfaction. For knowing and feeding the inputs to the research, the surveys are conducted among the Serbian population. To sum up, it has been concluded that the various aspects of customer satisfaction under categorized these 7 classes are enlisted. This will help to know the customer better and act accordingly in the coming period of time to ensure the rise in online shopping trend.

Internet usage has increased to greater extent in the current period of time. The total internet subscribers are 34.3% of the total population of the world as per the survey conducted in 2012 (San Lim et al., 2016). The exponential rise in internet users has helped many services like online shopping, entertainment, information surfing, online gaming, and knowledge transfers from the various corners of the world to be utilized with great ease by the common population. In this research paper, the researcher is trying to elaborate on how the rising use of the internet has made it possible to ease online clothing shopping, various payment methods inline to shop online and related aspects the same. In this paper, the researcher studied the customer's opinions about online shopping of clothing based on the below four parameters. Viz satisfaction on the website: usability, credibility, service quality and transaction cost. As per the researcher, the research conducted in Malesia with the 200 participants, the various aspects of customer satisfaction under these four parameters are studied and critically analyzed. It was found that these four aspects over which customer satisfaction is mapped are strongly interdependent. With this said there is a continual need for research into these aspects to accommodate the effects of changing expectations of the customers along with the fast-moving and changing markets.

There is a revolutionary change that the use and spread of the internet has brought to human life. The change has made human life much easier and has got many effects that comfort human beings for the well-being the same (Dr. Durga wati Kushwaha), In this research paper, the researcher has studied the increasing trend in online shopping and its good effects in line with the customer's satisfaction. The study is carried out in concern with clothing shopping online through various applications and websites. The whole research is carried out based on the primary data which is being recorded by the responses recorded. The participants were related to the population that does online shopping for clothing regularly. The critical analysis and evaluation of the data is done and found that the customers find it easy to do online shopping. Customers are satisfied with the increasing trend of online shopping.

Though online shopping saves a lot of time to visit markets and explore various shops & malls, the actual product feel cannot be taken while shopping online (Lin, 2008). In this research paper, the researcher has studied the various good and bad aspects of online shopping and how brand name plays an important role in online shopping. The research is majorly related to the clothing industry. While shopping online, it is impossible for customers to try the cloth, to feel it, and to check the texture of the cloth that strongly influences the choice of the customer as in conventional market shopping. In such conditions, to remove confusion while choosing a particular model of the cloth, the brand name comes into the picture and plays an important role in selecting the said product having confusion with. Good brand name in the market helps in freezing the sale of that product at that moment and prevents further search by the consumer online. To sum up, the brand name too plays an important role in selection of the product while shopping online.

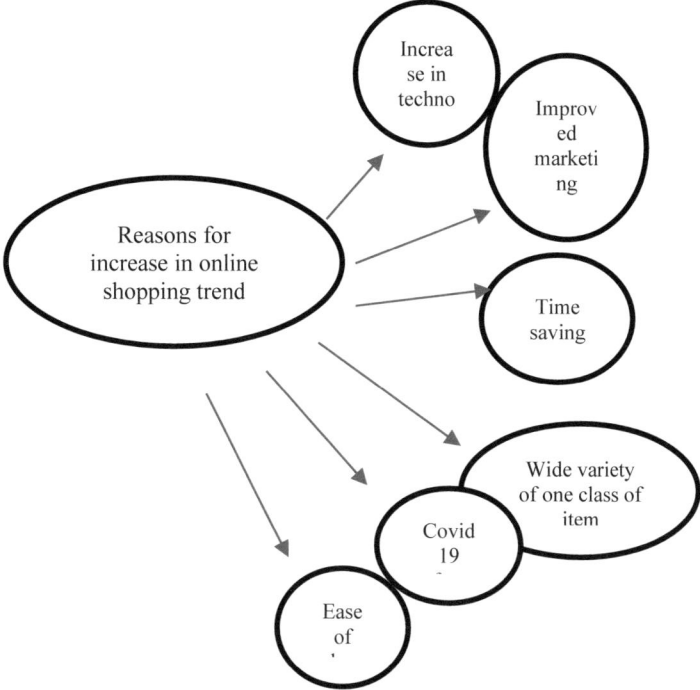

Figure 1. *The reasons for the rise of online shopping.*

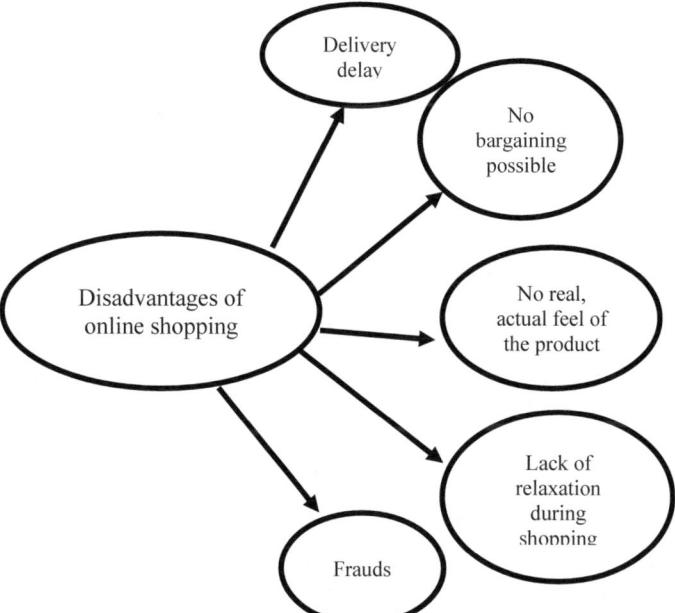

Figure 2. *Disadvantages/dissatisfaction of online shopping.*

4. RESEARCH QUESTIONS

R1. Are customers dissatisfied with online shopping for apparels?

R2. What are the disadvantages of online shopping for apparels?

The main factors that affect the satisfaction of the customer are quality, price, security, responsiveness, and delivery time. We have used an online survey on the internet to take the responses of the sample of people. The people then answered yes or no to various questions asked in the survey. The data is then analyzed, and an end conclusion is made. According to the results, we have to a hypothesis which we will present in the continuation (Vasić et al., 2019).

5. METHODOLOGY

With the detailed study of online and conventional market shopping, advantages and disadvantages ofboth individuals, we have to dig out the research objectives. The research is done purely based on primary data. The data which has been recorded with the responses provided by the participants to the fixed set of questions that has been asked to the individual participants inline with the aim to reach the outcome. The conclusion is drawn based on the critical analysis and evaluation of the data provided by the participants which is purely based on their past and current online shopping experiences. The research is mainly targeted to the online shopping of the clothing items.

Below are the set of questions being asked in the survey to the target participants.

Table 1. Questionnaire About the Online Shopping Habits of the Participants.

Sr. No.	Questions
1	What is your age?
2	How often do you shop online for clothes?
3	Do you shop online for apparels?
4	Is the website secure, from where you usually buy apparels?
5	Do you see accurate information in the online shopping website?
6	Are you satisfied with the shipping facility?
7	Do you get the exact quality of apparels which you expect before buying from an online shopping website?
8	Do you see correct pricing at online shopping websites?
9	Does online shopping for apparel save you time?

Out of 51 participants, 3 answered that their age is between 26 and 50. Between ages 51 and 75, there was 1 participant. Up to 25 years, there were 47 participants. Most participants replied that they shop online a few times occasionally. Sixteen people answered that they shopped for clothes occasionally. People who participated once a month is 14. People who participated once a week is 2. No. of participants shop online for apparel were 80%. Thirty-three participants replied yes that the website from which they shop is secure. Fifteen participants replied maybe the website is secure. Only 3 people replied that the website is not secure. When participants were asked if they saw accurate information on the online shopping website: 61% replied that they thought it was accurate; 25% said that maybe it was accurate; 14% said it was inaccurate. When asked are you satisfied with the shipping facility apparel 90% replied yes and 10% replied no. (Question 6) When asked. Do you get the exact quality of apparel you expect before buying from an online shopping website 16 people replied yes, 11 replied no and 24 answered maybe. Customer satisfaction with pricing is 63% of people answered that they saw the correct pricing of online shopping websites. 18% answered maybe and 20% answered no. 39 participants said online shopping for apparel saved their time. 10 said maybe and 5 participants replied to no.

Table 2. Consumer Tendency to Buy Online Apparels.

Question Item	Cronbach Alpha	Factor Loading			
		1	2	3	4
RATING REVIEW	0,576				
I believe that people who left a review about product is trustworthy					0,740
The reviewers gave detail information about particular premium cosmetic product					0,797
FREESHIPPING PROMOTION	0,733				
If an e-commerce site offers free shipping, I want to shop more				0,905	
If an e-commerce site offers free shipping, I will give positive feedback				0,752	

Question Item	Cronbach Alpha	Factor Loading			
		1	2	3	4
DISCOUNT PROMOTION	0,861				
I shop from e-commerce because the store gives cheaper price rather than the physical store's price.			0,863		
I shop from e-commerce that offers affordable prices			0,821		
PURCHASING DECISION	0,736				
E-commerce has a good ability to secure transactions		0,799			
E-commerce application provides enough information when a user tries to make a transaction		0,787			

Table 3. Coefficients of the Various Parameters.

Model		Unstandardized Coefficients		Standardized Coefficients	T	Sig.	Collinearity Statistics	
		B	Std. Error	Beta			Tolerance	VIF
1	(Constant)	1.889	1.133		1.667	0.0097		
	Rating Reviews	0.318	0.067	0.317	4.754	0.000	0.521	1.921
	Freeshipping promotion	0.022	0.072	0.018	0.305	0.761	0.656	1.523
	Discount Promotion	0.566	0.067	0.535	8.420	0.000	0.574	1.742

6. RESULTS

To find if the customers dissatisfied with online shopping for apparels we reach out to the answers of above research objectives, results achieved from responses to the below questions by the participants are tabulated in one table to have a final result.

Table 4. The Questions Asked to the Participants.

Sr. No.	Question	Response	
		Yes	No
1	Do you see accurate information in the online shopping website?	31	7
2	Are you satisfied with the shipping facility?	46	5
3	Do you get the exact quality of apparels which you expect before buying from an online shopping website?	16	11
4	Do you see correct pricing at online shopping websites?	32	10
5	Does online shopping for apparel save you time?	39	2

According to a survey, it has come up that most of the people feel that the information mentioned on the website regarding the product is accurate and satisfactory to trust over the product before buying it without actually physically experiencing the product rather than cloth more specifically in regard with the research objective. Majority of the population who is doing it firmly believe that the packing quality is quite good and satisfies them. Positive majority exists when the question Do you get the exact quality of apparels which you expect before buying from an online shopping website? Is asked to the target population of participants. Major chunk of the participants feel that the prices are accurate on the online shopping websites. They do feel that it saves a lot of time when they choose to do shopping online instead of conventional market shopping. **With this data put up, it can be summed up that customers are not dissatisfied with online shopping, rather they are satisfied with the same.**

To find the disadvantages of online shopping for apparels we have used secondary data.

Table 5. Disadvantages of Online Shop.

Sr. no.	Disadvantages of online shopping	Description
1	Delivery delay	Many times, it has been observed that there exists a greater delay in the delivery than the ideal said period of delivery. This doesn't happen in the conventional shopping in the markets. The customers can take up the product/service immediately once they procure.

Sr. no.	Disadvantages of online shopping	Description
2	No bargaining possible	As the prices are fixed on the online shopping platforms, no bargaining is possible at this platform which causes the customer to suffer. Though the discounts are given, the magnitude of discount is fixed over which no reduction is possible on the online platforms which can be done in the market depending on the volume of shopping from the single shop. Even many times, a discount is given only if the customers pay through the particular credit card or pays through the particular medium of pay app that restricts the customer to take that advantage many times as due to non- availability of that particular method of pay.
3	No real and actual feel of the product	When shopping online, the actual feel of the product cannot be taken. Especially when the thing comes to the clothing. In the actual market, we can actually feel and see the products. In case of clothing, we can try the various items and take the actual appearance feel of the clothes being chosen which helps in actually choosing the items that really suit the personality. This may be not really possible in online shopping.
4	Lack of relaxation during conventional shopping	Most people enjoy shopping by going to the market and exploring different places of shopping. This is the hobby of many people and they enjoy doing so and get relaxed by doing the same. In online shopping, this experience is missing.
5	Frauds	Many times it has been observed, the online shown product is something different, and actual delivered looks a bit different. This makes the customer totally disappointed and dissatisfied with the customers. This is really not possible in the market shopping.

7. DISCUSSION

During the survey, we interviewed 74 people to know about their experiences with online shopping. They responded according to a scale from 1 to 10. Their responses are measured, and we arrived at a consensus regarding their experiences with online shopping portals. On further observations, we divided the people according to their ages. People who are between 20 and 30 are the largest contributors to orders in shopping. People who are more tech-savvy are more likely to use online shipping websites.

8. CONCLUSION

With the detailed study of the advantages of online shopping and critical analysis and evolution of the data recorded by the participants in the survey, the majority population are on the positive side of online shopping. With the due disadvantages mentioned in the methodology experienced by some of the participants and based on the secondary data, it is analyzed that some of the efforts must be taken up to feel the target customer of online shopping satisfactory in the long run.

REFERENCES

Aaker, D. A. (1991). *Managing brand equity: Capitalizing on the value of a brand name.* New York: The Free Press.

Aaker, J. L. (1997). Dimension of brand personality. *Journal of Marketing Research, 34*(3), 347–356.

Abercrombie and Fitch. (n.d.). *A & F Careers.* Retrieved July 30, 2007, from http://www.abercrombie.com.

Alba, J. W., & Hutchinson, J. W. (1987). Dimension of consumer expertise. *Journal of Consumer Research, 13*(1), 454–511.

Alba, J., Lynch, J., Weitz, B., Janiszewski, C., Lutz, R., Sawyer, A., & Wood, S. (1997). Interactive home shopping: Consumer, retailer, and manufacturer incentives to participate in electronic marketplaces. *Journal of Marketing, 61*(3), 38–53.

Alam, S. S., Bakar, Z., Ismail, H., & Ahsan, M. N. (2008). Young consumers online shopping: An empirical study. *Journal of Internet Business, 5*, 81e98.

Casalo, L. V., Flavian, C., & Guinalíu, M. (2008). The role of satisfaction and website usability in developing customer loyalty and positive word-of-mouth in the e-banking services. *International Journal of Bank Marketing, 26*(6), 399e417.

Casalo, L. V., Flavian, C., & Guinalíu, M. (2007). The role of security, privacy, usability and reputation in the development of online banking. *Online Information Review, 31*(5), 583e603.

Chang, W., & Chang, H. (2011). A dynamic system of e-service failure, recovery and trust. In *PACIS 2011 proceedings* (pp. 1–13). Brisbane, Australia.

Cheesman, D. (2012). Don't call generation Y 'cheap': They're conscious, creative, but coddled. Retrieved 28 December 2012, from http://generationalguru.com/2012/10/dont-call-generation-y-cheap-theyre-conscious-creative-but-coddled/.

Cho, H., & Fiorito, S. S. (2009). Acceptance of online customization for apparel shopping. *International Journal of Retail & Distribution Management, 37*(5), 389–407.

Lin, H. L. (2008). *Consumer Satisfaction/Dissatisfaction in Apparel Online Shopping at the Product-Receiving Stage: The Effects of Brand Image and Product Performance* (Doctoral dissertation, Virginia Tech).

San Lim, Y., Heng, P. C., Ng, T. H., & Cheah, C. S. (2016). Customers' online website satisfaction in online apparel purchase: A study of Generation Y in Malaysia. *Asia Pacific Management Review, 21*(2), 74–78.

Shopping, A., Dotts, K., Dewitt, D., & Karuppaiyah, S. (2021). *Advantages and disadvantages of online shopping.* Retrieved 12 April 2021, from https://accountlearning. com/advantages- disadvantages-online-shopping/

Vasić, N., Kilibarda, M., & Kaurin, T. (2019). The influence of online shopping determinants on customer satisfaction in the Serbian market. *Journal of Theoretical and Applied Electronic Commerce Research, 14*(2), 70–89.

9. Effect COVID-19 on Customer Satisfaction in Hotel Industry Using Kano's Model

Rajamreddy Guru Sandeep Kumar Reddy, Dalavayi Reddy Niveditha, Mohammed Huzaifa and Dr. Subhadarshini Khatua

Woxsen University, Hyderabad, Telangana, India

ABSTRACT: A key pivot around which the success of any business rotates is the customer satisfaction. Customers must be satisfied with the product or service served to the customer by any company. The same is very much true for the hotel industry. Post-COVID-19 situation lockdown, the hotel's business again has opened, but the fear still exists in the customers' mind, which causes the few customers in the hotel. In order to attract more customers, the hotels must carry various actions to ensure high volume of customers gets attracted towards them. In this chapter, the Kano model is applied to dig out the scenario of customer's satisfaction before and after COVID-19 on the hotel industry. An online survey is carried out in which a fixed set of questions are asked to the participants about the post-COVID-19 conditions in the hotel, what they feel about the hoteling post and prior lockdown. The responses are recorded, a critical analysis and evaluation of recorded responses lead to the outcome required against the research gap found out with the complete study of the topic.

KEYWORDS: Kanos Model, Hotel Industry, Satisfaction, Pandemic

1. INTRODUCTION

COVID-19 has bought an exponential change in every step of life and has impacted the various sectors in the world that contributes to the income source of many families and indirectly to the economy of the particular country. One of the sectors that contribute to the economy to a greater extent is the hotel sector and its contribution to the economy varies depending on the country. This hotel sector is badly affected by the COVID-19 as the complete lockdown has restricted the tourism business. People also didn't travel from one place to the other resulting in no usage of hotel. Post-complete lockdown, during partial lockdown, the fear of COVID-19 was still there in the minds of people and which

Chapter 9 DOI- 10.4324/9781003397175-9

was really needed to prevent the spread of this pandemic disease. But with all these the hotel industry was badly affected, with the survey conducted, it is put forth that without government assistance, it is really hard for hotels to keep its existence in the market. Many hotel employees lost their jobs during COVID-19 setback. During complete lockdown, there was complete closure of hotels. during the said span of time zero income is incurred whereas the expenditure required for the static hotel set up was continually on. All this has impacted the hotel industry badly and a great fall is observed during the said span of time in the hotel industry.

2. KANO'S MODEL

This model was invented by Professor Kano who was a professor in Tokyo University in 1978. This model interlinks the customers' satisfaction against the customer requirements. In simple words, what level of requirement will satisfy the customer in what way is simpler elaborated by the inventor in this model. The inventor mentioned five characteristics of the requirements generally asked by the customers against which the satisfaction level of the customer is measured.

Characteristics of customer requirement as per Kano				
Threshold attributes	Performance attributes	Excitement attributes	Indifferent qualities	Reverse qualities.

- *Threshold Characteristics*: These are the very basic amenities that the customer must get and are generally taken up for granted as these are made available everywhere in order to keep the existence of any business in the competitive market. Eventually when these basic needs are not fulfilled, this may lead to strong dissatisfaction of the customer which in turn the business may lose the customer forever. Customers always consider to get at least these basic needs fulfilled by the business.
- *Performance Characteristics*: These characteristics make the customer feel satisfied and fulfilled. When these characteristics are fulfilled continuously over the period of time companies grow with greater extent in the long run. If these characteristics are not fulfilled, the customer feels unsatisfied, which results in the slow or no growth of the company over the period of time in the long run.
- *Excitement Characteristics*: These are the characteristics if fulfilled, customers feel delighted as these are far away from the basic requirements. Customers don't feel bad even if these are not fulfilled. This doesn't have that much impact on the growth of the company in the long run. These are

unusually not promises to the customer, hence when fulfilled by the company, the customer gets surprised and gets delighted towards the company. This in turn keeps the customer abided with the company for a long time.

- *Indifferent Characteristics*: These aspects are neither good nor bad. In simple words it has got no positive and no negative impacts on the customers satisfaction
- *Reverse Characteristics*: These have a very negative impact creating parameters on the customer satisfaction. These are the parameters that create trouble to the customer. This leads to the dissatisfaction of the customer, which in turn hampers the business to greater extent in the long run.

3. IMPORTANCE OF KANO MODEL

1. Various means to delight and satisfy the customer is identified with the help of this model with great ease which in turn helps the business to grow.
2. Must have characteristics required to grow the company with the fair compaction with the competitor is achieved with the aid of Kano model.
3. Identifying the acts that delight the customer that comes under excitement characteristics are identified with the help of Kano model to ensure the periodic business growth.
4. The Kano model helps in identifying the indifferent characteristics that have neither positive nor negative impact. This will help the company to focus more on the acts that add some value to the customer in the period of time.
5. The Kano model helps in identifying the reverse characteristics that add negative impacts to the customer's satisfaction. Identifying those acts and removing them can help the business to maintain the brand in the market in the long term.

Customer's Satisfaction: Any business model runs in better way when it aims to have customers delight and satisfaction. Customers are satisfied when he/ she is in position to get the expected basic and advanced amenities required for the requirement at the time. In the hotel industry customer attraction is one of the term which majorly affected by the customers satisfaction. When the customers are satisfied with the first visit to the hotel, he will attempt to come again to the hotel which in turn will impact the business growth of the hotel in a very positive manner.

Quality of Services Provided Measurement of Service Quality: It is very important for the hotel industry to provide quality which is much higher than the average required quality in order to sustain their existence in the competitive market in the long run. To provide the better quality , it is required to have the best possible ways that the market is offering to the customers for their ease and comfort. For this purpose , it is necessary for businesses to examine and measure their services time to time and brainstorming must be carried out to improve the existing services to sustain in the market. There are various means to measure the quality of the services that the industry is providing. One of it can be inclusion of customer feedback for the services being offered to have continuous improvement.

Importance of service quality measurement (as shown in Figure 1) shows the majors factors as:

1. Customers long lasting bonding
2. Better value to customer
3. Customers satisfaction
4. Customers delight

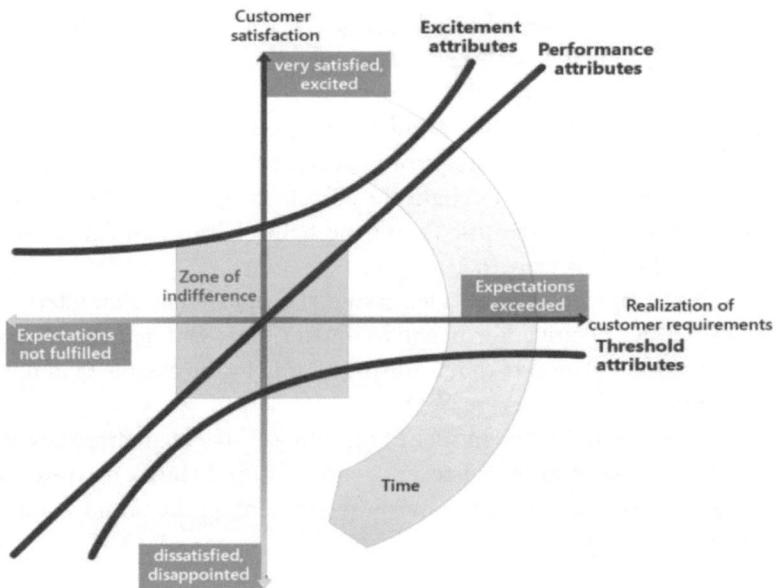

Figure 1. Service quality measurement.

4. LITERATURE REVIEW

It is very important to have the contact of the potential customers of the hotel industry that helps to have suitable marketing strategies. This strategy keeps on changing as per the need of the changing scenarios of the market. In this research paper, the researcher has applied the Kano model to analyse the customer contact possessing the potential to bear the services being provided by the industry. In this research paper, QFD that is quality function deployment is too applied to know the exact improvement points in the existing scenario to constantly improve the customers service in the long term. With the mentioned approach in the paper, brand new technical efforts are evolved to satisfy the customer and make the business of the hotel industry stand unique in the market. This output of the experiments will make the business of hotel stand better than the competitors and will enhance the customers' attraction and thus the overall profitability of the hotel industry (Chang and Chen, 2011)

The customer's satisfaction is the key about which the success of the hotel industry revolves. In order to have a high volume of customers' attraction and good profitability every hotel must focus on the single tool of customer satisfaction. A constant search must be done by the industry to invent various ways through which they can delight the customer in the long term. In this research paper, the researcher has considered 20 hotels in Sicily, Italy. These were most of the three-star hotels in Italy. Researchers have set a fixed set of 600 questions for the survey to be conducted. The purpose of the survey was to establish the key drivers to the customers' satisfaction that the aggregate population of the particular country expects from the hotel industry. Along with the key drivers to customer satisfaction, it is also discussed what offers the managers need to offer to the customer to have better customer satisfaction. In order to achieve the desired results, the researcher has applied the Kano model to the scenario mentioned (Dominici and Palumbo, 2013).

Identifying the customers thought about the satisfaction in the service and act accordingly is the key in the hotel industry to have the high volume of customers attraction and thus to have good profitability. In this research paper, the researcher has applied the Kano model to identify the service quality attributes required by the customer. Identifying the same will make it easy for the hotel managers to offer the similar kind of deal to the customers to have more satisfaction of the customer and more delight to the customer.

The results of this model applied to the scenario is presented in the form of mobile value-added service quality. This has enabled the hotel manager to identify the particular customer and provide him the best possible service quality within the possible sources of the particular hotel (Ying-Feng Kuoa, 2012).

Hospitality and tourism industry is growing to greater extent since last decade, with growing volume of tourists, the requirements of the customers have also started becoming stagnant. This allows the increasing competition in the hotel industry. It is the need of the hour that the hotels must keep on updating the ways they serve the customers to keep their existence in this competitive market, for the purpose they need to have a watch over the requirements of customers and the innovative ways the competitors are serving the consumers to set their own strategy to attract more customers. In this research paper, the researcher has considered the Russian tourists occupying the European hotels in Europe. In this research, the Kano model is applied by the managers of the hotels to find the key driving characteristics that dominantly drive the customer's satisfaction and delight. Customers' experience about the service quality and the trip was also demonstrated with the aid of the Kano model (Zobnina and Rozhkov, 2018).

5. RESEARCH QUESTION

1. What is the customer satisfaction before COVID-19 in the hotel industry with the help of Kano's model?
2. What is the customer satisfaction after COVID-19 in the hotel industry with the help of Kano's model?
3. How to improve customer satisfaction level?

6. METHODOLOGY

With the detailed data studied in introduction and literature reviews, a standard method is designed based on the primary data to dig out the answers to the research questions mentioned. The method to reach the conclusion based on the real time data consists of setting a fixed set of questions which will be asked to the various peoples in the country. The data collection is done on the online platform by Google's forms. The responses to the questions are recorded. Based on the responses provided, the critical evaluation and analysis is done to reach out the conclusion.

Below are the fixed set of questions asked to every participant in the survey to conclude as shown in Table 1.

Table 1. The Questionnaire Asked to the Customers.

Sr No	Questions
1	Have you visited any hotel after COVID-19 Unlocking?
2	Did you find any difference in Bed Sheets, Towels, and Room Cleanness?
3	Did you find any difference in TV, Coffee making facilities, cupboards, etc.?
4	Did you find any difference in WIFI, Swimming Pool, Laundry, and Dining?
5	When are you more satisfied with the hotel industry?
6	Have you seen any changes in hotel charges before COVID-19 lockdown and after COVID-19 unlocking?

7	Have you seen social distancing in the hotel industry after COVID-19 unlocking?
8	Do you see changes in hotel staff behaviour after COVID-19 unlocking?
9	Are you comfortable visiting the hotel after COVID-19 unlocking?

7. ANALYSIS

When the customers are asked if they had visited any hotel after COVID-19 unlocking, 42 participants responded yes, and 11 participants responded no, which implies 79.2% of participants are visited and 20.8% of participants have not visited.

When customers are asked if they find any difference in Bed Sheets, Towels, and Room Cleanness, 35 participants responded yes and 15 participants responded no, which implies 66% of participants found a difference in Bed Sheets, Towels, and Room Cleanness and 34% participants are not.

When customers are asked if they find any difference in TV, Coffee making facilities, cupboards, etc., 20 participants responded yes, and 31 responded no, only 2 are not provided any responses, which implies 38% of participants found differences in TV, Coffee making facilities, cupboards, etc, and 58.4% participants didn't find any differences TV, coffee making facilities, cupboards, etc.

When customers are asked if they find any difference in WIFI, swimming pool, laundry, and dining, 31 participants responded yes, and 21 responded no, which implies 58.4% found a difference in WIFI, swimming pool, laundry, and dining, and 41.5% are not.

When the question "When are you more satisfied with the hotel industry" is posed to customers, 24 participants responded after COVID-19 lockdown, 14 participants said before COVID-19 lockdown and 1 participant mentioned there is no change and 1 not responded, which implies 45.25% participants are more satisfied after COVID-19 lockdown and 26.4% participants are more satisfied before COVID-19 lockdown, 26.4% participants mentioned that there is no change in satisfaction, before and after lockdown.

When customers are asked if they have seen any changes in hotel charges before COVID-19 lockdown and after the COVID-19 unlocking, 40 participants responded yes, 11 participants responded no, and 2 participants did not respond, which implies 75.4% of customers seen changes in hotel charges before COVID-19 lockdown and after the COVID-19 unlocking, whereas 20.7% customers are not.

When customers are asked if they have seen social distancing in the hotel industry after COVID-19 unlocking, 33 participants responded yes, 17 participants responded no, and 3 participants not provided any response, which implies 62.2% of participants had seen social distancing in the hotel industry after COVID-19 unlocking, whereas 32% participants are not.

When customers are asked if they have seen changes in hotel staff behaviour after COVID-19 unlocking, 36 participants responded yes, 13 participants responded no, and 4 participants not provided any responses, which implies 70% of participants had seen changes in hotel staff behaviour after COVID-19 unlocking whereas 24.5% participants are not.

When the question "Are you comfortable visiting the hotel after COVID-19 unlocking?" is posed to customers, 40 participants responded yes, 10 responded no, and 3 are not responded, which implies 75.4% of participants are comfortable visiting the hotel after COVID-19 unlocking, 18.86% participants are not comfortable visiting the hotel after COVID-19 unlocking.

Out of total of 53 participants, 43 participants visited the hotel post-COVID-19. The data collected through them are elaborated in the form of graphs which itself speaks a lot.

7.1. What is Customer Satisfaction Before and After COVID-19 in the Hotel Industry With the Help of Kano's Model?

With respect to the various attributes by the Kano model, the customers satisfaction is mapped based on the responses recorded.

Threshold Attributes: 35 people feel that there is positive change in the basic amenities that hotel must provide post-COVID-19. Hence there is better satisfaction in the threshold characteristics that is the most quality aspect of the hotel which in turn conclude to have better customer satisfaction post-COVID. Basic amenities are improved to greater extent considering prevention of Covid in the society.

Performance Attributes: 20 people out of all noticed that there is positive change in the performance attributes of the Kano model. The positive change is observed in the parameters like TV, coffee making facilities, cupboards.

Excitement Attributes: 31 people felt that there is positive change in the excitement attributes post-COVID-19 in the hotel premises that elevates the customers' satisfaction and delight.

These characteristics include WIFI, swimming pool, laundry and dining.

7.2. How to Improve Customer Satisfaction Level?

- *Social distancing*: Customers must be made to follow the social distancing in the hotel to ensure the prevention of Covid.
- *Temperature check*: Temperature must be checked at the entrance of the customer in order to ensure no Covid patient should enter the hotel. This will eventually spread the further spread of the pandemic.

- *Provision of sanitizer*: Providing sanitizer at the entrance and in the room will ensure the customer to sanitize themselves periodically to ensure the safety towards good health.
- *Extreme cleanliness*: Cleanliness of the hotels must be increased to certain extent more as compared to the prior COVID-19 conditions.
- *Provision of basic medicines*: Basic medicines must be provided at every room to have customer ease.
- *Sanitization of room*: Every room must be sanitized after a single use.

The data collected are sorted and arranged in the form of basic attributes mentioned by the Kano model. This sorting will help to reach the final results which can be seen in Table 2.

Table 2. Results of the Research.

Sr. no.	Question	Kano Characteristics	Response	Count of Participants
1	Did you find any difference in Bed Sheets, Towels, and Room Cleanness?	Threshold attributes	Yes	35
			No	15
2	Did you find any difference in TV, Coffee making facilities, cupboards etc?	Performance attributes	Yes	20
			No	31
			No response	2
3	Did you find any difference in WIFI, Swimming Pool, Laundry and Dining?	Excitement attributes	Yes	31
			No	21
			No response	1
4	When are you more satisfied with the hotel industry?	Indifferent characteristics	Before Covid	14
			After Covid	24
			No change	14
			No response	1

Sr no	Research question	Solution	Explanation
		Table 2: Results of the Research.		
1		Social distancing	customers must be made to follow the social distancing in the hotel to ensure the prevention of Covid.
2		Temperature check	Temperature must be checked at the entrance of the customer in order to ensure no Covid patient should enter the hotel. This will eventually spread the further spread of the pandemic
3	How to improve customer satisfaction level?	Provision of sanitizer	Providing sanitizer at the entrance and in the room will ensure the customer to sanitize themselves periodically to ensure the safety towards good health.
4		Extreme Cleanliness	Cleanliness of the hotels must be increase to certain extent more as compared to the prior COVID-19 conditions
5		Provision of basic medicines	Basic medicines must be provided at every room to have customer ease.
6		Sanitization of room	Every room must be sanitized after a single use

8. CONCLUSION

Considering the need of the hotel by the population for stay and meals, its importance cannot be denied even after the COVID-19 situation. With the application of Kano s model to find out the customer's state of satisfaction post-COVID-19 situation and what are the means by which the hotels can gain more customer satisfaction and delight are tried to put forth. With the improving situation, the means that may satisfy the customer are identified like cleanliness, sanitation of rooms, temperature check at the entrance. With the survey conducted it has been established that the hotels have become cautious with the customers health and care post-COVID-19. The responses proved that the basic and advanced amenities have drastically improved post lockdown in order to have more satisfaction to the customer.

REFERENCES

Base, K. (2021). What is the Kano Model? – Knowledge Base. Retrieved 7 April 2021, from https://www.microtool.de/en/knowledge-base/what-is-the-kano-model/

Chang, K. C., & Chen, M. C. (2011). Applying the Kano model and QFD to explore customers' brand contacts in the hotel business: A study of a hot spring hotel. *Total Quality Management, 22*(1), 1–27.

COVID-19'S IMPACT ON THE HOTEL INDUSTRY | AHLA. (2021). Retrieved 7 April 2021, from https://www.ahla.com/covid-19s-impact-hotel-industry

Dominici, G., & Palumbo, F. (2013). The drivers of customer satisfaction in the hospitality industry: Applying the Kano model to Sicilian hotels. *International Journal of Leisure and Tourism Marketing, 3*(3), 215–236.

Kuo, Y. F., Chen, J. Y., & Deng, W. J. (2012). IPA–Kano model: A new tool for categorising and diagnosing service quality attributes. *Total Quality Management & Business Excellence, 23*(7–8), 731–748.

Trilyo, T., & Trilyo, T. (2018). Service Quality & Customer Satisfaction In The Hotel Industry | Trilyo Blog. Retrieved 7 April 2021, from https://www.trilyo.com/blog/service-quality-customer-satisfaction-in-the-hotel-industry/

Zobnina, M., & Rozhkov, A. (2018). *Listening to the voice of the customer in the hospitality industry: Kano model application.* Worldwide Hospitality and Tourism Themes.

14 Ways to Improve Customer Loyalty During COVID-19. (2020). Retrieved 7 April 2021, from https://www.singlegrain.com/customer-retention/customer-loyalty-covid-19/

10. Effect of Distraction on Cognitive Processes during the Task Performance

Arijit Sinha[1], *Diya Chatterjee[1], *Dr. Anindita Majumdar[2]

[1]Post-graduation student, the University of Calcutta

*[1]Post Graduate, Department of Psychology, *[2]Associate Professor *[1] The University of Calcutta, *[2]Woxsen University

[1]arijitsinha.off.21@gmail.com, *[1]diyachatterjee80@gmail.com

ABSTRACT: Cognitive processes focus on human thought and interpretation of their environment. On the basic level, these include sensation, perception, and attention. Whereas memory, learning, language, problem-solving, decision-making, reasoning, and intelligence are some of the complex cognitive processes. While performing these functions induced stress and distraction may impact accuracy and speed. The main objective of this study is to intervene at the different levels of complexity of the tasks at different cognitive levels. The tasks provided were primarily based on attention, language, logical-mathematical, bodily-kinesthetics, and visuospatial ability. Data collection was done by conducting an experiment on young adults, followed by a structured interview. Nowadays art therapy set-ups are being used frequently in counseling, yet it has not been precisely studied empirically. Even though the levels of stress and anxiety are introduced at the same intensity while performing other tasks, it is easier to complete the bodily-kinesthetic and visuospatial tasks compared to the tasks of attention, language, and logical-mathematical. The result of the study suggested that there were distractions in each cognitive functioning. Although among all the tasks provided, the participants felt least uncomfortable during the Mandala coloring task which is based on bodily kinesthetics and visuospatial abilities. Whereas the most difficult and disliked task was the Anagrams which involved critical thinking and language. Therefore, these findings of the present study can be used in the implementation of bodily-kinesthetic and visuospatial tasks in the therapeutic level of art and sensorimotor integration to reduce stress and anxiety in individuals.

KEYWORDS: Cognitive processes, distraction, stress, visuospatial, bodily-kinesthetic

1. INTRODUCTION

Cognition involves mental events and processes. It can be referred to as all forms of knowing and awareness, such as perceiving, conceiving, remembering, reasoning, judging, imagining, and problem-solving. Along with affect and conation, it is one of the three traditionally identified components of the mind (*APA Dictionary of Psychology*). Cognition may include a spectrum of mental processes like storage, acquisition of information, retrieval, and manipulation (Subedi, 2022). The cognitive approach has seen a surge in fields like clinical and social psychology since the significance of cognitive biases and communication has been studied. (Eysenck and Keane, 2015). Cognitive psychology studies the representation of human knowledge and its use in human action. This knowledge can be declarative or procedural, episodic, or semantic. The system of declarative memory has been associated with learning, representation, and the use of knowledge about facts which are semantic knowledge, and events which are episodic knowledge. (Ullman, 2001). Some of the most significant and accepted methods are attention, perception, memory, pattern recognition, organization, language, reasoning, problem-solving, classification, and concepts (Best, 1986).

The Theory of Multiple Intelligences was first presented in 1983 by Howard Gardner when he published his book *Frames of Mind*. This approach was new to understanding human intelligence (Kurt, 2021) which described each human being as capable of eight relatively independent forms of information processing. Here, Intelligence was explained more as abilities (Gardner and Hatch, 1989) namely, intelligence is linguistic, logical/mathematical, bodily-kinesthetic, spatial, musical, interpersonal, intrapersonal, and naturalist. Generally, in academic settings, most significance is given to linguistic and logical–mathematical intelligence, and it is important to develop various skills and abilities based on the strength of a particular individual. This helps in the building of self-accomplishment and self-esteem of an individual (Brualdi, 1996).

Attention is very useful in everyday life when an individual is voluntarily attending to any stimulus. The process of actively attending anything involves effort (Gaddes and Edgell, 1994). On the contrary, during passive attention, the individual may not be attending to the stimulus voluntarily yet the stimulus may stand out. The other divisions could be "focused" and "divided" (Eysenck and Kene, 2015), where focused attention is when there is selective attention. When there is focused attention, an individual chooses to attend to one stimulus among all the stimuli present whereas in divided attention the individual attends to more than one stimulus at a time. Based on the concept of multiple spotlights theory (Awh and Pashler, 2000), there can be another type of attention known as "split attention" where the attention is split and directed to more than one region of space. For human beings, life is incomplete without language and communication as they are considered the most superior social animal. It is believed that language can provide a powerful tool for social cognition (Fitch *et al.*, 2010). Human beings are considered to know a lot of words, around 60,000 separate words (Solso *et al.*, 2005) which are in an individual's verbal memory known as a lexicon. Linguistics focuses on the study of language which has an essential role in many aspects of human life such as communication, perception, thought, etc. Logical reasoning is the cognitive functioning of a human being where a person is performing goal-directed thinking, which involves reasoning and problem-solving. It has been found in cognitive psychology, that there is an Impact on an individual's knowledge of problem-solving (Wang *et al.*, 2010).

Bodily kinesthetics is to develop certain skills that help in physical activities. As per the Theory of Multiple Intelligence, the category of bodily kinesthetics focuses on certain parameters such as movement, balance, flexibility, dexterity, etc. (Gardner and Hatch, 1989). The functioning of bodily kinesthetics was found to be stimulated inside classrooms via physical training, dancing, crafts, etc. (Bartolomei-Torres, 2022).

Visuospatial understanding is quite important in daily life. These skills mainly focus on cognitive processes required in identifying, integrating, and analyzing space. Therefore, this understanding is required for proper movement along with having a deep and distant perception. The right hemisphere of the brain primarily controls visuospatial functioning (Sternberg and Sternberg 2017). Thus, damage in these portions can cause impairment in visuospatial skills.

While attending to any stimuli an individual may face certain disturbances which can distract them from performing a task. There can be any vibrant stimuli that can lead the individual to get distracted. These could be unnecessary noise, bright colors, use of cell phones, automobiles, pedestrians, etc. (Nemeth *et al.*, 2014). Distraction can be external as well as internal which may be related to the ongoing task or may not be related to the same (Forster 2013). Stress can become an important factor to affect the cognitive functioning of a person (Sandi 2013). As per the optimal level theory by Yerkes and Dodson it is found that when a person is under optimal stress, cognitive performance is best while it is impaired below or above (Hamilton 2019).

In order to understand the effect of distraction on cognitive processes like attention, language, and problem-solving while performing tasks in the presence of induced stress brought by instructions, it is important to review and analyze some of the past research regarding these topics.

There are mainly two types of conditions that require discussion when it comes to distraction and its effect on cognitive processes. The classification given by Craik in 2014 describes the first being if there exists a general attention resource that gets used to create a barrier against stimuli that a person does not want to willingly perceive i.e., distracting unwanted stimuli. This might leave fewer cognitive and attentional resources required for the completion of the original task. The second type of condition elaborates if a distracting stimulus only acts as an interruption when it is qualitatively like the original task given to the person to perform i.e., an undesirable auditory stimulus will be distracting and have a deteriorating effect on a task that is of auditory in nature.

In a research, Mastroberardino and Vredeveldt (2014) worked with 120 children, 60 females and 60 males where they were asked to recall the events of a 5-minute video clip that showed a theft scenario. During the recall phase, there were four conditions, two being distraction-free and two being added with distraction. In the distraction-free condition, the recalling had to be done while viewing a blank screen or keeping their eyes closed. In the distraction condition, a visual or auditory distraction was brought in during the recall. It was seen that the recalling of the visual responses of the video clip was more correct with fewer errors during the distraction-free condition in comparison to the condition where distractions were added. This research was replicated on adults by Vredeveldt *et al.*, (2011) that showed similar results.

Research by Craik and McDowd (1987), Hasher and Zacks (1988), Hasher *et al.*, (1999), and Wais and Gazzaley (2014) suggest that age is a factor when it comes to distraction and cognition. Age acts as a vulnerability factor and there is a decrement in performance in older adults than younger adults. Although older adults were less efficient in blocking the undesirable stimuli of distraction compared to the younger adults; the older adults were able to use the information gathered from the distracting stimuli in a later task where this information became relevant.

Calvo and Gutiérrez-García (2016) showed in their research that when environmental factors exceed an individual's potential and capacities for coping, he is likely to perceive stress. Lack of individual cognitive resources in managing stress enhances the vulnerability one feels in a stressful condition. In a study by Jerath *et al.*, 1993 state-anxiety scores were measured for a problem-solving task before and after stress instructions were given. It was seen that there was a substantial increment in state anxiety scores after stress was induced via instructions during the problem-solving task.

One research done by Rajanil *et al.*, (2021) on adults in healthcare or nursing centers emphasizes the activity of expressive art among these adults in a group. Later, a proper narrative taken from them indicated that art activities could be linked to building and acknowledging new skills and strengths. These skills could help the patients in several aspects of their lives such as personal growth, personal alleviation, communication, etc.

2. METHODS

2.1. Objective

- To understand the effect of distraction on different levels of cognition.
- To understand among all the tasks in which the participant got least distracted after introducing a stressor.

2.2. Hypothesis

- There is an effect of distraction on different levels of cognition.
- Among all the tasks', coloring mandala was the task where participants get the least distracted even after a stressor was introduced

2.3. Sample

Convenient sampling method was used with a sample size of 30. The age range of people who can participate was 18–30 years of age. Anyone below or above was not considered.

2.4. Materials Used

To perform this study, four sets of cognitive tasks were used which were based on attention, vocabulary, logical mathematics, and visuospatial performance. In addition, the participants were provided with pen/pencil and colored pens. Lastly, external noise was used to distract the participants.

2.5. Procedure

The participants were provided with a data sheet and a pen. Then they were instructed that they are provided with 4 tasks, which need to be completed as early as possible by committing the least number of errors. Then the time was recorded with the help of a stopwatch. In the meantime, an extremely disturbing noise was being played in front of the participants at certain intervals. Then after completing the first three tasks, they were provided with color pens to color the mandala. After completion of all the tasks, the participants were asked certain questions in the form of a structured interview to understand which task was most relaxing. Then with the final data, the mean of time taken and error for each task were computed.

3. RESULTS

Result Table 1.

Cognitive Task	Mean Time Taken (min)	Mean Error	Total Correct Response	% Error
Cancellation sheet	5.56	17	146	11.64
Anagrams	4.02	4	10	40
Mathematics	2.97	1	10	10
Mandala Art	13.62	—	—	—

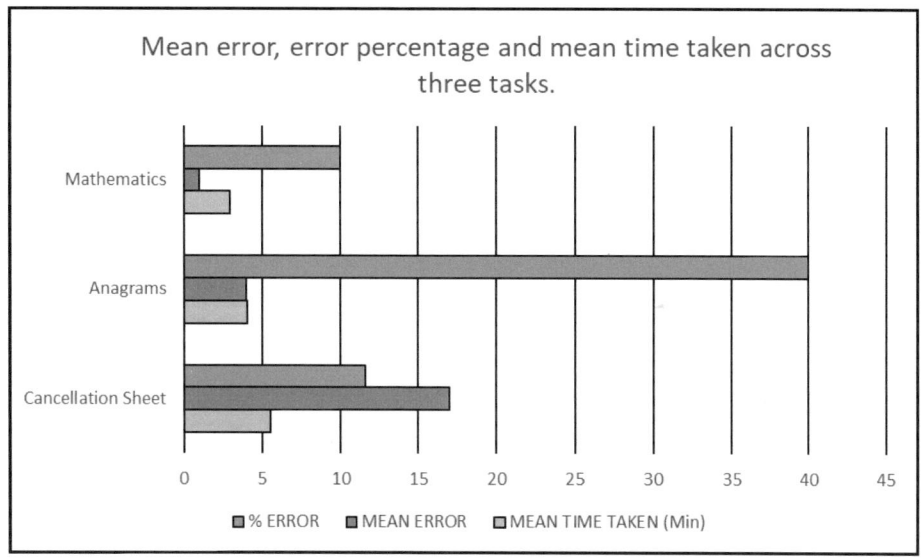

By observing the mean, (Result Table 1), it is seen that the most time has been taken to complete the Mandala Art. It is followed by the Cancellation task, Anagrams, and then the Mathematics task. From the structured interview at the end of the tasks, it was gathered that the participants were enjoying themselves and continued doing the Mandala Art longer as they were not time-bound and there was no strict ending to the art. From the mean, it can be observed that the maximum errors have been committed in the Cancellation task. It is followed by the Anagrams and then the Mathematics task. So, it can be inferred that the participants were most distracted by the noise stimuli during the Cancellation task. The cancellation task, which was used to determine the attention and concentration of the participants, was most affected by the noise.

Result Table 2.

Most Liked	Most Disliked	Easiest	Most Difficult	Task Most Affected by Distraction	Task Least Affected by Distraction
Mandala Art	Anagrams	Mandala Art	Anagrams	Cancellation	Mandala Art

The Mandala Art was the most mentioned task in the "most liked" and "easiest" category of performance tasks, as seen in Result Table 2. The Anagram task was mentioned most in the "disliked" and "difficult" task category, also found (Result Table 2).

When the participants were asked about the distracting noise they described it as "annoying", "bothering", "weird", "frustrating", "awkward", and "disturbing". In the interview, participants explained that it was difficult to focus and concentrate on their tasks when the noise was being delivered. They were affected and felt bothered when the noise was being delivered. They felt most disturbed during the Anagrams and Mathematical task. For some participants, it was challenging to continue the Cancellation, Anagrams, and Mathematics during the noise. It was considerably easier to continue the Mandala Art even when they were made to hear the noise.

When the participants were questioned about the Mandala Art, they described it by using words like, "creative," "relaxing", "peaceful", "soothing", and "artistic". Some of them mentioned that the Mandala Art reminded them of their childhood art works. The participants told it was enjoyable to do patterns and choose color combinations while doing this task. The participants explained how it was less mentally taxing to fill colors or sketch the Mandala Art in comparison to the previous task that they did. Some of the mentioned it was much easier to focus and continue this task even amidst the distracting noise.

4. DISCUSSION

There is an effect of distraction on cognitive task performance. The first hypothesis for this study has been accepted, as there were distractions observed and noted among the participants in different levels of cognition. Wyss *et al.*, in 2013 did research to understand the effect of distraction on cognitive task performance, it was also observed from their study that distractors can have varied effects on performances. Sandi, 2013, saw that stress can affect cognition in numerous ways and the outcome can either be facilitating or impairing which depends on a number of factors of both the stress and the cognitive function that is under study. Mainly the problem-solving tasks were most affected. Mandala art was found to be the least difficult task, it seems that Mandala art is the most preferable task. From this, the second hypothesis of this study which stated that Mandala will be the task least distracted has been accepted. Ashmi in 2019 has inferred that using Mandala art can reduce the level of anxiety along with elevating the mood. Mandala Art therefore can be used as a therapy to intervene in state anxiety and stress in daily life. Vennet and Serice (2012) from their study also found that coloring a mandala was helpful in reducing stress. Sandmire *et al.*, in 2012, observed that for a specific span of time using mandala art therapy can be helpful to reduce state anxiety and stress, mainly among college students. Schouten *et al.*, (2014) worked on the effectiveness of Mandala art therapy and found it helpful in reducing trauma and induced stress among adults.

Kim *et al.*, (2018) observed that mandala art therapy has been found to increase hope, resilience, and well-being among psychiatric patients. Muthard and Gilbertson, (2016) studied the effect of mandala coloring on reducing negative affect, state anxiety, and psychological stress. The results supported that mandala art as a therapy can be used to reduce state anxiety, stress, and negative affect. Dilawari and Tripathi (2014) portrayed that art therapy can be used to help mental health patients to overcome stress, communicate in a better way, and have increased the well-being of a person suffering from mental health problems. Yvette in 2021 inferred that expressive art therapies such as music, dance movement, art, etc. can be helpful in reducing stress and helping in the process of treatment of mental health issues. Mohammad and Mohammad (2013) in their study mentioned that expressive art therapy is helpful in the case of counseling as this can help the participants to express and share their feelings such as anger, regret, stress, disappointment, etc. Kheiberi *et al.*, (2014) in a study on orphan girls, found that the state anxiety score has decreased from what was the score before the involvement of art activity. Thus, it was concluded that an episode of expressive art therapy can reduce state anxiety in a person.

The Mandala Art which focused on designing and coloring used fine motor skills for its performance. Dewi *et al.*, (2019) proved that painting application has an effect on the fine motor skills of children through coloring activities. Another study by Basa *et al.*, in 2020 found that finger paint activities helped in the development of gross motor and fine motor skills which results in a child's more creative and artistic thinking. There was focused attention on the Mandala Art task which was divided between visuospatial and bodily-kinesthetic and sensory functioning. Boonaphatjarern *et al.*, (2020) concluded from their study that visuospatial activity along with visual attention may increase by using Mandala-based images. A study by Lerner in 2016, showed that visual-spatial learning improves through art and design learning. In a research by Sternberg and Sternberg in 2017 showed that the right hemisphere of the brain primarily controls visuospatial functioning. Vartanian and Goel (2004) observed that brain areas like the right caudate nucleus, bilateral occipital gyri, left cingulate sulcus and bilateral fusiform gyri get activated when an individual does some creative activity like painting. It is considered that people with bodily-kinesthetic intelligence can learn things to do easily by exploring and discovering (Frothingham, 2020). Bartolomei-Torres in 2022 found that bodily kinesthetics functions could be stimulated in the classroom through physical training, dancing, crafts, etc.

In the research of O'Neil, Spielberger, and Hansen (1969) when participants were given difficult material, the state anxiety and errors were higher than when the material was easier. Anagrams that focused on Language skills and critical thinking were found to be more difficult and disliked compared to mathematics, which focuses on problem-solving. Although all of them have gone through the standard academic system, there seems to be a poor transfer of training in the case of language and critical thinking. Nauman (2017) did a study on the lack of critical thinking among students in developing countries. She mentioned that it is not an easy task to introduce the knowledge of critical thinking in the classroom as the teachers themselves were lacking in such skills. Students can portray proper critical thinking skills in one context but they may lack in another (Willingham

2007). Liang and Fung, 2021 in research inferred that there is a requirement for the implementation of cultivating critical thinking in English classrooms. So the experience and training at the academic level in the case of mathematics seem to be better than in the case of language and critical thinking.

5. CONCLUSION

It has been observed that when the stressor was introduced repeatedly while performing the tasks, there was a significant disturbance in performance and concentration. From the study followed by a structured interview, it can be concluded that the participants took the most time in performing the mandala coloring. Yet, this task was the most relaxing by the maximum number of participants, and most of them mentioned enjoying coloring even if there was a stressor. Thus, it can be inferred that stress and anxiety can be relieved by filling colors in a Mandala as expressive art therapy.

5.1. Limitations

Due to time constraints, a small sample size was considered which may not have high generalization. The results are based on descriptive statistics alone. Currently, a short-structured interview was taken from the participants based on words. To implement these findings in an anxiety or stress reduction model an elaborated qualitative verbatim needs to be taken. This study did not aim to understand the lack of critical thinking among young adults. The study also showed a lower transfer of training in the case of Anagrams but considerably better for Mathematics. This could be studied further.

5.2. Implications

The findings of this study can be used to understand the implementation of the concept of art therapy among young adults to reduce stress and anxiety. On further research, it can be observed whether there can be an implementation of coloring and other art-based therapy on adults in reducing life and work stress. Findings portrayed that there was a lack of critical thinking among the students. This can be analyzed further to understand the lack of critical thinking exercises in an academic curriculum.

REFERENCES

Anat. J., 2010 Feb; 216(2): 177–183. Published online 2009 May 28. doi: 10.1111/j.1469-7580.2009.01099.x

Ashmi, M. S. (2019). Efficacy of mandala creation on anxiety, mood, and self-healing.

Awh, E., and Pashler, H. (2000). Evidence for split attentional foci. Journal of Experimental Psychology: Human Perception and Performance, 26(2), 834.

Bartolomei-Torres, P. (2022, May 19). Bodily-kinesthetic intelligence: Definition, characteristics, and activities for its development. Learningbp. Retrieved November 7, 2022.

Basa, F. L., Sutarto, J., and Setiawan, D. (2020). Finger Painting Learning to Stimulate Motor Development in Early Childhood. *Journal of Primary Education*, 9(2), 193-200.

Best J. B. (1986). *Cognitive psychology*. West Pub.

Boonaphatjarern, N., Panthong, K., and Juntapremjit, S. (2020). The Effects of Applying Graphic Images Based on the Mandala structure for Increasing Visual Attention and Visuospatial ability for Children at Risk of ADHD. *Dhammathas Academic Journal*, 20(3), 173-186.

Brualdi Timmins, A. C. (1996). Multiple intelligences: Gardner's theory. *Practical Assessment, Research, and Evaluation*, 5(1), 10.

Craik FIM (2014) Effects of distraction on memory and cognition: a commentary. *Front. Psychol.* 5:841.

Craik, F. I. M., and McDowd, J. M. (1987). Age differences in recall and recognition. *J. Exp. Psychol. Learn. Mem. Cogn.* 13, 474-479. doi: 10.1037/0278-7393.13.3.474

Dilawari, K., and Tripathi, N. (2014). Art therapy: A creative and expressive process. *Indian Journal of Positive Psychology*, 5(1), 81.

Eysenck, M. W., and Keane, M. T. (2015). Cognitive psychology. A student's handbook. Hove, East Sussex, UK.

Eysenck, M. W., and Keane, M. T. (2015). Cognitive psychology: A student handbook. Psychology Press.

Fink, G. (Ed.). (2016). *Stress: Concepts, Cognition, Emotion, and Behavior: Handbook of Stress Series, Volume 1* (Vol. 1). Academic Press.

Fitch, W. T., Huber, L., and Bugnyar, T. (2010). Social cognition and the evolution of language: constructing cognitive phylogenies. Neuron, 65(6), 795-814.

Forster, S. (2013). Distraction and Mind-Wandering Under Load. Frontiers in Psychology, 4. doi:10.3389/fpsyg.2013.00283

Frothingham, S. (2020, February 28). Are you a bodily-kinesthetic learner? Healthline. Retrieved November 7, 2022.

Gaddes, W. H., and Edgell, D. (1994). Attention Deficit Disorder. In Learning Disabilities and Brain Function (pp. 253-303). Springer, New York, NY.

Gardner, H., and Hatch, T. (1989). Educational implications of the theory of multiple intelligences. Educational researcher, 18(8), 4-10.

Hamilton, H. (2019). Examining the bipolarity of the MMPI-2-RF behavioral/externalizing dysfunction (BXD) scale using a laboratory measure of impulsivity. (Doctoral dissertation, Central Michigan University).

Hasher, L., and Zacks, R. T. (1988). "Working memory, comprehension, and aging: a review and a new view," in *The Psychology of Learning and Motivation*, Vol. 22, ed. G. H. Bower (New York: Academic Press), 193–225.

Hasher, L., Zacks, R., and May, C. (1999). "Inhibitory control, circadian arousal and age," in *Attention and Performance XVII*, eds D. Gopher and A. Koriat (Cambridge, MA: MIT Press), 653–675.

Jerath, J. M., Hasija, S., and Malhotra, D. (1993). A study of state anxiety scores in a problem-solving situation. *Studia Psychologica*, 35(2), 143.

Kheibari, S. Z., Anabat, A. M., Largany, S. F. H., Shakiba, S., and Abadi, M. E. H. (2014). The effectiveness of expressive group art therapy on decreasing anxiety of orphaned children. *Practice in Clinical Psychology*, 2(3), 135-14.

Kim, H., Kim, S., Choe, K., and Kim, J. S. (2018). Effects of mandala art therapy on subjective well-being, resilience, and hope in psychiatric inpatients. *Archives of psychiatric nursing*, 32(2), 167-173.

Kurt, D. S. (2021, April 29). *Theory of multiple intelligences - Gardner*. Educational Technology. Retrieved November 8, 2022, from https://educationaltechnology.net/theory-of-multiple-intelligences-gardner/

Lee, Y. (2020) Expressive art therapy and mental wellbeing. *Motifs: A Peer Reviewed International Journal of English Studies*, 6(1), 1-7.

Liang, W., and Fung, D. (2021). Fostering critical thinking in English-as-a-second-language classrooms: Challenges and opportunities. *Thinking Skills and Creativity*, 39, 100769.

Mastroberardino, S., and Vredeveldt, A. (2014). Eye-closure increases children's memory accuracy for visual material. *Front. Psychol.* 5:241. doi: 10.3389/fpsyg.2014.00241

Muthard, C., and Gilbertson, R. (2016). Stress Management in Young Adults: Implications of Mandala Coloring on Self-Reported Negative Affect and Psychophysiological Response. *Psi Chi Journal of Psychological Research*, 21(1).

Nauman, S. (2017). Lack of critical thinking skills leading to research crisis in developing countries: A case of Pakistan. *Learned Publishing*, 30(3), 233-236.

Nemeth, C., Papautsky, L., Grome, A. and Fallon, C. (2014). Computer-Based Training in Human-Systems Integration. Technical Report. Federal Railroad Administration. U.S. Department of Transportation. Washington, DC. 20590

Oktavia, D., Bali, M., Rahman, H., Umar, U., Syakroni, A., and Widat, F. (2019, June). Exploration of Fine Motor Skills through the Application of Paint. In *Proceedings of 1st Workshop on Environmental Science, Society, and Technology, WESTECH 2018, December 8th, 2018, Medan, Indonesia*.

Sandi, Carmen (2013). *Stress and cognition. Wiley Interdisciplinary Reviews: Cognitive Science, 4(3)*, 245–261. doi:10.1002/wcs.1222

Sandmire, D. A., Gorham, S. R., Rankin, N. E., and Grimm, D. R. (2012). The influence of art making on anxiety: A pilot study. *Art Therapy*, 29(2), 68-73.

Schouten, K. A., de Niet, G. J., Knipscheer, J. W., Kleber, R. J., and Hutschemaekers, G. J. (2015). The effectiveness of art therapy in the treatment of traumatized adults: a systematic review on art therapy and trauma. *Trauma, violence, & abuse*, 16(2), 220-228.

Solso, R. L., MacLin, M. K., and MacLin, O. H. (2005). Cognitive psychology. Pearson Education New Zealand.

Spielberger, C. D., and Hansen, D. N. (1969). Effects of state anxiety and task difficulty on computer-assisted learning. Journal of Educational Psychology, 60(5), 343.

Sternberg, R. J., and Sternberg, K. (2017). Cognitive psychology. Cengage Learning.

Subedi, K. (2022). Cognition in the Psychological Perspectives.

Ullman, M. T. (2001). The neural basis of lexicon and grammar in first and second language: The declarative/procedural model. *Bilingualism: Language and Cognition*, 4(2), 105-122.

Vaartio-Rajalin, H., Santamäki-Fischer, R., Jokisalo, P., and Fagerström, L. (2021). Art making and expressive art therapy in adult health and nursing care: A scoping review. *International journal of nursing sciences*, 8(1), 102-119.

Van Der Vennet, R., and Serice, S. (2012). Can coloring mandalas reduce anxiety? A replication study. *Art therapy*, 29(2), 87-92.

VandenBos, G. R. (Ed.). (2007). *APA Dictionary of Psychology.* American Psychological Association.

Weeks, J. C., and Hasher, L. (2014). The disruptive – and beneficial – effects of distraction on older adults' cognitive performance. *Front. Psychol.* 5:133. doi: 10.3389/fpsyg.2014.00133

Willingham, D. T. (2007). Critical thinking: Why is it so hard to teach? American Educator, 8–19.

Wyss, N. M., Kannass, K. N., and Haden, C. A. (2013). The effects of distraction on cognitive task performance during toddlerhood. *Infancy, 18*(4), 604-628.

11. Knowledge Management for Employees

Chekhilla Bharathi

ABSTRACT: Knowledge management (KM) is the discipline of creating a thriving work and learning atmosphere that fosters the continuous creation, aggregation, use and reuse of both organizational and personal knowledge in search of new business value. In the 21st century, KM has emerged as one of the most important management methodologies in the organizations across the globe. This methodology has enabled organizations to access and share hidden knowledge(s) among all levels of employees within the organization. Moreover, KM incorporates all possible information, description, analysis and synthesis of various phenomena concerning an enterprise/organization. Knowledge asset as well as managing knowledge is a crucial strategic tool to overcome environmental challenges in achieving a competitive advantage. A successful KM implementation in an organization is directly proportionate to its performance. KM focuses on the hard resources of the organization. This largely requires fostering learning environments that encourage people to share, create and acquire knowledge, enable socialization, externalization and transfer of knowledge. Since the last decade of the 20th century, knowledge has become the most important factor of production after labor, land and capital. In today's rapidly growing economic world, KM has become a serious factor in many fields of knowledge such as education, cognitive science, etc. With successful KM, many organizations have seen less repetitive efforts, significant performance improvements, new developments, better decision making, and increased efficiency. This chapter highlights the importance, principles of KM and also highlights the challenges/issues of KM.

KEYWORDS: Knowledge Management, Knowledge Sharing, Organizations, Learning, Intellectual Knowledge, Competitive Advantage, Corporate Advantage

Chapter 11 DOI- 10.4324/9781003397175-11

1. INTRODUCTION

Knowledge management (KM) is a new concept, where each company has its own key knowledge assets, which is unique to itself. KM relies on the knowledge base of the organization; knowledge gained from years of interaction with its customers and other stakeholders; the economies of the business; and the employees it hires. Knowledge is important for an organization to carry on any business and, KM plays a key role for a successful performance of an organization.

In addition, knowledge is something that can be understood and utilized by the people at their workplace. All the information is collected through Information Technology (IT) to build create support system, management information systems and data warehousing.

Knowledge does not have a universal definition. Similarly, KM does not have a universal definition. KM is often considered in a broader context. It is a process that helps organizations create value from their intellectual and knowledge-based assets. It is a method of harnessing and using people's intellectual capabilities to gain competitive advantage and customer engagement through efficiency and innovation. It also helps in quick decision making.

KM is defined as "a multi-disciplined approach to achieving organizational objectives by making the best use of knowledge. KM focuses on processes such as acquiring, creating and sharing knowledge, and the cultural and technical foundations that support them. The aim is to align knowledge processes with organizational objectives" (Standards Australia, 2001, HB-275).

In today's expanding universe of knowledge, KM comprises of strategies and wide range of practices that are used in an organization to discover, innovate, signify, distribute, and adopt insights into its valuable experiences. Significant growth of KM in modern and dynamic working environment allowed knowledge to completely utilize data, combined with skills of individuals, ideas, innovation, motivation, commitment and experience.

Organizations usually use previous data and strategies in formulating new innovations to sustain in highly competitive environment. Knowledge will provide confidence to anorganization in decision making process; and also help organization in guiding people to arrive at the decision.

2. OBJECTIVES

- To identify the importance of Knowledge Management
- To understand the Digital Knowledge Management
- To identify the challenges/issues of Knowledge Management
- To understand how Knowledge Management is related to the training and development

2.1. Literature Review

After numerous researchers providing descriptions of KM, the single globally accepted definition of KM does not exist, as different standpoints or conservatories of KM can yield different dimensions and meaning (AI-Hawary, 2015).

Armstrong (2006) said KM as a process or practice of generating, gaining, encapsulating, sharing, and using knowledge, wherever it belongs, to enhance performance and learning in organizations.

KM is the processes of identifying, capturing, sharing, circulating, applying, and gathering of knowledge (Allameh et al., 2011).

Ravanpykar et al. (2014) defined KM to be a process that is implemented by organizations combined with organizational learning. All the organizations use this collective knowledge while producing and distributing knowledge in order to build their organizational knowledge and will use it accordingly.

It is not possible to confine KM cannot to a singular definition and it is interpreted differently in different line of industries. It can be simply put as a process for forming, organizing, transferring, sharing and leverage explicit knowledge and tacit knowledge toward the successful progression of an organization.

Macintosh (1997) said that KM is "The identification and analysis of available and required knowledge, and the subsequent planning and control of actions to develop knowledge assets so as to fulfil individual and/or organizational objectives."

Moreover, several organizational research streams look upon the significant role of organizational knowledge such as, organizational design (Sanchez and Mahoney, 1996), strategic alliances (Inkpen and Beamish, 1997), and international acquisitions (Bresman et al., 1999)

Parag Sanghani (2009) articulated that KM implementation framework recompenses of, learning culture, training, and technology as a common influencer on organizational and individual KM; and organizational KM is specifically influenced by leadership strategy, and structure. Parag Sanghani stressed that individual KM is more influenced on individual's attitude and personality.

As per Martensson (2000), KM is a significant and essential component for any organization to endure and sustain competitive eagerness. Hence, it is compulsory for organizational leaders to consider KM as a prerequisite for enhanced productivity and flexibility in all sectors. Based on this explanation, KM is defined as a process of creating, acquiring, storing, sharing, and using knowledge, wherever learning and performance enhancement is required in an organization.

2.2. What is Knowledge?

The logical way of beginning KM should begin with the definition of knowledge itself, but the concept of knowledge is defined in many different versions and definitions, where each of it has its own valid context.

As per Webster's dictionary, knowledge is "the fact or condition of knowing something with the familiarity gained through experience or association." One of the most frequently used definition of knowledge is "the ideas or understandings, which an entity possesses that are used to take an effective action to achieve the entity's goal(s)." This definition of knowledge is specific to the instance at which it was created.

There are numerous kinds of knowledge in an organization, but they can be briefly categorized as following

1. **Explicit Knowledge:** Any articulated knowledge that is often recorded pictographically, moreover, in the form of tables, writings, product specs and so on. "The knowledge creating company," an article from Harvard business review states that, explicit knowledge is a "formal and systematic." Therefore, explicit knowledge will be methodically documented and, thus, knowledge can be effortlessly obtainable by every individual in the organization.

2. **Implicit Knowledge:** Any form of learned skill that essentially teaches know-how of an event is called as implicit knowledge. During this process an explicit knowledge applied during the implementation, in any specific situation will result in implicit knowledge. Organizations will record the experience, learning outcomes and will synthesize it with other information to solve other similar situations.

3. **Tacit Knowledge:** Most knowledge forms can easily be transferred from one stake holder to another. However, with respect to tacit knowledge, it is hard transfer knowledge from one stake holder to another through oral or written communication. The major challenge with tacit knowledge is it cannot be articulated. A great example of tacit knowledge is soft skills such as how to communicate, innovate, leadership skills, body language, time management, and emotional intelligence.

4. **Propositional Knowledge:** Propositional knowledge, literally this form of knowledge will only be expressed in propositions; mostly in declarative sentences. For instance, a mathematical equation is an example of propositional knowledge.

5. **Procedural Knowledge:** A demonstration of operational procedures, that focuses mainly on "how" will operate, especially sequence of work will fall under procedural knowledge.

6. **Posteriori Knowledge:** Any kind of subjective knowledge that comes from personal experience is a posteriori knowledge. Although this knowledge is not recorded in the company's knowledge base, it still plays an important role in the team's success. This type of knowledge allows individuals to know their strengths and weaknesses based on their experience and helps companies diversify their team's skill sets.

7. **Priori Knowledge:** A priori knowledge is the opposite of a posteriori knowledge, and is acquired independently of experience or evidence,

and a priori knowledge is often shared through logical reasoning and the ability to think abstractly.

3. IMPORTANCE

The 21st century economy is heavily relied upon knowledge workers who are keys to high productivity making it a high greed resource. Ever more organizations are creating their knowledge repositories for several day-to-day activities of the organization. These repositories play vital role in the product development, customer service, human resource (HR) management, and many more. Several new positions such as, such as knowledge developer, knowledge facilitator, and corporate knowledge officer are also emerging.

1. **Key Management Challenges:** The key competitive advantage for any successful business today is knowledge and it pioneers land, capital, raw materials, and technology in any organization.
2. **Key to Change and Growth:** Intelligence and knowledge in an organization and among individuals is usually driven through change, and, for it to be on positive node, organizations and people must adapt to change.
3. **Build competitive strength:** An effective use of knowledge is an add value in an organization's drive competitive advantage, which will extensively absorb knowledge, and share then among its several verticals effectively to build for cutting edge product in the market; thus, building a competitive strength.
4. **Knowledge Demands Constant Renewal:** The requirement for new knowledge in this fast-moving technological world is enormous to make this possible knowledge should be constantly renewed and updated in an organization.
5. **Improves Learning Capability:** KM creates a bundle of learning opportunities which provide stronger learning foundations. Thus, motivating many individual to improve their implicit and explicit knowledge.
6. **Turns the Brainpower into Profitable Products:** Organizational leaders should thrive to build good teams because, for any organization the collective intelligence of individuals is its intellectual capital.
7. **Strategic and Integrating Force:** Chief knowledge officer represents the importance of KM as a strategic and integrating force
8. **Knowledge Explosion:** The information is estimated to be tripled every seven years, so, Organization should develop plans to prioritize and manage the information flow to stay updated with the market conditions.
9. **Improvement in Decision Making:** To improve the creation, decision making and usage of knowledge, KM has been developing new ways, tools, processes systems, etc.

4. CHALLENGES AND ISSUES

a) **Getting employees on-board:** The HR teams of any organization will consider people's cultural issues because they are expected to cause problems in KM. Moreover, the employees will participate solely disregarding the quality of knowledge. However, to avoid it HR teams motivate employees to work as teams and encourage team building activities to uphold the quality of knowledge.

b) **Difficulty in turning knowledge into action:** Many organizations face major huddles in learning how knowledge is captured, handled, and acted on. As the Knowledge is captured, organizations struggle with the stockpile of information and how to act upon it.

c) **Pace of Change:** In the fast-moving world, knowledge is growing exponentially so organizations are struggling to keep up with the pace in change of knowledge and how it is to be shared.

d) **Technology and People:** Technology plays a critical role in storing and sorting knowledge and people are using it more as a toll of communication in sharing knowledge.

e) **Difficult to Deal With Tacit Knowledge:** Tacit knowledge gained through experience becomes difficult to express. Henceforth it's a challenge to codify, transfer or share the knowledge.

f) **Continuous Research Required:** Knowledge has no end point, there should always be updation in the knowledge which needs to have continuous research in process and the problem is it needs an extensive dedicated team and resources.

g) **It is Not Static:** As knowledge gets enhancing every now and then, the value of knowledge gets diminished overtime. It should be improvised, modified, and deleted continuously.

h) KM is regarded as the still neglected area of collaborations.

5. DIGITAL KNOWLEDGE MANAGEMENT

5.1. What is Digital Knowledge?

The process of detecting, gathering, recording, combining, and centralizing organizational knowledge, specifically in digital format. And, especially the term "Organizational Knowledge" itself has a wide scope because it has knowledge on both the customer and internal organizational information.

Nowadays organizations are storing information on customer preferences, habits and demographical information of target locations to serve their customers better. This information is stored digitally and used along side specific algorithms to gain competitive edge over rival organizations.

1. Customer-facing information includes:
   - Geographical/Location information
   - Product/Service information
   - Transactional information
   - Brand/company information

2. Internal information includes :
   - Company's vision, mission statements
   - Staff information
   - Standard operations procedures
   - Competitive metrics and industry trends

6. COMPONENTS OF DIGITAL KNOWLEDGE MANAGEMENT

Though digital KM may not sound complicated together information in current digital era there are six major components, which each an every organization usually follows.

Strategy: It is important for every organization to work for a single goal. Outlining digital KM's purpose will clearly identify needs of a business system. Once when the strategy is put in place, it will be easy for an organization to monitor and progress the goals of the organization.

People: The entire digital KM revolves around the people in an organization. Access to digital KM for teams and individual within the organization will allow stakeholders from various parts of the organization to plan and implement the strategy.In this process professionalswill contribute knowledge for trainers to access and use of information.

Processes: People of the organization will employ best practices and resources to gain knowledge, most organizations will optimize the knowledge sharing process while knowledge is being shares. Organizations also need to put processes in place for identifying, collecting, documenting and sharing knowledge in the future.

Technology: Organizations should be aware of limitations of technology in place to supports knowledge sharing. Though budget or IT limitations restrict the knowledge sharing method and process, organizations use all their ability to grow knowledge base, categories and sharing systems in the future.

Content: All transferable knowledge that is to be delivered will reside in form of some content format. It is critical that this contentcreation should be engaging, understandable formats will, or the knowledge transfer will be obsolete. A well-streamlined content creation and delivery process will effectively engage people of the organization during knowledge transfer.

Culture: It is critical for an organization to break down barriers within the organization to manage knowledge effectively, whether they are management hierarchies, team silos, or political or social cliques. The work culture of an organization will play a prominent role here. it should encourage information sharing across every corner of the organization, The knowledge sharing culture should be such that it benefits both individuals and the organization. Knowledge sharing should be cultivated and nurtured just like any other part of your company culture.

7. BEST KNOWLEDGE MANAGEMENT SYSTEMS AVAILABLE IN INDIA

KM is a multi-disciplinary strategy that needs to be shared, created, used, and managed effectively by an organization to achieve organizational goals through which, every individual in the organization can gain and share the knowledge.

The need for KM in emerging economies like India should be well managed. This is done effectively by employing KM system. Following are a few KM systems serving in India.

1. **monday.com:** Monday.com is an open framework that is designed to help businesses in the development their essential solutions. This frame works has several visual and editable templates from which, work operating system allowsseamlessintegration of tasks from a single workspace. Employees of Stack Roger Solutions Private Limited, Hyderabad have mentioned some of the best features of Monday.com.

2. **ProProfs Knowledge:** ProProfs Knowledge is an effective and a straight forward tool. A stack overflow member mentioned it to be a "thoughtfully created to improve internal team collaboration and client assistance." Another member of stack overflow community said "It aids in developing both an inner body of Knowledge for your staff and a self-service information base for your customers."

3. **ClickUP:** Managing projects through ClickUP allows KM, processes, tasks, and time very efficient. This platform is loaded up with many features like, extension tools, including documents and web links integration, collaboration and reporting knowledge bases. For example, documents can all be made on ClickUP allowing all Teams members to post comments and work together immediately.

   In KM industry ClickUP is known for its exquisite formatting across documents that are commonly shared between teams, allowing teams to linkworkflows, making the execution of ideas simpler.

4. **Scribe:** Scribe is a modest KM solution to manage articles in a knowledge base. It is often used as a health care management system. Adithya Excel Hospital in Dammaiguda, Hyderabad is using Scribe for last 3 years. They mentioned following as their advantages.

- Seamless procedure to edit and copy instructions.
- Take photos while you work through a procedure.
- Easy to generate stepwise guidance.

5. **Zendesk:** Zendesk is an adaptable, expandable, and opensource platform for KM, because of its highly adaptable in Customer service segments.

8. HOW KNOWLEDGE MANAGEMENT IS RELATED TO TRAINING AND DEVELOPMENT?

Organizations in today's fast-growing economies are investment of enormous time, capital, and resources in KM, it has become essential more than ever for organizations to make their training sessions as effective as possible. And this is where it is important for organizations to develop a strategy to make use of KM more efficiently.

Organizations preserve their intellectual capital and takes special care in capturing and preserving knowledge. Organizations encourage knowledge distribution and collaboration, and this methodology makes learning more efficient and makes a ton of difference in training new hires. Thus, it is evident that KM is crucial.

A pool of intellectual capital sorted and stored by organizations is used alongside better KM systems, which enables effective training and development in the organization. It can be overwhelming for a recruit. A well-organized KM repository plays a key role in fostering collaboration during training and learning sessions. All employees are encouraged to work to their fullest potential by learning knowledge from their co-workers.

KM and training and development are intertwined in important things such as developing skills and behaviours. Traditional KM techniques are used to implement many ways to carry forward and transform informal learning

The raise of informal knowledge learning and sharing is making path for new roles in the segment of learning and development communities like social learning strategist, curator, community manager and collaboration specialist.

From the start of previous decade, KM practitioners across the world have endeavoured to capture best KM practices. They practices made sure others will learn, adapt, and even evolve their practices to enhance worker performance of every individual in the organization. Over the years, many learning and development professionals realized that, formal classroom learning, or e-learning, is merely a procedure to impart knowledge for advanced skills. However, the elective data flow and informal learning methods can significantly influence performance and engagement in an exceedingly a lot of just-in-time, economical manner.

There are four ways in which, learning professionals will invest their effort in cognitive content management strategies:

1. **People**: Getting people together face-face or virtually to exchange the best practices and resolve the challenges and go on.
2. **User-Generated Content**: Set up the governance structure around people who share their best content and define the process to support user-generated content for legal education programs
3. **Content-Curation**: Develop learning tools to include external content and farming to support the selection of resources, information or learning.
4. **Project-Debriefs**: Carefully review the lessons to be learned, what to avoid and what to change to continually improve results. KM and learning and development use many different skills to achieve common goal: equipping specific employees with knowledge and skills to improve performance technologies such as communities, user-generated content, content improvement, and project reporting will serve the educational purpose more. These KM techniques, along with other techniques such as knowledge objectives, training, and data collection, have been replicated, particularly in learning and development approaches where technology and collaboration are more popular.

9. CONCLUSION

Companies have understood that same as physical assets, intellectual possessions and corporate knowledge can be managed effectively to improve performance. The main attention of KM is to link people, process and the technology in an effort to stowing and exploiting the corporate knowledge. An integral role played by data organizers (future knowledge managers) is absolutely crucial in making these connections possible.

Organizations are concerned with the management of executing KM in their organization as they face a number of challenges in emerging sound methods for this still evolving area of management practice. Though it is difficult to execute all the different kinds of KM practices, the organizations get benefited only when KM in implemented. Implementing the best practices of the organization is one of the method to overcome this concern.

The main concept in KM is learning. Learning revolves around interaction with number of sources in a useful way to build new ideas, knowledge and understanding. KM also helps in effective learning and in gaining competitive advantage for an organization in their respective industries.

REFERENCES

Al-Hawary, S. I. S. (2015). Human resource management practices as a success factor of knowledgemanagement implementation at health care sector in Jordan. *International Journal of Business and Social Science, 6*(11), 83–98.

Allameh, M., Zare, M., & Davoodi, M. (2011). Examining the impact of KM enablers on knowledge management processes. *Procedia Computer Science, 3*, 1211–1223.

Boadu, F., Dwomo-Fokuo, E., Boakye, J. K., & Kwaning, C. O. (2014). Training and development: A tool for employee performance in the district assemblies in Ghana. *International Journal of Education and Research, 2*(5), 130-146.

Bresman, H., Birkinshaw, J., & Nobel, R. (1999). Knowledge transfer in international acquisitions. *Journal of International Business Studies, 30*(3), 432–462.

Chaudhury, N. B., & Achrya, P. (2003). Knowledge management. In *Paper presented at the MANLIBNET 5th Annual National Convention.* Xavier Labour Research Institute, Jamshedpur, March 6–8, 2003.

Chawla, D., & Joshi, H. (2010). Knowledge management initiatives in Indian public and private sector organizations. *Journal of Knowledge Management*, 811–827

Cabrera, E., & Cabrera, A. (2005). Fostering knowledge sharing through people management practices. *International Journal of Human Resource Management, 16*(5), 720–735.

Choy, C. S., & Suk, C. Y. (2005). Critical factors in the successful implementation of knowledge management. *Journal of Knowledge Management Practice, 6*(1), 234–258

Gupta, S., Tuunanen, T., Kar, A., & Modgil, S. (2022). Managing digital knowledge for ensuring business efficiency and continuity. *Journal of Knowledge Management.* Epub-ahead-of-print. https://doi.org/10.1108/JKM-09-2021-0703

https://prowly.com/magazine/digital-knowledge-management-strategy

https://www.thecloudtutorial.com/digital-knowledge-management

Inkpen, A. C., & Beamish, P. W. (1997). Knowledge, bargaining power, and the instability of international joint ventures. *The Academy of Management Review, 22*(1), 177–202.

Martensson, M. (2000). A critical review of knowledge management as a management tool. *Journal of Knowledge Management, 4*(3), 204–216.

Ravanpykar, Y., Fyzi, J., & Pashazadh, Y. (2014). Examine the relationship between knowledge management with organizational learning and employee empowerment in national companies of south oil producing. *Indian Journal of Science Research, 5*(1), 284–295

Sanghani, P. (2009). Knowledge management implementation: Holistic framework based on Indian study. In *Pacific Asia Conference on Information Systems (PACIS), Association for Information Systems, PACIS 2009 Proceedings.*

Sanchez, R., & Mahoney, J. T. (1996). Modularity, flexibility, and knowledge management in product and organization design. *Strategic Management Journal, 17*, 63–76.

Singh, M. D., & Kant, R. (2008). Knowledge management barriers: An interpretive structural modeling approach. *International Journal of Management Science and Engineering Management, 3*(2), 141–150.

Sachan, D. (2002). Knowledge management: Challenges for the information professionals/librarians. In *Paper presented at the MANLIBNET 4th Annual National Convention.* National Institute of Financial Management, Faridabad, April 3–5, 2002.

Wells, J. (2021, March 8). *KM World 100 companies that matter in knowledge management.* KM WORLD.

12. Alcohol Dependence Syndrome (Continuous Use)—Psychotherapeutic Intervention

Hospital No: 4014/9/2021/Manasa
Date: 19/9/2021
Consultant Name: Mr. Srikanth Goggi * Dr. Saroj Arya**

Sociodemographic data:

Name	:	Mr. S
Gender	:	Male
Age	:	32 years
Marital Status	:	Married
Mother Tongue	:	Telugu
Education	:	Intermediate
Occupation	:	Carpenter (unemployed currently)
Religion	:	Hindu
Residence	:	Urban
Family Type	:	Nuclear
Number of Family Members	:	4 (Including Patient)

Referral:

The Referral Source	:	Psychiatrist's Note
Referral Reason	:	Interventional Therapy

The client and his wife served as confidential informants. The data was sufficient and credible.

Chapter 12 DOI- 10.4324/9781003397175-12

1. MAJOR GRIPES

As stated by the Client

- Marital issues that have persisted for almost two years
- Not able to maintain employment; received a warning letter from my superior 7 months ago
- 7 months of sleeplessness and exhaustion
- He disregarded his responsibilities as a family man.

The Source Claims That:

- Heavy drinking during the last six months
- Eighteen months ago: Getting arrested for driving under the influence of alcohol or drugs
- Aggressive behavior, agitation, exhaustion, and a lack of domestic responsibilities

Nature of illness:

- **Onset:** Insidious
- **Course:** Continuous
- **Progress:** Deteriorating

2. CASE SUMMARY

The index client was admitted to the Manasa psychiatric hospital seven months ago (in the year 2021) for binge drinking. The customer's wife was brought in to provide some further background on the customer. They tied the knot in 2011, and she has since followed him to the hospital for treatment, where she has been residing with their two children at her parents' house. Mr. K was a skilled carpenter at an architectural business. Client started drinking alcohol when he was 14 years old, when several of his classmates brought beer to school and forced him to try it.

Following this encounter, he began consuming alcohol socially once every three months while out with friends as a way to celebrate becoming adulthood. He had always be able to keep his parents in check at home so long as he kept his breath minty. His behavior remained same during Intermediate. Then he became an apprentice to the town carpenter and learned his trade from the ground up. At the age of 21, he moved to Hyderabad in order to establish himself professionally as a carpenter, a trade he really enjoyed. He has been accepting contracts for close to four years.

He found work in the field of architecture when he was 25 years old. He liked his job working on corporate mega-projects. Over time, he and his coworkers all became heavier drinkers. At 27, he found himself in an arranged marriage, something he said he was thrilled with. For next two years, everything went swimmingly. As his alcohol use became more frequent and excessive over time, his wife began nagging him to sober up and take responsibility for the family. Because of this, he often became irritable and agitated, and the pair fought often. He started drinking more, and he said that his coworkers often invited him to join them at bars after a long day of work.

He steadily increased his alcohol use, resulting in everyday drinking and physical dependency. He dodged accountability in numerous areas. When his wife made unpleasant requests, he disregarded her. This practice evolved regularly: client had fun with coworkers, and then felt team spirit at work the next day. The customer favored beer and vodka, drinking a maximum of 32 drinks each day. He usually had 10–12 drinks every day.

After asking his wife about his drinking habits, investigators learned that he had been arrested three times, had fought with neighbors, brought alcohol to work and received numerous warnings for his behavior, had relationship issues with his wife, had financial difficulties, etc. Several blackouts and signs of physical dependence (drinking before work, intense cravings when he could not drink, skipping work to drink all day) were present in his life.

She said the customer was not putting in the necessary effort at work. Even when she asked for his help monetarily, he refused. The financial support from her parents was minimal. The client's wife claims her husband had a six-month drinking problem. The day he and three friends were arrested for DUI (driving under the influence) and noise violations was the one day he did not drink. To convince him to stop drinking, she returned to her parents' house, denied sex, contacted the police, and did more. She also worked hard to promote his sobriety by doing lovely things for him, talking positively about him, and doing good things when he did not drink.

She had concerns about their relationship and his lack of responsibility, which she believed stemmed from his incapacity to hold down a job, care for children, etc. She was worried about him since his drinking had gotten out of hand in the previous six months. There was a noticeable uptick in his wrath, irritation, and restlessness. Despite all of this, with the aid of her parents' financial support, she joined him at Manasa Hospital so that he could become sober via treatment. There were no signs of a brain damage, seizures, hallucinations, paranoia, suspicion, intrusive or unwanted thoughts, compulsions, repetitive actions, depersonalization, or derealization.

The client's premorbid character traits point to his having been an optimistic, fun-loving individual. He is anxious and irritated when given responsibilities since he does not want to take charge. He has a positive outlook on life and is good at making friends. He is easily offended and has a low tolerance for criticism. His dietary habits were unpredictable, and he also sometimes smoked and imbibed. Restless and distracted in the psychomotor activity; speech was relevant, coherent, and goal directed; no abnormality in the stream, form, and possession of thought; worried, guilt feelings appeared in the content of thought; subjective and objective tensed, sad; no pity. There was no discernible impairment to memory, reasoning, or perception, and Level IV Insight was determined to be present.

Working Diagnosis: ICD-10 criteria for F1x.25: Alcohol Dependence Syndrome -Continuous usage are met based on the client's presenting symptoms, medical history, and minimum systemic evaluation.

3. ASSESSMENTS FOR THE PURPOSE OF THERAPY

- **Michigan Alcohol Screening Test**—A measure of alcoholism's severity, it was utilized..
- **Bender Gestalt Test (BGT)**—Evaluation of visual-motor perception.
- **Behavior Analysis**—For the purpose of analyzing one's typical behaviors.

4. RELEVANT TEST FINDINGS

1. **Michigan Alcohol Screening Test**—Client has a drinking issue, as shown by his score of 9.
2. **Bender Gestalt Test (BGT)**—The client has shown satisfactory levels of both visuomotor and perceptual coordination. There was no evidence of overwriting or repetition, and he was successful in reproducing six of nine different patterns. That means his memory is good enough.
3. **Behavior analysis**
   a) **An first assessment of the circumstance:**
      Absurd Acts of Behavior:
      i. Distress, rage, agitation, and exhaustion
      ii. Drinking in excess
      iii. Fights often with neighbors.
      Deficiencies in Behavior:
      i. Regular Activities
      ii. Ignoring domestic duties
      iii. Low sense of self-worth.
      Behavioral Strengths:
      i. Wife, mother being supportive
      ii. Knowledge and skills to earn and live independently.
   b) **Elaboration on the Nature of the Issue**
      Client consumes excessive alcohol. In spite of several warning from his professional place, he could not control and lost the job. He was idle, not taking care of basic responsibilities at home financially and that lead to interpersonal conflicts between couple. He had many debts and often fights with near and dear ones and created havoc in their life.

The client was arrested for drinking and drives case but still unable to control taking alcohol.

c) **Analyzing What Drives People**

The customer was determined to quit drinking and was admitted to the hospital as a result. He felt terrible about his actions and wished he could live a regular life, supporting his loved ones.

d) **Analysis of Change**

The client grew up with a father who was forceful and stern, and a mother who doted on him and protected him. He was too terrified of his father to show any emotion at home. His upbringing was normal in every other way. He is more obstinate and often displays tantrum-like behavior. He gets along well with his brothers, sisters, and cousins.

e) **Evaluation of Self-Regulation**

After completing therapy, the client feels in charge of his life and is prepared to engage in treatment in order to permanently abstain from alcohol.

f) **Examining Pertinent Social Connections**

The patient's marriage is in trouble. Client often yelled while inebriated, making children hesitant to approach him. The client's mother blames his wife for turning him into an alcoholic and a spoilt brat, yet she still stands by him no matter what. Their relationship remains amicable. He gets along well with his brothers and sisters.

g) **Evaluation of the Physical and Cultural Setting**

The client's personal and professional relationships were tense. His record of arrests and violent altercations gave him a reputation as a violent drunk. He saw alcohol use as a normal part of his culture and a sign of maturity, and this attitude toward alcohol persisted throughout his life. His perspective gradually led to his addictive behavior. After he departed, his relationships with his friends and family were strained, and he lost his job and his extended family.

3. THERAPEUTIC FORMULATION IN PSYCHOTHERAPY

Client index is a 32-year-old married Hindu man from an urban background who has only completed an intermediate level of education. He was brought in by his mother and wife with the chief complaints of excessive alcohol consumption, including being arrested for DUI, exhibiting aggressive behavior, restlessness, fatigue, not taking responsibility at home, and marital issues. After 7 months of increasing behavioral problems, hospitalization was finally necessary. Adolescence brought out his obstinate, belligerent nature, and he also began drinking at that time. In addition, his father was quite stern and terrified of him, so when the client saw his father drinking toddy, he assumed that drinking was commonplace in his village and that he, too, could do so without consequence.

He started off as a sociable drinker who was able to keep his drinking under control until he reached Intermediate. After completing his apprenticeship, he moved to Hyderabad to pursue a career in the furniture industry, where he worked and made money by collecting contracts from local businesses. There, he met new people and began drinking with them. The client's drinking habit rose to 10–12 drinks per week when he found a job and felt comfortable in at an architectural business; this was reinforced gradually and was providing him a sense of fulfillment. It was great up until marriage, but then the couple started having problems in many areas. This included communication problems, aggressive attitude, a drop in professional success, conflicts with the neighbors, and more. Lack of control over domestic and professional obligations led to a rise in alcohol use. When he got jailed, lost his job, and refused to accept responsibility for his actions, his ability to operate in society and the workplace declined. In addition to regret and concern for the future, he has recently sought assistance in order to kick his drinking habit. Alcohol Dependence Syndrome, Continuous Use (F1x.25) according to ICD-10 criteria was the preliminary diagnosis based on the patient's clinical history and test results.

Premorbid stability, a lack of debilitating symptoms, and a strong desire to undergo therapy all point to a positive prognosis. Psychological psychotherapy and medication were designed to assist patients cut down on and eventually quit drinking, as well as learn and practice abstinence-sustaining skills such dispute resolution within marriage, problem-solving, coping, and relational enhancement.

Therapeutic Objectives

Short-term goals:

- Inform the patient about their current health status
- To strengthen and sustain the drive for improvement
- Introducing effective coping methods.

Long-term goals:

- Continuing sobriety
- Relating to others and fostering strong families
- Methods to avoid relapse.

Methods of Treatment

Mind training

- Training to improve motivation
- Behavioral and cognitive therapies
- Preventing relapse.

4. SESSION 1—PSYCHOEDUCATION

Psychological counseling concerning alcoholism and its treatment was provided to the client. Members of the family were given the tools they would need to work together effectively and to better communicate with one another. The client was also taught about the dynamics of triggers and how to recognize them. She was instructed in effective cue control tactics for optimizing therapeutic results. Interventions were discussed in terms of their potential to aid people at each of the four phases of the change model: (1) initiation, (2) maintenance, (3) cessation, and (4) relapse. This helped the client understand the roots of her habit and the factors that keep it going.

5. SESSION 2—ENHANCING AND MAINTAINING MOTIVATION TO CHANGE (MET)

The client's desire to alter their conduct was evaluated after the first session. The therapist and patient then collaborated to implement the following steps discussed in the session:

- Setting objectives for health, interests, hobbies, relationships, and financial security.
- Understanding the positives and negatives of alcohol misuse is crucial for therapists to grasp its impact on the client's life.
- Include consequences of continuing alcohol use to persuade clients to reduce or quit consumption and prevent long-term impacts.
- Motivators for quitting include retaining family, career, independence, and physical wellness.
- Effective tactics for reducing or ending drug abuse include generating non-abusive social chances and reconnecting with hobbies and interests.
- Handling dangerous circumstances (e.g., associating with alcoholics, solitude, boredom, and bad family relationships)

Presented a wrap-up of the meeting's findings.

6. SESSION 3—COGNITIVE BEHAVIOR THERAPY

In session three, the client was taught how to utilize the stimulus control approach to reorganize or avoid potentially dangerous circumstances. This strategy encourages the client to spend more time engaging in activities that do not include alcohol.

The therapist has pinpointed both the initial and subsequent causes of the client's conduct.

Primary triggers were:

- Bars
- Friends in the workplace
- The festivities
- Celebrations of all kinds

Secondary causes include:

- To be bored
- Arguments with the spouse
- A desire for
- Anxiety

Inner excitement is a common secondary trigger that leads to substance abuse. The goal of therapy was to help the patient develop strategies for dealing with the triggers that had been identified, such as systematic extinction, learning alternative responses or solutions, increasing life skills like decision-making, expressing emotions, saying no, using time wisely, rejection, social skills, etc., through role-playing and group work.

7. SESSION 4—MINDFULNESS

Client was taught urge control strategy to cope with impulses and notice and adjust thoughts, emotions, and plans that lead to alcohol usage after reviewing prior session. This mindfulness method, "**Urge Surfing**," helps us avoid avoidance and conflict. The rationale to practice mindfulness and how it helps was explained as "when we use mindfulness we stay exposed to the thoughts feelings or urges for their natural duration without feeding or repressing them. In fact, if we just let an urge be—non judgmentally—without feeding it or fighting it (Fighting it is just another way of feeding it anyway) then it will crest subside and pass. Of course they come back again but over a period of time. However each time you overcome a bout of cravings they become less intense and less frequent if we do not feed the urges and if we do not give in to the addiction. Moreover our mindfulness technique of urge surfing improves".

7.1. Technique Procedure

Purpose: To "ride out" urges in a different manner until they go. Preparation:

- Remember that impulses fade on their own.
- In my imagination, cravings are like ocean waves that come and go. They start tiny, flourish, and eventually disappear.
- Regularly practice mindfulness and observe impulses.
- Mindfulness prepares us to surf these waves without succumbing.

Urge surfing:

- Practice mindfulness.
- Be mindful of your breath. Never change it. Let the air breathe.
- Reflect on your ideas.
- Gently return your attention back to the breath without judging or battling cravings. Observe how cravings influence the body.
- Focus on the location where the desire is felt and observe:
- Observe sensation quality, location, bounds, and intensity.
- Observe that these changes occur throughout in and out breaths.
- To concentrate, repeat the technique with each body component involved.
- Be inquisitive and observe changes over time.

Replace the anxious desire that wanting would go away with curiosity in our reality. When we do this, appetites vary, peak, and subside like ocean waves. It gets more bearable.

Day's work done; participants were encouraged to maintain regular mindfulness practice until therapy sessions were completed.

8.   SESSION 5—COPING STRATEGIES

The goal of the workshop was to teach participants effective methods of managing with negative affect and emotional issues, as well as correct any erroneous beliefs they may have had regarding alcohol.

The following are craving-coping strategies:

- Distraction
- Discussing cravings
- Adhering to cravings
- Recalling bad effects of cocaine misuse
- Using self-talk.

Therapists cautioned that the patient's desire to learn these techniques alone may not be enough to permanently alleviate the patient's cravings. One may lessen the frequency and severity of cravings and find less frustration and distress from them with practice.

8.1. Distraction

Distraction, particularly physical activity, may help with conditioned cravings for alcohol or other substances. Prepared a list of diverting activities with patients in preparation of future cravings, such as walking, playing basketball, and relaxing. Client behavioral interaction was used to create a list that minimizes alcohol cravings.

8.2. Discussing cravings

Explained how talking about desire in a supportive atmosphere might help the person stay abstinent and lessen anxiety and susceptibility. It may also help patients recognize signals.

Close family members were trained since they may feel concerned when clients speak about desire because they anticipate it to lead to use.

8.3. Adhering to cravings

This strategy lets cravings come, peak, and dissipate without resisting or giving in. Imagery helps acquire control by avoiding resistance.

You should rehearse steps in sessions and at home before cravings. Tell client that the goal is not to eliminate desires, but to make them less anxiety-provoking and harmful, making them simpler to manage. It involves paying attention to the urge, concentrating on the location where it happens, and remembering unpleasant outcomes.

When desiring, many individuals recall only the favorable effects of alcohol and ignore the bad ones. Thus, while desiring, reminding oneself of the advantages of abstinence and the risks of using might help. Patients might remind themselves that using will not help.

8.4. Recalling bad effects of cocaine misuse

Many people have unconscious automatic ideas that accompany desire. Automatic desire ideas sometimes include urgency and exaggerated repercussions, such as "I must use now," "I'll die if I do not use," or "I cannot do anything else until I use." Recognizing and combating automatic thinking helps with desire. Therapists pointed out cognitive distortions during sessions (e.g., "A few times") to assist clients notice automatic thinking.

You've mentioned you need to utilize today. When you think, do you notice? ").

After identifying automatic thoughts, positive self-talk makes it simpler to face them. Challenge the concept ("I won't really die if I don't have cocaine") and normalize desire. (e.g., "Craving is uncomfortable, but a lot of people have it and it's something I can deal with without using").

The session ended with the participant pledging to practice mindfulness and coping techniques and calling his wife to discuss their marital issues.

9. SESSION 6

The purpose of the meeting was to assess the client's method implementation. The family was also counseled on the value of partner participation in therapy and the need of open lines of communication in maintaining a positive bond between partners. Is it crucial to attend AA meetings and other support organizations after leaving treatment to stay sober?

10. SESSION 7: RELAPSE PREVENTION

The client was instructed on relapse prevention. This was crucial to alcoholism therapy. Consider Marlatt and Gordon's RP model. Model indicated that immediate determinants (high-risk events, coping abilities, outcome expectations, and abstinence violation impact) and covert antecedents (lifestyle variables including impulses and cravings) might cause relapse. The RP model includes several particular and global intervention methods to help therapists and clients handle each relapse stage.

As soon as the client was released from the hospital, the therapist ended the session.

Evaluation: The client was asked to come for evaluation after 3 months.

11. OUTCOME

After intervention, client's mental and physical health improved significantly. Therapist told client to practice all strategies as homework.

12. FOLLOW-UP

Clients and families were urged to follow up.

Mr. Srikanth Goggi*

Consultant clinical psychologist & Asst. Professor - University of Technology, Jaipur, Rajasthan, India

Dr. Saroj Arya**

Professor & Head, Department of Psychology - University of Technology, Jaipur, Rajasthan, India

Corresponding Author: Mr. Goggi Srikanth, Email:srikanthgoggi@gmail.com Mobile: 9849137095

13. Social media and aggressive behavior in youth

Aseema Misra, Ishwari Pednekar, and Sai Sushma Ganna
Woxsen University, Hyderabad, Telangana, India

ABSTRACT: Social media empowers the flow of information, enables connectivity, and ensures the exchange of ideas. In today's changing world, we are dependent on it because of the convenience it provides. It is widely used to surf data related to various fields. In the corporate world, it is used to conduct virtual meetings and conferences. Young teens use it too to get an insight into the current trends. However, its continuous usage becomes a cause of distraction. The youth tend to be obsessed with it as one gets to live a virtual life that is different from reality.

The youth, being vulnerable by nature, fall prey to social media and become highly dependent on it. This results in low self-esteem. The research paper aims to understand the negative correlates of social media on youth and the implementation of certain measures which could improve their lifestyle. This could help in understanding their thought process and aid them.

KEYWORDS: Social media, youth, networking, adolescent

1. INTRODUCTION

Social media is an open platform that includes networking sites. It enables people to voice their opinions on a large scale. As it is convenient, it has replaced electronic media like television and radio (Shetty et al., 2015). Lucrative electronic devices such as mobile phones, iPad, Laptops, etc. assist the usage of social networking sites at any point in time (Siddiqui and Singh, 2016). The innovation of these handy devices compels everyone to be a part of the social media circle. It involves apps such as Snapchat, Instagram, LinkedIn, Twitter, Pinterest, etc.

In the past, newspapers ensured the flow of information in a day's time from various parts of the world. Social media such as Twitter aims at the same but at a faster pace (minutes after the occurrence of the event). Messenger and WhatsApp replaced the communication channels of the past. Conversation on such apps is now in the form of emojis, wherein the emojis hold different meaning to every individual. This increases the gap in understanding. Thereby, the messaging apps tend to be misleading though they focused to be a medium

Chapter 13 DOI- 10.4324/9781003397175-13

which assists interaction. Infotainment is also a part of social media. The most popular infotainment today is YouTube.

Social networking is more engaging as it involves the use of visuals while communicating. It saves time and energy, which would be otherwise used in personal interaction or finding data through books and newspapers. It also saves the resources used. For example, LinkedIn reduces the human effort of submitting resumes at various places. Instead, one could directly contact the recruiter.

The usage of social media impacts the field of education as well as business (Siddiqui and Singh, 2016). It is used by:

a) Teacher: A teacher could use it to extract ideas for teaching aids. It could also be used to spread information to students. It enables an online mode of teaching.

b) Students: In order to interact with fellow students, it is used to form groups and take part in research projects. Students could use it to grasp information in less time.

c) Business: To increase the marketing of a product or service businesses use social media. It could also be used for surveying, which reduces the time involved in gaining feedback.

d) Customers: A customer could use it for complaining, demanding, and fulfilling their needs. It provides a variety of choices to customers.

e) Society: It helps society in networking, connecting, and contributing to bringing positive change.

Screen exposure begins at an early age among youth. US children aged below two spend 42 minutes per day on screen media (Nesi, 2020). American statistics show that over 60% of kids aged 13–17 have at least one profile on a social networking site (Richards et al., 2015). In India, about 53% of the children spend at least 2 hours daily on social media. Due to increased exposure to social media, children get habituated to it by the time they reach adolescents.

Research suggests that social media exposure to violent activities and behaviours has a detrimental psychosocial effect on children and adolescents (Marcum et al., 2010; Ybarra et al., 2007a, 2007b). In addition, youth has normalized aggressive behaviour and are more likely to endorse it too (Hinduja and Patchin, 2013; Williams and Guerra, 2007).

The present research paper aims to understand the negative correlates of social media – cyber-bullying, phishing, and cyber suicide. It provides applicable measures to control the aggression among youth due to social media.

2. YOUTH AND SOCIAL MEDIA

As young adults are influenced the most by social media, they tend to construct identities that are different from their offline personas. The portrayal of a happy and happening life (which includes tattoos, designer clothes, expensive cars, etc.)

is done on social media. This provides them with a sense of accomplishment as the youth is insecure by nature. Their insecurities include body image, social circle, and relationships.

To meet these insecurities, they follow the current trends. The adaptability of the new age fashion, usage of slang, owning updated technological devices, etc. are parts of following trends. By doing this, the youth feels accepted by their peers, which provides them with a sense of superiority.

This mindset of youth is well understood by a few people, especially scammers, who lure them to gain benefits of any kind. Scammers do it to loot money, personal gain, promote malicious software, etc. The youth fall prey to various activities which are as follows:

a) Cyberbullying: Cyberbullying pertains to bullying through instant messaging apps, e-mails, or any other related electronic communication technology (Hinduja and Patchin, 2012; Kowalski et al., 2012, 2014). Any form of misbehavior or bullying at school is more likely to be associated with cyberbullying (Hay and Meldrum, 2010; Hinduja and Patchin, 2007; Nixon, 2014).

b) Dating: It is the second most popular platform for paid online content (Matthews et al., 2018). A lot of dating apps are available today which provide a variety of features. This includes adding pictures, enabling location, and other personal information. This form of networking could be risky and involves the formation of unsafe social contacts (Sumter et al., 2017). The data provided could be misused and lead to victimization experiences. Using tactics such as coercion, fraud, force, and bogus job offers, the criminals scour dating platforms in an attempt to exploit the personal situations of down on their luck individuals by promising to help them out. They groom their victims, establish a false sense of trust, and ultimately meet them in person. Before long, they force the targets into sex work or forced labour.

c) Honeytrap: This is a modus operandi that uses an online platform by providing the details of a popular figure, usually a government servant or a popular personality. They gain the trust of the youth by emotionally connecting with them. On obtaining their image or personal information, the youth is exploited through blackmail.

d) Phishing: Scammers use this method to gather information and later use this as a tool to trap users. As building fake profiles is easy, social media is used to scam users to gain monetary benefits. This includes extracting bank details, passports, or other personal details.

e) Gaming: Due to the increased aggression in gaming culture, the youth tend to be following it regularly. This impacts the thought process of adolescents as well as young adults. They get influenced by the nature of the game which is mostly based on a destructive mindset. Online gaming involves shooting, looting, or killing opponents which promotes bullying, violence, and group violence.

3. IMPACT OF SOCIAL MEDIA

It is found that social media usage among youth could have both positive and negative impacts. It enhances communication and plays a vital role in networking. However, the academic performance as well as the attention span of the youth declines with the increased usage of social media. It leads to psychological distress which in turn causes anxiety, depression, and low self-esteem. The lifestyle portrayed on social media, especially via scrolling causes inferiority among youth. This creates cognitive biases. Cognitive bias relates to the habit of opening and scrolling social media apps which forms a part of the unconscious part of the brain. The youth is attracted by persuasive technology. This term defines the social influence caused due to social media. Influence implies acting a certain way by indulging in video gaming, online dating, or following current trends.

Youth tend to become dopamine deficient as they feel they cannot stop scrolling. This causes the fear of being left behind or missing out. Due to this youth indulges in prolonged use of social media. This turns into an addiction. However, addiction is not determined by frequent usage alone but also includes the way an individual reacts to abstaining from it (American Society of Addiction Medicine, 2016).

Using a qualitative approach, secondary data was evaluated which was collected from newspapers and journals. This estimated that the total amounts of registered users of social media are 2.80 billion around the world. According to the recent reports, Facebook is one of the most popular social media that has achieved 2.23 billion active users in a month followed by YouTube, Twitter, and Instagram (Statista, 2018). It was also found that 95 percent of teenagers have full access to use smartphone and 45 percent of them stay online constantly (Monica and Jingjing, 2018). Thirty-one percent teens believe that effect of social media is mostly positive, 24 percent teens describe the negative affect and the largest share of 45 percent believe social media has neither negative nor positive affect on them (Monica and Jingjing, 2018). As understanding the impact of social media on teenagers by surveying teenagers themselves was not relevant, the research was further elongated.

Teenager benefited in various aspects when using social media including education, improving relationship, job opportunities, keeping in touch with the world, influence on exercise and social media as e-commerce (Anderson and Jiang, 2018). At present, teacher sends student's result, attendance sheet, homework, and exam schedule through email. E-commerce marketing has become popular among teenagers to buy and sell product easily without wasting the time (Jeniffer, 2014). E-commerce marketing is mostly social media-based.

Employees and freelancers found networking through professional platforms like LinkedIn and increased their chances for seeking jobs with more opportunities.

The most dangerous negative impact of social media on teenagers is addiction. Among all of the negative impacts, addiction has deteriorated amid teenagers. A survey conducted by Mediakix revealed that it is estimated 210 million people are now addicted to social media and Internet directly or indirectly (Mediakix, 2018; NCBI, 2016). Social media can destroy someone's prestige and dignity through fake account in social media and spreading fake and vogues news. According to the Digital Security Bill-2018, whoever tries to spread out false propaganda through social media will be punished ranging from 3 to 10 years imprison terms (Dailystar, 2018).

4. AGGRESSION DUE TO SOCIAL MEDIA

Throughout the adolescent years, cyber aggression is found to be increasing (Tokunaga, 2010). Violence-related acts include bullying, rape, group violence, harassment, etc. Initially, violence is triggered through social media comments, stalking popular social media figures, and sending threatening or insulting messages (Perren et al., 2012). Such activities on the internet turn into a daily practice. Online dating too forms a part of violence which is differentiated as sexual violence. Involvement in sexually risky behaviour amounts to violence, especially on adolescent girls (Perren et al., 2012).

Risky behaviour leading to violence tends to pave the way for habits of consuming alcohol, cigarette, or drugs. These behavioural changes in young adults go unnoticed. It is found that adolescents' behaviour online has a connection with their offline attitudes (Subrahmanyam et al., 2006, 2009). This might lead to involvement in gang violence. Gang violence relates to a group wherein criminal or destructive activities are carried out for profit. The gang members use networking websites like Twitter, Instagram, Facebook, etc. It has been found that 45% of internet users engage in at least one form of online offending which includes selling drugs, harassment, or uploading violent videos (Pyrooz et al., 2013).

5. IMPACT OF STREAMING MEDIA

An Indian case survey analysed the psychosocial impact of web series and online streaming content amongst youth on online platforms. A questionnaire was carried out for the study. The respondents included students from various colleges and universities. Entertainment series or movies streamed on either television or theatres. Currently, the trend has changed with the introduction of Over-the-top (OTT). OTT, originally used in western countries, is now adapted in India too wherein streaming is done via platforms like Netflix, MX Player, ALT Balaji and other video streaming platforms.

These online based platforms follow the binge-watching model as they are monthly or yearly subscription based. The target audience of these streaming platforms is 18–30 years audience. They provide the convenience to be accessed on smartphones, laptops, or TV. It includes various genres like comedy, action, romantic, sci-fi, etc. which provides a variety of options to choose from along with the control to watch as per one's personal choice, unlike television.

It is found that the binge-watching habit is closely related to negative feelings (Sung et al., 2005). The rhythmic pattern of binge-watching makes one less social. Teenagers tend to feel lonely and try to imitate the characters of the show they watch. They fail to discriminate the reality and fantasy (as portrayed in OTT platforms). The influence brought in by western culture affects the youth's lifestyle as a majority of the youth follow western culture.

The content streamed on OTT platform contains the usage of abusive language, vulgarity, rebellious nature against family, country, or religion, and other taboos in the society. The presence of nudity and obscenity in online streaming platforms tends to promote unnecessary sexual activity in youth. Apart from this, it was found that Indian youth spends around 9 hours on OTT platforms whereas in other countries it is 7 hours.

The research used a quantitative method. The sample was students of colleges and universities in Haryana. A total of 250 responses were received by asking close-ended questions. 48% of the respondents were 18–22 years old. The age of the respondents was 18 to 30. The key findings were as follows:

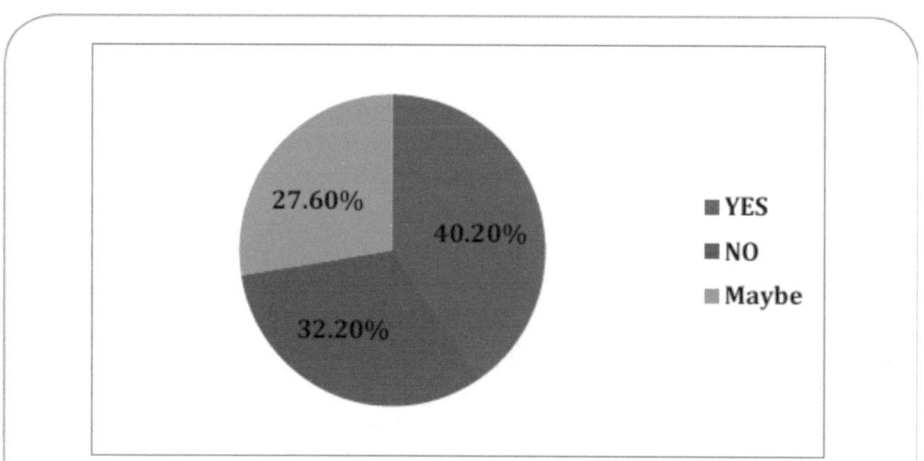

Figure.11 *Do you think that web series shows enhance criminal activity in society.*

The observation drawn from the first question implied that 40.20% of the youth agree that web shows enhance criminal activity in society. 32.20% disagree with the mentioned statement whereas 27.60% are doubtful about the impact of web series.

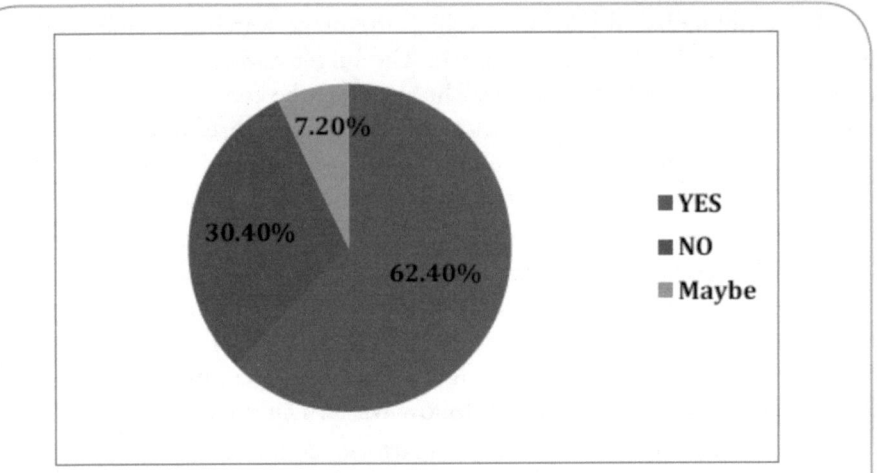

Figure. 12 *Do you thing that web series are changing the language and behavior of the youth/.*

The information received in the survey was that 62.40% of the respondents agreed that the language and behaviour of the youth are influenced by watching web series. However, 30.40% disagreed and 7.20% of the students were uncertain.

6. CONCLUSION

It is evident that teenagers who engage in social media tend to spend more than the stipulated time on social media in adolescents. This leads to a decline in performance, loss of memory and inability to concentrate. Being trapped in the loop of social media, youth tend to lose their people skills.

Indulging in games that are based on problem-solving and the usage of critical thinking instead of games involving shooting can be implemented. This would not hinder negativity in the subconscious mind of the gamer. OTT platforms' focus should be on delivering realistic stories for limited hours. This would prove to be less impactful as the youth would not indulge in binge-watching.

Limiting screen time and introducing "no tech day" in schools and at homes are the best ways to limit social media usage. This encourages the youth to have face-to-face communication and use books for the purpose of research. Apart from this, avoiding compulsive checking of messages or e-mails is an effective measure and encourages socializing. It has been found that social media has increased the number of individuals' social connections while decreasing the meaningfulness of these connections. This lack of social participation "decreases social capital circulating in the community and weakens trust relations among citizens. Individuals are becoming more isolated and inward looking" (Parigi and Henson, 2014).

Structural differences were highlighted between online and offline interactions. These structural differences lead to systematic psychological and behavioural consequences that are functionally changing the landscape of social life. The traditional or offline interaction involves spontaneous and unplanned flow of communication. They need not be formal in nature. These are the ones used in day-to-day communication. This improves the presence of mind and develops ones' behaviour. Gossiping too forms a major part of offline communication.

Online interaction, on the other hand, provides:

1. Fewer nonverbal cues: Reading, versus hearing, ones' opinions make the communicator appear less mentally capable. This suggests that text-based interaction may be dehumanizing, since derogating a person's mental capacities is a form of dehumanization.
2. Greater potential for anonymity: Greater anonymity is associated with disinhibition and aggressive behaviour – potentially because anonymity reduces accountability and thereby licenses bad behaviour. Online environments – which cater to anonymity and weakened social norms – may serve as a breeding ground for such behaviours.
3. Forming and maintaining ties online: Facebook users found that despite social media's explosion, the number of significant contacts that people report remains similar to before the rise of social media. It was found that every individual had around 5 intimate friends, 15 close friends, 50 general friends, and 150 acquaintances. Therefore, though forming social ties was convenient, maintaining them continues to be difficult.

The recommendations to reduce the time spent on social media by youth are as follows:

a) Turning off electronic device notification: Unnecessary notification kills teenager's valuable time and turns irritating. Statistic reveals a report which discovered that every people check their smartphone within 15 minutes or less than this duration, if they do not mute their smartphone notification option (Phobe, 2016; Picard, 2018).
b) Indulging in a new hobby: Teenagers should do something, which they have wished to do but never have enough time to do it. New leisure activities will keep their brain and hands distracted from wishing to use the social media. Going for long drive is one of the best options for teenagers. Participating in indoor and outdoor activities could also assist them get rid of social media addiction. Apart from this, getting a pet and watching it grow can emotionally develop a teenager.
c) Make and follow proper schedule: The teenagers are normally very sensitive and are easily distressed when asked to restrict their daily schedule. They prefer enjoyable schedule, which is easy to follow. They unintentionally end up browsing social media for hours. Focusing on

daily activities is the prime solution for teenager to concentrate in their study. Teenagers' life styles can be altered by having a regular wake up and sleeping time.

d) Spending time with beloved persons: Teenage in general need to be nurtured on a regular basis as they are more vulnerable to indulge in wrong doings. For this reason, teenagers must surround themselves with people who are optimistic and have a goal-oriented approach. They must be encouraged to be around people they trust or feel comfortable to spend time with. These people could be friends, siblings or grandparents. On sharing one's ideas and beliefs with people an individual is believed to turn more mature.

The usage of social media to a limited extent is favourable and provides a promising future to the youth. Using it as per the requirement maintains a positive spectrum for improving personality. It stabilizes and makes one realize their nature.

REFERENCES

Adachi, P. J., & Willoughby, T. (2011). The effect of violent video games on aggression: Is it more than just the violence? *Aggression and Violent Behavior*, 16(1), 55–62.

Anderson, M., & Jiang, J. (2018). Teens, social media & technology 2018. *Pew Research Center*, 31(2018), 1673–1689.

Dhiman, B., & Malik, P. S. (2021). Psychosocial impact of web series and streaming content: A study on Indian youth. *Global Media Journal*, 19(46), 1–7.

Donlevie, K. (2018). Screen time versus face time: How social media usage affects time spent face to face.

Ferguson, C. J., Olson, C. K., Kutner, L. A., & Warner, D. E. (2014). Violent video games, catharsis seeking, bullying, and delinquency: A multivariate analysis of effects. *Crime & Delinquency*, 60(5), 764–784.

Hay, C., & Meldrum, R. (2010). Bullying victimization and adolescent self-harm: Testing hypotheses from general strain theory. *Journal of Youth and Adolescence*, 39(5), 446–459.

Hinduja, S., & Patchin, J. W. (2007). Offline consequences of online victimization: School violence and delinquency. *Journal of School Violence*, 6(3), 89–112.

Hinduja, S., & Patchin, J. W. (2012). Cyberbullying: Neither an epidemic nor a rarity. *European Journal of Developmental Psychology*, 9(5), 539–543.

Hinduja, S., & Patchin, J. W. (2013). Social influences on cyberbullying behaviors among middle and high school students. *Journal of Youth and Adolescence*, 42(5), 711–722.

Huang, J., Stringhini, G., & Yong, P. (2015). Quit playing games with my heart: Understanding online dating scams. In *Detection of intrusions and malware, and vulnerability assessment: 12th international conference, DIMVA 2015, Milan, Italy, July 9–10, 2015, proceedings 12* (pp. 216–236). Springer International Publishing.

Kobiruzzaman, M. M., & Yaakup, H. (2021). Impact of social media towards society: A case study on teenagers. *International Journal of Education and Knowledge Management*, 1(3), 1–12. https://doi.org/10.37227/IJEKM-03-2018-14.

Kowalski, R. M., Giumetti, G. W., Schroeder, A. N., & Lattanner, M. R. (2014). Bullying in the digital age: A critical review and meta-analysis of cyberbullying research among youth. *Psychological Bulletin*, 140(4), 1073.

Kowalski, R. M., Limber, S. P., & Agatston, P. W. (2012). *Cyberbullying: Bullying in the digital age*. John Wiley & Sons.

Kutner, L., & Olson, C. (2008). *Grand theft childhood: The surprising truth about violent video games and what parents can do*. Simon and Schuster.

Lieberman, A., & Schroeder, J. (2020). Two social lives: How differences between online and offline interaction influence social outcomes. *Current Opinion in Psychology, 31*, 16–21.

Machmutow, K., Perren, S., Sticca, F., & Alsaker, F. D. (2012). Peer victimisation and depressive symptoms: Can specific coping strategies buffer the negative impact of cybervictimisation? *Emotional and Behavioural Difficulties, 17*(3–4), 403–420.

Marcum, C. D., Higgins, G. E., Freiburger, T. L., & Ricketts, M. L. (2010). Policing possession of child pornography online: Investigating the training and resources dedicated to the investigation of cyber crime. *International Journal of Police Science & Management, 12*(4), 516–525.

Matthews, J. A., Winkler, S., Wilson, P., Tomkins, M. D., Dortch, J. M., Mourne, R. W., ... & Vater, A. E. (2018). Small rock-slope failures conditioned by Holocene permafrost degradation: A new approach and conceptual model based on Schmidt-hammer exposure-age dating, Jotunheimen, Southern Norway. *Boreas, 47*(4), 1144–1169.

Mitchell, K. J., Ybarra, M. L., Jones, L. M., & Espelage, D. (2016). What features make online harassment incidents upsetting to youth? *Journal of School Violence, 15*(3), 279–301.

Monica, A., & Jingjing, J. (2018). Teens, social media & technology 2018. *Pew Research Center, 31*(2018), 1673–1689.

Nesi, J. (2020). The impact of social media on youth mental health: Challenges and opportunities. *North Carolina Medical Journal, 81*(2), 116–121.

Nixon, C. L. (2014). Current perspectives: The impact of cyberbullying on adolescent health. *Adolescent Health, Medicine and Therapeutics, 5*, 143.

Parigi, P., & Henson, W. (2014). Social isolation in America. *Annual Review of Sociology, 40*, 153–171.

Pyrooz, D. C., Sweeten, G., & Piquero, A. R. (2013). Continuity and change in gang membership and gang embeddedness. *Journal of Research in Crime and Delinquency, 50*(2), 239–271.

Richards, S., Aziz, N., Bale, S., Bick, D., Das, S., Gastier-Foster, J., ... & Rehm, H. L. (2015). Standards and guidelines for the interpretation of sequence variants: A joint consensus recommendation of the American College of Medical Genetics and Genomics and the Association for Molecular Pathology. *Genetics in Medicine, 17*(5), 405–423.

Shetty, A., Rosario, R., & Hyder, S. (2015). The impact of social media on youth. *International Journal of Innovative Research in Computer and Communication Engineering, 3*(7), 379–383.

Siddiqui, S., & Singh, T. (2016). Social media its impact with positive and negative aspects. *International Journal of Computer Applications Technology and Research, 5*(2), 71–75.

Sumter, S. R., Vandenbosch, L., & Ligtenberg, L. (2017). Love me Tinder: Untangling emerging adults' motivations for using the dating application Tinder. *Telematics and Informatics, 34*(1), 67–78.

Tokunaga, R. S. (2010). Following you home from school: A critical review and synthesis of research on cyberbullying victimization. *Computers in Human Behavior*, *26*(3), 277–287.

Williams, K. R., & Guerra, N. G. (2007). Prevalence and predictors of internet bullying. *Journal of Adolescent Health*, *41*(6), S14–S21.

Ybarra, M. L., Diener-West, M., Markow, D., Leaf, P. J., Hamburger, M., & Boxer, P. (2008). Linkages between internet and other media violence with seriously violent behavior by youth. *Pediatrics*, *122*(5), 929–937.

14. Metaverse: The new reality of entertainment and business world

Nikitha Kommareddy, Pinnamaraju Sujithvarma, and Koushik Kuma
Woxsen University, Hyderabad, Telangana, India

ABSTRACT: This study focuses on the impact of the metaverse on the entertainment and business industry, which has become one of the most relevant fictional universes that has affected the industry. The study analyzes the vital factors associated with the metaverse's core concept and explores the incorporation of techniques used by the entertainment and business industry. The study shows how these techniques have had a positive impact on the economic arena. Additionally, the study examines the core concepts of virtual and augmented reality and the tools incorporated into the metaverse's framework. Theoretical background and methodology are used to understand the perception of the metaverse. The study analyzes the key factors that facilitate the immersive virtual world associated with the metaverse. The study also examines the incorporation of business specifications in the entertainment industry, and relevant resources are identified. These prominent aspects of the study help to analyze the core concepts of the metaverse. The study aims to provide insights into the metaverse's real evaluation and its potential impact on the entertainment and business industry. It is important to understand the key factors associated with the metaverse's core concept to effectively utilize it in the entertainment and business industry.

1. INTRODUCTION

Metaverse is a concept that combines technology with science fiction, representing a universal, single, and immersive virtual world facilitated by Virtual Reality (VR) technology and Augmented Reality (AR). The metaverse has a significant impact on society, especially in the entertainment and business industries. Companies can leverage the platform to provide new experiences to consumers, potentially transforming their business performance. Experts believe that the metaverse will become essential for all companies to expand their businesses in the future.

This assessment aims to understand the significance of the metaverse in the entertainment and business sectors, as well as the challenges it faces. The research objectives are to understand the importance of the metaverse in the business and entertainment sectors and to identify the challenges the technology is facing.

In the entertainment industry, the metaverse is accessed through VR headsets, providing users with immersive experiences. However, the technology faces problems such as potential bullying, harassment, and assault. Privacy concerns also affect the business performance of the technology, particularly with eye-tracking technology.

This assessment highlights the significance of the new reality in the entertainment and business world, where the metaverse is a prominent player. Through this research, various concepts have been clarified, and the problems the technology faces have been identified. To gather information on the topic, keywords such as "augmented reality," "virtual reality," "metaverse," and "usage of metaverse in the entertainment industry and business sector" were used. This resulted in a plethora of information, and relevant articles were selected after careful cross-checking.

In conclusion, the metaverse is an essential technology for the entertainment and business industries. It enables businesses to deliver new and immersive experiences to users, potentially transforming their business performance. However, the technology faces several challenges, including bullying, harassment, assault, and privacy concerns. Understanding the significance of the metaverse in the entertainment and business sectors and identifying its challenges is crucial to the successful implementation and growth of the technology.

2.   THEORETICAL BACKGROUND

2.1. Introduction

The literature review chapter is essential in providing the theoretical background of the research topic and analyzing previous research on the subject of the metaverse. This chapter will identify the gaps in current knowledge and provide a better understanding of the main factors that have been studied by authors in the past. By utilizing previous literature, this chapter will contribute to the theoretical perspective of the study and remain effective in the long run.

2.2. Previous Literature

The metaverse has brought a new level of integration between computing technology and virtual reality, leading to a completely new approach to the internet. This has been made possible using augmented and virtual reality technologies. The metaverse has the potential to optimize human experience in a sustainable manner and has become a significant aspect in various industries worldwide. Augmented reality, followed by virtual reality, remains a standard aspect in deriving internet usage. The importance of the metaverse in different industries and sectors cannot be overstated, and it has become an exciting development that brings virtual reality into the realm of human experience like never before.

2.3. Gaps in Literature

This study aims to address the gap in the literature by exploring the impact of metaverse on the business and entertainment industry, which has not been thoroughly discussed by previous researchers.

2.4. Theoretical Perspective

Theoretical aspects are crucial in understanding the perception of the metaverse in relation to the internet and related technologies. The incorporation of psychological factors is also considered to develop a virtual world for users. The selected context evaluates relevant theories to understand the theoretical perception of the metaverse, which includes the consideration of technology-driven theories and psychological factors. These theoretical aspects aid in comprehending the metaverse's different attributes and provide a foundation for the course of action in the study.

2.5. Metaverse and its Significance

Metaverse is a virtual world that is created through the integration of virtual reality and augmented reality technologies that make it seem comfortable for users to interact with. It is an evolution of the internet that has no physical existence, and its technical specifications have become highly connected to users. The metaverse is accessed through niche applications, mainly for gaming and entertainment purposes, and its functionality and accessibility are controlled by computing ecosystems. The significance of the metaverse is largely due to two main technologies:

Virtual reality that stimulates the 3D evaluation of the metaverse, and augmented reality that enhances interaction with the real environment. The dynamic nature and technology specifications that are associated with the metaverse are typically evaluated by the internet of things, extended reality 3D modelling, edge computing, blockchain technology, and videos among other technologies.

The metaverse is a hypothetical interaction specific to the internet, and its conceptualization is all about a single virtual world facilitated by virtual and augmented reality technologies. Users can interact with this virtual world, and it has become one of the most reliable and highly connected aspects of the internet.

The metaverse is typically associated with gaming and entertainment purposes, but its scope can be extended to other industries, such as business. It has become an important technology that can be used to optimize human experiences in a virtual setting. Its dynamic nature and technology specifications make it a promising platform for the future.

2.6. Understanding the Perception of Metaverse in the Business World

The perception of the metaverse has become increasingly important for the business industry in recent years, as it has the potential to significantly improve the user experience by incorporating exclusive virtual reality technologies. The technical factors associated with the metaverse also make it a sustainable aspect that can be easily incorporated into business operations to enhance communication and target market audience accomplishment, allowing for a more in-depth approach towards digital insights and developing business sustainability.

The optimization of real-life evaluation using the metaverse is considered essential for digital transformation and enhancing transaction and shopping experiences in every sector of business. The virtual environment provided by the metaverse allows individuals to socialize and evaluate trading purposes that can be virtually accessed, leading to a sustainable impact on profit and revenue generation. Businesses of all sizes can access opportunities that come with the incorporation of the metaverse, allowing companies and brands to focus on business matters while improving communication and collaboration.

Furthermore, the metaverse has the potential to enable digital collaboration and a new perspective on monetary transactions, making it a prominent factor in enhancing business operations and their sustainable outcomes. The incorporation of the metaverse will also allow businesses to enhance the scope of customer satisfaction and ensure that individuals can have an in-depth approach towards digital insights.

Overall, the metaverse has become a vital aspect of the business industry, and its incorporation will lead to sustainable outcomes for companies and brands alike. The incorporation of the metaverse will enable businesses to focus on their core competencies while enhancing communication, collaboration, and customer satisfaction, making it an asset to businesses of all sizes.

2.7. Importance of Metaverse in the Entertainment Industry

The metaverse is a game-changing concept that has revolutionized the entertainment industry by integrating different digital tools to create a virtual world. This has enhanced the entertainment experience for viewers by providing a more immersive and satisfying experience. Virtual and Augmented Reality have played a crucial role in enabling the sustainable incorporation of the metaverse. These technologies have provided opportunities for innovation and creativity, which has contributed to market stability and the overall sustainability of the industry. The metaverse has transformed the functions of the entertainment industry by providing a highly advanced and transformative virtual world that offers various opportunities for satisfying audiences. It has opened new avenues for the industry, which can be explored and exploited to enhance the entertainment experience for viewers. The incorporation of the metaverse has been one of the most positive developments in the entertainment industry and has the potential to bring about even more significant changes in the future.

2.8. Issues Associated with Metaverse

While the metaverse is an important concept with potential benefits for various industries, there are also several significant issues associated with its implementation. One of the major challenges is the high cost of the equipment and technical tools required for successful integration. Moreover, the metaverse requires highly skilled and trained individuals for maintenance and operation. The complexity of the metaverse also makes it difficult for many companies to implement it successfully, and automation may pose challenges to data security.

The impact of the metaverse on mental health and addiction is also a concern. Implementing the metaverse requires time and experience, and it is subject to complex rules and regulations that can affect human access to it. These issues affect not only the business and entertainment industries, but also other aspects of society. Overall, while the metaverse has potential benefits, its implementation requires careful consideration of the associated challenges and risks.

3. SUMMARY

In this chapter of the literature review, it has been observed that the different researchers and theoretical backgrounds that have been evaluated in association with the metaverse have been taken into consideration. This eventually helps to focus on the importance of the metaverse and its significance and the way it has impacted the entertainment and business industry as a whole. Moreover, through this chapter, the previous literature has been evaluated based on the selected topic and at the same time, the gaps of literature have also been accessed that remain extremely reliable for the research to carry forward the research in the most sustainable manner.

4. METHODOLOGY

The passage discusses research methodology and its various components, including research philosophy, research approach, research choice, research strategy, research design, data collection, and data analysis. The study aimed to evaluate the concept of metaverse effectively with appropriate research methodology. The research philosophy used was interpretivism, which is based on the socially constructed reality. The research approach chosen was the inductive approach, while the research choice was secondary qualitative data. The research strategy used was the case study analysis, and the research design chosen was exploratory research design. The data collection method used was secondary data collection through various authentic sources such as news articles, blogs, and social media. Finally, the data analysis method used was thematic analysis, which is an effective process used to analyze qualitative data.

5. RESULTS

The selected study highlights the significance of the metaverse in the industrial framework and its evaluation in terms of core aspects. The appropriate usage of the metaverse is reliant on industrial specifications and incorporated technology. The advantages associated with the metaverse include sustainable enhancements of business specifications, customer acquisition, and increased satisfaction. The metaverse has developed 3D and virtual specifications that offer impossible experiences for human users. The entertainment industry has incorporated metaverse technology to enhance user and viewer experiences. The social media industry is heavily facilitated by the core practices of metaverse, enhancing society's specifications and expectations. The growth of the metaverse has changed the overall perception of users and businesses towards it. However, there are consequent disadvantages related to privacy and data management issues and identity hacking, which require regulations and sustainable individuals well integrated with the metaverse before implementing its technology-related attributes. The study highlights the relevance of the metaverse's core concepts and its methodological factors, including augmented and virtual reality, to understand the information given attributes and communication factors.

In summary, the selected study demonstrates the significance of the metaverse and its evaluation in the industrial framework. The metaverse's appropriate usage depends on industrial specifications and incorporated technology, offering sustainable enhancements to business specifications, customer acquisition, and increased satisfaction. The metaverse has developed 3D and virtual specifications that offer impossible experiences for human users. The entertainment industry has incorporated metaverse technology to enhance user and viewer experiences, and the social media industry is heavily facilitated by the core practices of the metaverse. However, there are consequent disadvantages related to privacy and data management issues and identity hacking, which require regulations and sustainable individuals well integrated with the metaverse before implementing its technology-related attributes.

The study highlights the relevance of the metaverse's core concepts and its methodological factors, including augmented and virtual reality, to understand the information given attributes and communication factors.

6. DISCUSSION

The metaverse is an important concept for consistent growth in various industries. A study analyzed the factors that help understand the metaverse and its integration into business activities. It found that a sustainable approach is necessary for monitoring and management of the latest trends. The entertainment industry has benefited from the metaverse as it enhances media diversification and improves viewer engagement. The global market size of the metaverse in the entertainment industry is valued at around $14 billion in 2022. The metaverse is not limited to the business and entertainment industry but also facilitates the banking and finance, healthcare, real estate, automation, retail, and manufacturing industries. However, the study also identified negative impacts associated with the metaverse, which must be evaluated to ensure sustainability.

7. CONCLUSIONS

The study on the metaverse has evaluated various factors associated with it and its impact on the business and entertainment industry. The study has analyzed the positive and negative aspects of the metaverse, and the economic benefits it offers. The research methodology adopted for the study included systematic data collection and the consideration of secondary data. It is important to understand the perception of the metaverse as it has a significant impact on the functioning of the business and entertainment sectors.

To enhance the sustainability of the metaverse, it is recommended to ensure that society has sustainable knowledge of the metaverse. This will ensure that the metaverse remains accessible to all. Sustainable technology needs to be incorporated into the metaverse to develop time management skills and reduce the scope of addiction for some users. The technical specifications of the metaverse require highly technical approaches and tools that need to be well-integrated to ensure that the overall activities are accomplished successfully and without redundancy.

The study has highlighted the future scope of research in the conceptualization and accessibility of the metaverse. The topic remains one of the trending factors that will have a sustainable impact on various industries operating globally. Incorporating the core elements of the metaverse will help enhance the sustainability of the topic and its discussion in the future.

In conclusion, the study on the metaverse has been significant in analyzing the various factors associated with it and its impact on the business and entertainment industry. The positive and negative aspects of the metaverse have been evaluated, and the economic benefits it offers have been analyzed. It is important to understand the perception of the metaverse as it has a significant impact on the functioning of the business and entertainment sectors.

To enhance the sustainability of the metaverse, society needs sustainable knowledge of the metaverse, and sustainable technology needs to be incorporated into it. The technical specifications of the metaverse require highly technical approaches and tools that need to be well-integrated. The future scope of research in the conceptualization and accessibility of the metaverse is significant, and incorporating the core elements of the metaverse will help enhance the sustainability of the topic and its discussion in the future.

8. AUTHOR CONTRIBUTIONS

In this assessment, the contribution of the author is very important. The author is the only person who identifies this topic which is not in highlight for a long time. However, during this research, the author faced so many problems as the information regarding this is not available that much. However, it can be said that with the utmost interest of the author, the research got a good output and a balanced output too. Therefore, it can be said that the author had to take an important role to make sure that all the subjects have been covered properly.

9. CONFLICT OF INTEREST

The conflict of interest is something that sometimes plays a key role to understand what can be done and what could have been done for the betterment of the research. In this case, it can be said that as the topic is comparatively new, therefore, the researcher could take the information after conducting extensive research. However, it can be said that the researcher could also take interview process to understand the topic more effectively.

REFERENCES

Akour, I. A., Al-Maroof, R. S., Alfaisal, R., & Salloum, S. A. (2022). A conceptual framework for determining metaverse adoption in higher institutions of gulf area: An empirical study using hybrid SEM-ANN approach. *Computers and Education: Artificial Intelligence, 3,* 100052. Retrieved from: https://www.sciencedirect.com/science/article/pii/S2666920X22000078

Al-Ghaili, A. M., Kasim, H., Al-Hada, N. M., Hassan, Z., Othman, M., Hussain, T. J., … & Shayea, I. (2022). A review of metaverse's definitions, architecture, applications, challenges, issues, solutions, and future trends. *IEEE Access.* Retrieved from: https://ieeexplore.ieee.org/stamp/stamp.jsp?arnumber=9966605

Almarzouqi, A., Aburayya, A., & Salloum, S. A. (2022). Prediction of user's intention to use metaverse system in medical education: A hybrid SEM-ML learning approach. *IEEE Access, 10,* 43421–43434. Retrieved from: https://ieeexplore.ieee.org/stamp/stamp.jsp?arnumber=9761199

Alvarez-Risco, A., Del-Aguila-Arcentales, S., Rosen, M. A., & Yáñez, J. A. (2022). Social cognitive theory to assess the intention to participate in the Facebook metaverse by citizens in Peru during the COVID-19 pandemic. *Journal of Open Innovation: Technology, Market, and Complexity, 8*(3), 142. Retrieved from: https://www.sciencedirect.com/science/article/pii/S2199853122007430

Anderson, J., & Rainie, L. (2022). The metaverse in 2040. *Pew Research Center.* Retrieved from: https://www.pewresearch.org/internet/wp-content/uploads/sites/9/2022/06/PI_2022.06.30_Metaverse-Predictions_FINAL.pdf

Bansal, G., Rajgopal, K., Chamola, V., Xiong, Z., & Niyato, D. (2022). Healthcare in metaverse: A survey on current metaverse applications in healthcare. *IEEE Access, 10,* 119914–119946. Retrieved from: https://ieeexplore.ieee.org/stamp/stamp.jsp?arnumber=9940237

Bauer, T. (2016). Research philosophy and method. In *Responsible Lobbying: Conceptual Foundations and Empirical Findings in the EU* (pp. 69–84). https://doi.org/10.1007/978-3-658-15539-1_3

Bojic, L. (2022). Metaverse through the prism of power and addiction: What will happen when the virtual world becomes more attractive than reality? *European Journal of Futures Research*, *10*(1), 1–24. Retrieved from: https://eujournalfuturesresearch. springeropen.com/articles/10.1186/s40309-022-00208-4

Buhalis, D., Lin, M. S., & Leung, D. (2022). Metaverse as a driver for customer experience and value co-creation: Implications for hospitality and tourism management and marketing. *International Journal of Contemporary Hospitality Management* [ahead-of-print]. Retrieved from: https://eprints.bournemouth.ac.uk/37717/1/METAVERSE%20 paper_IJCHM_R2_clean%201sept22.pdf

Cha, S. S. (2022). Metaverse and the evolution of food and retail industry. *The Korean Journal of Food & Health Convergence*, *8*(2), 1–6. Retrieved from: https://koreascience. kr/article/JAKO202210951903144.pdf

Cheng, R., Wu, N., Varvello, M., Chen, S., & Han, B. (2022). Are we ready for metaverse? A measurement study of social virtual reality platforms. In *Proceedings of the 22nd ACM Internet Measurement Conference* (pp. 504–518). Retrieved from: https://dl.acm. org/doi/pdf/10.1145/3517745.3561417

Damar, M. (2022). What the literature on medicine, nursing, public health, midwifery, and dentistry reveals: An overview of the rapidly approaching metaverse. *Journal of Metaverse*, *2*(2), 62–70. Retrieved from: https://dergipark.org.tr/en/download/article-file/2495651

Dufour, I. F., & Richard, M. C. (2019). Theorizing from secondary qualitative data: A comparison of two data analysis methods. *Cogent Education*, *6*(1), 1690265. https:// doi.org/10.1080/2331186X.2019.1690265

Dwivedi, Y. K., Hughes, L., Baabdullah, A. M., Ribeiro-Navarrete, S., Giannakis, M., Al-Debei, M. M., ... & Wamba, S. F. (2022). Metaverse beyond the hype: Multidisciplinary perspectives on emerging challenges, opportunities, and agenda for research, practice and policy. *International Journal of Information Management*, *66*, 102542. Retrieved from: https://www.sciencedirect.com/science/article/pii/S0268401222000767

George, A. H., Fernando, M., George, A. S., Baskar, T., & Pandey, D. (2021). Metaverse: The next stage of human culture and the internet. *International Journal of Advanced Research Trends in Engineering and Technology (IJARTET)*, *8*(12), 1–10. Retrieved from: https:// www.researchgate.net/profile/A-Shaji-George/publication/357354932_Metaverse_The_ Next_Stage_of_Human_Culture_and_the_Internet/links/61c9f701b6b5667157ac7b69/ Metaverse-The-Next-Stage-of-Human-Culture-and-the-Internet.pdf

Golf-Papez, M., Heller, J., Hilken, T., Chylinski, M., de Ruyter, K., Keeling, D. I., & Mahr, D. (2022). Embracing falsity through the metaverse: The case of synthetic customer experiences. *Business Horizons*, *65*(6), 739–749. Retrieved from: https:// www.sciencedirect.com/science/article/pii/S0007681322000982

Gursoy, D., Malodia, S., & Dhir, A. (2022). The metaverse in the hospitality and tourism industry: An overview of current trends and future research directions. *Journal of Hospitality Marketing & Management*, 1–8. Retrieved from: https://www. researchgate.net/profile/Suresh-Malodia/publication/360356207_The_metaverse_ in_the_hospitality_and_tourism_industry_An_overview_of_current_trends_and_ future_research_directions/links/627778f12f9ccf58eb3703f0/The-metaverse-in-the-hospitality-and-tourism-industry-An-overview-of-current-trends-and-future-research-directions.pdf

Hammarberg, K., Kirkman, M., & de Lacey, S. (2016). Qualitative research methods: When to use them and how to judge them. *Human Reproduction, 31*(3), 498–501. https://doi.org/10.1093/humrep/dev334

Hwang, G. J., & Chien, S. Y. (2022). Definition, roles, and potential research issues of the metaverse in education: An artificial intelligence perspective. *Computers and Education: Artificial Intelligence*, 100082. Retrieved from: https://www.sciencedirect.com/science/article/pii/S2666920X22000376

Jiang, Y., Kang, J., Niyato, D., Ge, X., Xiong, Z., Miao, C., & Shen, X. (2022). Reliable distributed computing for metaverse: A hierarchical game-theoretic approach. *IEEE Transactions on Vehicular Technology*. Retrieved from: https://ieeexplore.ieee.org/stamp/stamp.jsp?arnumber=9880566

Jungherr, A., & Schlarb, D. B. (2022). The extended reach of game engine companies: How companies like epic games and Unity technologies provide platforms for extended reality applications and the metaverse. *Social Media+ Society, 8*(2), 20563051221107641. Retrieved from: https://journals.sagepub.com/doi/pdf/10.1177/20563051221107641

Liao, T. (2015). Augmented or admented reality? The influence of marketing on augmented reality technologies. *Information, Communication & Society, 18*(3), 310–326. https://www.tandfonline.com/doi/abs/10.1080/1369118X.2014.989252

Opie, C. (2019). Research approaches. In *Getting Started in Your Educational Research: Design, Data Production and Analysis* (p. 137). https://books.google.com/books?hl=en&lr=&id=ELuODwAAQBAJ&oi=fnd&pg=PA137&dq=Research+approach+&ots=SNqKvcj5PY&sig=gSopzzuJVLJgYazD2f-C0wKeXbc

Pandey, P., & Pandey, M. M. (2021). *Research methodology tools and techniques*. Bridge Center.

Patino, C. M., & Ferreira, J. C. (2018). Inclusion and exclusion criteria in research studies: Definitions and why they matter. *Jornal Brasileiro de Pneumologia, 44*, 84. https://doi.org/10.1590/S1806-37562018000000088

Patten, M. L., & Newhart, M. (2017). *Understanding research methods: An overview of the essentials*. Routledge.

Qin, Y. (2022, March). Investment potential analysis on Chinese stock market in metaverse-take VR industry as a sample. In *2022 7th International Conference on Financial Innovation and Economic Development (ICFIED 2022)* (pp. 1001–1007). Atlantis Press. Retrieved from: https://www.atlantis-press.com/proceedings/icfied-22/125971757

Terry, G., Hayfield, N., Clarke, V., & Braun, V. (2017). Thematic analysis. *The SAGE handbook of qualitative research in psychology* (Vol. 2, pp. 17–37). https://books.google.com/books?hl=en&lr=&id=   AAniDgAAQBAJ&oi=fnd&pg=PA17&dq=thematic+analysis&ots=don5nsDjM0&sig=xK__UBr_FNZMo3rQkYf1j9jiKAk

Tlili, A., Huang, R., Shehata, B., Liu, D., Zhao, J., Metwally, A. H. S., ... & Burgos, D. (2022). Is metaverse in education a blessing or a curse: A combined content and bibliometric analysis. *Smart Learning Environments, 9*(1), 1–31. Retrieved from: https://slejournal.springeropen.com/articles/10.1186/s40561-022-00205-x

Yemenici, A. D. (2022). Entrepreneurship in the world of Metaverse: Virtual or real? *Journal of Metaverse, 2*(2), 71–82. Retrieved from: https://dergipark.org.tr/en/download/article-file/2467111

15. Creator in Despair

Tinni Dutta
Assistant Professor, Department of Psychology, Muralidhar Girls' College, Kolkata, India

BACKGROUND: Human Development and Research Institute (HDRI) is a pioneer organisation, working in the field of drug addiction since the decade of 80s. Apart from drug addiction, the organisation is deeply involved in the prevention and treatment of STD/HIV/AIDS among marginalised groups. The organisation also stretches its hand regarding rural development in the district of Burdwan.

The present case was referred to me for diagnosis, treatment, rehabilitation and follow-up. Tears came to my eyes unknowingly when recapitalisation was going on. Beyond the confinement of day-to-day activity, sometimes, we all are overflowed by emotions and values. Though this present case was challenging, it had kept a positive note at the end.

Subhasis, a 35-year decent gentleman, came to the centre. He was suffering from alcohol consumption. This was coupled with unipolar depression. Symptoms of loss of interest in pleasurable activities, lack of energy, loss of appetite and sleep disturbances are pronounced. He had lost his father in childhood and was reared u by mother alone. He was admitted in English medium school. He had obtained high grades throughout the academic career. He became a graduate and had joined a company.

Apart from routine job, he had engaged himself in drawings and paintings. The unconscious processes of displacement and sublimation were reflected in his drawings and paintings. Thus on and often he was isolated and was absorbed in art. It provided him with great pleasure. Initially, he was a social drinker. He had maintained sobriety.

He was married, he had no issues in the relationship, and conjugal life was satisfactory during the first two years of his life. Then, he came to know that his wife was engaged in an extramarital relationship. He was disheartened and was absorbed in alcohol consumption. As he was good natured, his well-wishers had come forward to help him. Initial case history revealed that he had consumed alcohol too much due to depression. He had lost interest in life and enjoyable pursuit. After the detoxification process, his case history was taken again. Predisposing factors like loneliness, isolation and dominating attitude of mother got aggravated by marital dissatisfaction.

After completion of detoxification, detailed case history was taken, and short-term and long-time goals were identified. The short-term goal was to teach him to maintain sobriety, and the long-term goal was to reduce stress and anxiety levels, to function whole heartedly. To reach these objectives, psychological tests were administered to confirm the findings gained from case history and mental status examinations. He was found to be worried and apprehensive, but no disturbances were found regarding his judgement and thinking processes.

Bender-Gestalt test (BG) was given for screening. Draw-a-Person test (DAP), Rorschach inkblot test (RIBT) and thematic apperception tests had been selected for probing the psychodynamics of the present cases.

So, the tests that were administered are as follows: BG, DAP, RIBT and TAT. During the time of administering the psychological tests, he was found to be co-operative and attentive. He appeared to be a sober gentleman.

In BG test, no gloss psychopathology was found. Z score of '68' put him in a suspect category. Qualitative analysis revealed aggression, conflict with author figure and emotional disturbances. It is reflected in modern theory. Cellular engine of our body controls our energy levels. Fatigue issues stem from the environmental toxicants which is avoidable by positive life style. BG test was levelled as screening test he falls him in the co-called normal category. His mother appeared to be affectionate but since she had to play both the role of parents, sometimes she became dominating. But at the same time, she was caring for her only son. The son had the ambivalent attitude towards his mother which will be discussed later.

He was found to be little bit shaky, while he was instructed to draw a picture of human beings. At first, he drew the opposite sex which is indicative of the latent homosexual conflict. Both the figures drawn were of the ages of 20-25 which were reflective of regression, wish to return to youth and fantasy activity. Hair style on both the figures was indicative of inner sensitivity which was projected on figures. Pockets in both figures had indicated maternal dependency which was also confirmed by real-life situation. Drawings on the middle of the page reflected ego-centrism, and strong bold strokes had indicated aggressiveness towards himself and others.

Quantitative analysis was not done for this particular case as he was an adult, and it is usually applied on children. But face validity had indicated that he was intellectually mature.

Rorschach inkblot test was developed by Hermann Rorschach to make a diagnostic investigation of personality as a whole. Only 10 cards that differentiated between various psychiatric syndromes were needed. The Rorschach test consisted of 10 cards, of which six cards (cards I, IV, V, VI and VII) are made in shades of black and grey, two cards (cards II and IV) contain bright patches of red in addition to the shades of black and grey, and the remaining three cards contain several pastel sheds.

The present client was inhibited initially in visualising cards I, IV, V, VI and VII, and it is moreover reflected that he had great difficulties in card nos. IV and VI. He gave achromatic responses known as burnt child reaction. Scoring was based on three categories: location, determinant and content. He gave whole responses and D responses which are indicative of that he is capable of perception of the entire situation, but a low percentage of D response indicated maladaptive perception. He had shown the C$'$ response, the pure achromatic colour response scored as C$'$, the achromatic colour form response scored as C$'$F, and the form achromatic colour response scored as FC$'$. These responses were indicative of traumatic experiences and withdrawal tendencies. In tabular form, his responses were as follows:

			L	D	C
Card	I	Butterfly	(M	(F)	A
Card	II	Heads of the dogs	(D)	(F)	Ad
Card	III	Human figures	(M)	(M)	H
Card	IV	Animal skin	(M)	C$'$F	H
Card	V	Butterfly	(M)	FM	H
Card	VI	Animal skin	(M)	C$'$F	A
Card	VII	Head of the child	(P)	F	Hd
Card	VIII	Tree	(M)	FC$'$	Pl
Card	IX	Tree	(D)	FC	Pl
Card	X	Colour scattered	(M)	C	Obj

The thematic apperception test (TAT) was used in clinical and non-clinical settings. Here, 10 cards were used. The term 'thematic' has been derived from the term 'Thema' on which a client may think–which is based on the connotation 'need' and 'press' and the term apperception–recognising the particular situation. For the present client, the result was as follows: card nos. 1, 2, 3 BM; 4, 6 BM; 7 BM; 8 BM; 12 M; 13 MF; and 17 BM.

The instruction was given in the following manner: I shall show you some pictures, one at a time, and your task will be to describe a story. Your story must include what has led up to the event shown in the pictures, what is happening at the moment, and what the characters are feeling and thinking and the outcome.

The client perceived the environment sometimes congenial and sometimes non-congenial. He had shown difficulties in establishing bond. He had shown need for achievement and recognition, but it was marred with aggression. Aggression was pronounced in almost all the stories. He had revealed need and anxiety of loss of love, anxiety of separation and anxiety of being rejected. Conflicts which were pronounced were as follows: adequacy/inadequacy, activity/pleasure and activity/passivity.

Based on case history, mental status examination and psychometric assessment, the present client had undergone 45 days of psychotherapeutic programme which were discussed in the following:

The purpose of psychotherapeutic programme is to identify the root causes of alcoholism, to probe the psychodynamics of the client and to provide him with therapy that restore his

Detoxification	Day 1 to Day 5
Detailed case history taking along	Day 6 to Day 7 with mental status examination
Drug chart (alcohol)	Types of drug and quantity Chronologically Day 8
Twenty harmful	Areas where it has an impact – Day 9
Tool	Am I really recovering – Day 10
Lecture session	Disease concept Days 11 - 18
Lecture session	Quality of life Days 19 - 26
Lecture session	Relapse – 27 - 35 (warning sign and relapse prevention)
Lecture session	After Care- d Follow-up 36 - 44

Within this time frame, meditation, morning meeting, yoga, group discussion, encounter, emotional interview and Daily Feelings Journal were organised time to time.

Simultaneously, individual counselling, group counselling and family counselling were conducted as per schedule.

Findings: Drug chart reflected that both quantity and frequency of consumption had increased due to marital dissatisfaction. Harmful effects were chiefly on self and family. During psychotherapeutic programme, he had shown interest in meditation and morning meeting, which are based on activated concern therapy (ABCT). He was lucid and spontaneous in pulling up personal and interpersonal issues and as a whole providing acknowledgement. Emotional interview and encounters had revealed unconscious aggression and depression which was reconfirmed by his test.

Daily Feelings Journal had shown his loneliness, depression and brooding trend, but, however, it has improved gradually.

CONCLUSION

Case history, mental status examination, psychometric assessment and psychotherapeutic programme were indicative of alcoholism associated with unipolar depression. Organisation had extended hands of support and encouragement. He tried to support and encourage. He tried to lead a better quality of life in the labyrinth of human relationship.

16. The Power of Visual Merchandising: Investigating the Impact on Consumer Impulse Buying in the Apparel Retail Industry of Bangalore City

Ms. D. Maria Nirmal Preethi
Research Scholar Presidency University, Assistant Professor in Don Bosco college Bengaluru
Dr. Chithambar Gupta V
Professor School of Management Presidency University
Ms. Marina Sarah
Student Don Bosco College, Bengaluru

ABSTRACT: Consumers in this postmodern era are forced to engage in fierce, constant rivalry with businesses and retailers willing to pay for undifferentiated merchandising. Today's retailers use the merchandising tool to set themselves apart from rivals and stand out in the marketplace to draw in customers. This article will focus on one such instrument to gain a competitive edge in the market and draw clients into the firm. That tool is visual merchandising, which encourages shoppers to make spontaneous purchases. This education was founded on first-hand information that questioners compiled. The four hypotheses being explored are window display, forum display, floor merchandising, and brand name. Hypotheses are tested through Chi-Square analysis using Statistical Package for Social Science (SPSS) software, through which the study will relate visual merchandising to practical impulse-buying tools.

KEYWORDS: Visual merchandising, consumer impulse, impulsive buying behaviour

Chapter 16 DOI- 10.4324/9781003397175-16

1. INTRODUCTION

Understanding consumers are the key element to run a business successfully, their behaviours are a reflection towards their likes and dislikes of a product and the characteristics of influence. When we talk about likes and dislikes how consumer get influenced to these like and dislikes are the major study which researchers have to look into. There are many factors of influence such as culture, reference groups, social class, personal and psychological factors which act as a major element of buying decision of consumers. These different factors motivate the customer to buy the products and services according to their needs and wants. One such factor that stimulates the consumers is visual merchandising that can be described as 'All that the consumer sees in a retail store that creates a positive image for the business and results in attention seeking, interest that generates a desire to buy the products, is one such factor that stimulates consumers" (Bastow *et al.*, 1991).

Visual merchandising includes Window Display, Forum Display, Floor Merchandising and Brand names along with promotion signage (Mills, 1995). Presentation of goods often plays a very important role in decision making process of consumer as they show a better interest for goods which are attractive and well arranged (Oakley, 1990). Four dimensions of store atmosphere that is visual(sight), aural(sound), olfactory(smell) and tactile(touch) which are notable in customer preferences (McGoldrick, 2002). Thus visual merchandising is purely a marketing jargon which represents the important factors or elements which influences consumer buying decision process that specifies the direct and indirect communication of products to consumer through the usage of promotion indication like Bill boards, Banners, Posters, Pana flex, Shop Boards, lighting, space and layout allocation, Aesthetic design of the store and the brand names which the customers can envision during their visit to the store. It also indicates the shelving style of the stores, different sections, store atmosphere, different brands available in the stores and also the scent and aroma of the store. These visual indicators create an interest to the consumers that will generate sales to the retail stores. Visual merchandising is also a marketing strategy that has been used by many retailers online and offline to pull in many customers to their stores.

Impulsive behaviour involves making rash, unplanned purchases without considering a particular product category to satisfy particular needs. These behaviours occur following a consumer experience that results in an unforeseen phenomenon of a strong temptation to buy and being forced to make unsolicited purchases beyond any reasonable consideration. According to Jondry Adrin Hetharie (2019), a composite hedonic component that frequently energizes emotional conflict, which would be predicted from internal (psychological aspects) or external (inducement from promoters), is the impact to cause impulsive behaviour.

Visual merchandising plays a significant role in impacting consumer impulse buying behaviour, particularly in the apparel retail industry. Studies have found that the use of attractive visual displays, colour, lighting and in-store layout can influence consumers' perception of product appeal and increase the likelihood of impulse buying. Additionally, research has shown that the use of mannequins, window displays and other visual cues can create a sense of urgency and scarcity, leading to increased impulse buying. Furthermore, the use of technology such as digital displays, augmented reality and virtual fitting rooms have also been found to enhance the consumer's in-store experience and increase the likelihood of impulse buying. Most of the research points to the possibility that successful visual merchandising can favourably influence customers' impulsive buying behaviour in the retail garment sector.

One such psychological aspect being visual merchandising that leads the potential customer to consumers towards impulse purchasing. Impulse buying is known as unexpected buying of goods and services that impacts the shoppers' plan of purchase. Thus, this behaviour was recognized by shop-owners to create more business to attract and increase consumers intake to their stores.

The purpose of this study is to analysis the elements involved during impulse buying behaviour in visual merchandising that create a larger impact in consumers buying process.

1.1. Objectives of the Study

1. To understand window-shopping impulses.
2. To research how consumer impulse buying behaviour is affected by floor merchandising.
3. To examine and analyse the impact of forum display on impulsive purchasing
4. To highlight the influence of brands on impulsive purchasing

2. LITERATURE REVIEW

Prior to now, the classification of impulse buying included organized, unorganized and impulse gestures that result in purchases. A logical decision-making process and information searches are always time-consuming components of planned purchases. Unorganized purchases, on the other hand, refer to choices made without prior planning or references. Unplanned purchases are more common than impulse purchases when prompt purchasing decisions occur. In addition to unanticipated purchases, impulsive behaviour necessitates an immediate, compelling and desired urge to buy (Muruganantham and Bhakat, 2013) without any prior desire to shop (Beatty and Ferrell, 1998).

According to Bayley and Nancarrow (1998), the further definition of impulse buying is 'unexpected, enticing, impulsive, hedonistic shopping practises that prevent customers from thoughtfully weighing their options and information'. In contrast, pragmatic behaviour, which considers economics and looks for practical value among consumers, is pleasurable. The opposite of pragmatic behaviour, which experiences economics and looks for useful value among consumers, is hedonic behaviour, which is patented as pleasure.

The initial study on impulsive buying behaviour was uneasy with problems like rationality, distinguishing impulsive behaviour from non-impulsive behaviour, and attempting to rank the various types of impulsive behaviour into distinct categorizations (Kollat and Willett, 1969). The idea that impulsive behaviour is a characteristic of consumer behaviour was necessary for this strategy (Park and Lennon, 2009). Attributes are a propensity that alludes to the specific psychological characteristics of customers.

2.1. Consumer Behaviour in Apparel Industry

The study of consumer behaviour examines how individuals or groups select, get, use or discard goods and services that meet consumers' needs, wants and desires (Solomon, 2006). The phrase also refers to the investigation of organizational and personal behavioural patterns. According to Dadfar (2009), consumer analysis entails knowing what to buy, how to buy it, where to buy it and why. An attempt has been made to understand consumer behaviour as it relates to the garment industry, including the decision-making process that affects the movement of purchases, and this understanding has formed the body of knowledge currently available. Comprehending or studying people's purchases and how frequently and under what circumstances they make them is, therefore, part of comprehending consumer behaviour.

2.2. Visual Merchandising

The tool that businesses or shopkeepers use to entice customers or shoppers to make impulsive purchases is known as visual merchandising. When marketers and store owners set up their stores or engage in promotional activities that draw customers in by virtue of the store's layout, special offers, signboards, atmosphere, shelf arrangement, section divisions, cleanliness and numerous other factors that encourage impulsive or unplanned purchases.

According to Mehta (2012), window display plays a very important role with direct relation with impulse buying and he further did an investigation on floor merchandising which did not have direct relation. Impulse buying behaviour has various factors under it which directly or indirectly impacts the buying decision process of consumers one such important factor being visual merchandising. Visual merchandising further can be explained as exterior and interior design of a store which can attract the potential customers, one such element of attraction being window display which can attract the consumers into the store and also creates the first impression for the store. Physical attractiveness also plays a very important role in impressing customers in store selection (Omar, 1999).

Bashar and Irshad (2012) used a sample size of 250 Indian respondents using Pearson correlation to examine the effects of form display, window display, promotional signage and floor merchandising. According to his research, window displays and impulsive buying are positively correlated. Impulsive shopping has nothing to do with retail displays. Impulsive purchasing and floor selling often go together (Bashar and Irshad, 2012).

According to Sujata *et al.* (2012), impulse buying has an antecedent. He treated floor merchandising, form display and window display as independent variables. He used Pearson correlation on a sample of both males and females between the ages of 18 and 45. He comes to the conclusion that window display, impulse buying and forum display are all strongly correlated. The relationship between floor merchandising and impulsive purchases is weak.

Researchers have examined how visual merchandising affects consumer attitudes regarding women's clothing (Vinamra *et al.*, 2012). Visual merchandising is his dependent variable, and significant and neutral role in influencing the purchase are his independent variables. He collected data from 150 Indian ladies who frequented shopping centres. According to his research, visual merchandising has a significant influence over consumers' purchasing decisions. In some cases, visual merchandising also encourages impulsive purchases. The study discovered that customers' purchasing decisions were unaffected by visual merchandising.

Research was done on the influence of visual merchandising in mall fashion retailers (Maria *et al.*, 2010). His independent variables included aspects that shoppers valued prior to entering a business and characteristics that affect gender-specific buying decisions. Visual merchandising served as his dependent variable. On a sample of 334 respondents, he conducted statistical analysis using the mean standard deviation. His research indicates that gender variations in store window displays have a major impact on consumer purchasing behaviour but that less visible gender disparities exist in the attributes that shoppers value upon entering a mall.

3. RESEARCH METHODOLOGY

Identifying the numerous visual merchandising components that have a stronger influence on customer purchasing decisions is the study's main goal. The relevant study was conducted in Bengaluru, India, with a total of 200 questionnaires filled out by genuine shoppers, of which 189 were finished. As a result, the sample size was proportionally decreased. The variables were calculated using a Likert scale with five points. Customers were surveyed in a shopping centre while they were there to shop and wearing our study samples to gather information from them.

3.1. Statement of Problem

'How does visual merchandising affect consumers' impulse purchases'.?

3.2. Theoretical Framework

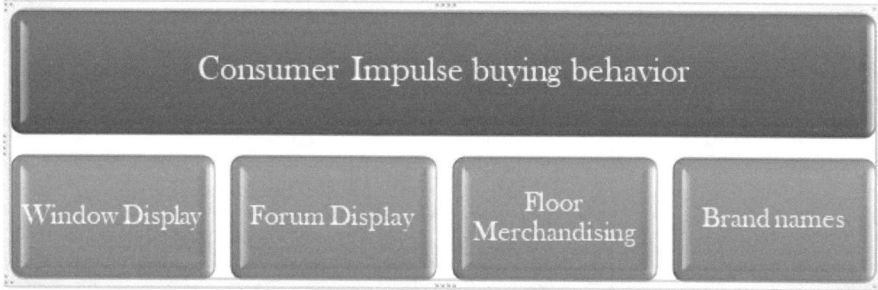

3.3. Data Collection

Since the information was gathered from the original source all at once, cross-sectional data collection was used for the study.

3.4. Analysis and Discussion of the Hypothesis

Chi-square test was used to analysis the data and to find out which among the variables had high impact on creating impulse buying among the consumers. Hypothesis testing was also done to test all the variables used for this study in detail to find out the significant relationship between the variables.

3.5. Hypothesis

- H0: Window display does not significantly influence impulse buying behaviour.
- H1: Window display does significantly influence impulse buying behaviour.
- H0: Forum display has no significant impact on impulsive purchasing.
- H2: Forum display does significantly influence impulse buying behaviour.
- H0: Floor merchandising does not significantly influence impulse buying behaviour.
- H3: Floor merchandising does significantly influence impulse buying.
- H0: Brand name does not significantly influence impulse buying behaviour.
- H4: Brand name does significantly influence impulse buying behaviour.

4. RESULTS AND DISCUSSION

Table 1 describes the study's demographic characteristics, including the gender of the respondents. It is mentioned that in this survey, 54.0% of respondents are women and 45.5% are men.

Table 1: Demographic profile of the cohort.

Variables	Categories	Overall ($N = 189$)
Age		26.62(10.98)
Sex	Male	86(45.5%)
	Female	102(54.0%)
	Other	1(0.5%)
Education	UG	128(67.7%)
	PG	40(21.2%)
	PhD	2(1.1%)
	Other	19(10.1%)

The first figure mentions the income level of the consumers which is determined per month salary of all consumers.

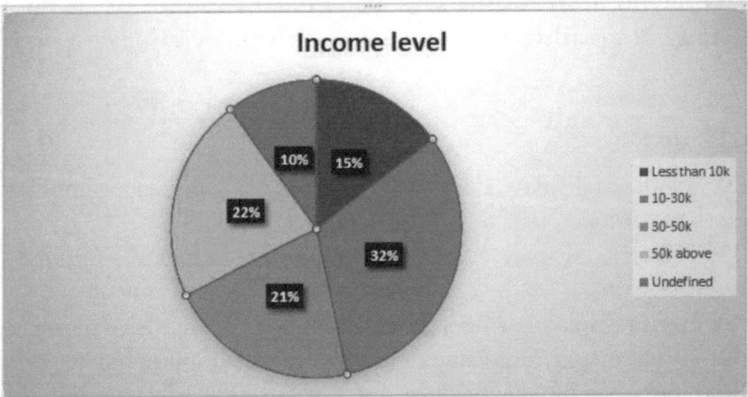

4.1. Chi-Square Analysis

The Chi-Square test of Independence examines whether categorical variables are connected or independent by looking for a correlation between them.

- H0: Window display does not significantly influence impulse buying behaviour.

Crosstab

Do you get attracted by what is displayed and does it prompt you to go inside the store?			Which of the following factors influences you to walk into a store?						Total
			Brand name and logo	Kind of merchandise on display	Outer view of store	Promotional offering at entrance (sales/discounts)	Store exterior graphic and signage	Window display setting	
Agree	Count		23	23	2	31	6	8	93
	%		51.1%	63.9%	18.2%	44.9%	54.5%	47.1%	49.2%
Disagree	Count		3	0	1	2	0	0	6
	%		6.7%	0.0%	9.1%	2.9%	0.0%	0.0%	3.2%
Neutral	Count		10	5	2	22	2	3	44
	%		22.2%	13.9%	18.2%	31.9%	18.2%	17.6%	23.3%
Strongly agree	Count		9	8	6	11	3	6	43
	%		20.0%	22.2%	54.5%	15.9%	27.3%	35.3%	22.8%
Strongly disagree	Count		0	0	0	3	0	0	3
	%		0.0%	0.0%	0.0%	4.3%	0.0%	0.0%	1.6%
Total		Count	45	36	11	69	11	17	189
		%	100.0%	100.0%	100.0%	100.0%	100.0%	100.0%	100.0%

Chi-Square Tests

	Value	df	Asymptotic Significance (2-sided)
Pearson Chi-Square	26.085a	20	.163
Likelihood Ratio	27.327	20	.126
N of Valid Cases	189		

a. 18 cells (60.0%) have expected count less than 5. The minimum except count is 1.7.

Crosstab

		Which of the following factors influences you to walk into a store?						Total
		Brand name and logo	Kind of merchandise on display	Outer view of store	Promotional offering at entrance (sales/discounts)	Store exterior graphic and signage	Window display setting	
Do you look for promotional sings or sale signs once you enter a store?	No Count	10	10	2	13	4	4	43
	%	22.2%	27.8%	18.2%	18.8%	36.4%	23.5%	22.8%
	Yes Count	35	26	9	56	7	13	146
	%	77.8%	72.2%	81.8%	81.2%	63.6%	76.5%	77.2%
Total		45	36	11	69	11	17	189
	%	100.0%	100.0%	100.0%	100.0%	100.0%	100.0%	100.0%

Interpretation: We do not reject the null hypothesis since the p-value above the significance level we determined to be significant (0.05). The window display has no significant impact on impulsive purchasing ($p > 0.05$)

Chi-Square Tests

	Value	df	Asymptotic Significance (2-sided)
Pearson Chi-Square	2.421a	5	.788
Likelihood Ratio	2.309	5	.805
N of Valid Cases	189		

a. 3 cells (25.0%) have expected count less than 5. The minimum excepted count is 2.50.

Interpretation: We do not reject the null hypothesis since the *p*-value exceeds our chosen significance level ($\alpha = 0.05$). Window display does not significantly influence impulse buying behaviour ($p > 0.05$).

H0: Forum display does not significantly influence impulse buying behaviour.

Crosstab

		6. Which of the following factors influences you to walk into a store?						Total
		Brand name and logo	Kind of merchandise on display	Outer view of store	Promotional offering at entrance (sales/discounts)	Store exterior graphic and signage	Window display setting	
10. Which of the following is the most influential factor for you in going inside a store?	Bold graphics	Count	0	1	2	2	1	1	7
		%	0.0%	2.8%	18.2%	2.9%	9.1%	5.9%	3.7%
	Creative product display	Count	34	28	4	47	8	9	130
		%	75.6%	77.8%	36.4%	68.1%	72.7%	52.9%	68.8%
	Mannequin styling	Count	2	5	2	9	0	4	22
		%	4.4%	13.9%	18.2%	13.0%	0.0%	23.5%	11.6%
	Music and lighting	Count	9	2	3	11	2	3	30
		%	20.0%	5.6%	27.3%	15.9%	18.2%	17.6%	15.9%
Total		Count	45	36	11	69	11	17	189
		%	100.0%	100.0%	100.0%	100.0%	100.0%	100.0%	100.0%

Chi-Square Tests

	Value	df	Asymptotic Significance (2-sided)
Pearson Chi-Square	22.123ᵃ	15	.105
Likelihood Ratio	22.823	15	.088
N of Valid Cases	189		

a. 13 cells (54.2%) have expected count less than 5. The minimum excepted count is 41.

Interpretation: We do not reject the null hypothesis since the p-value exceeds our chosen significance level ($\alpha = 0.05$). Forum display does not significantly influence impulse buying behaviour ($p > 0.05$).

Crosstab

		Which of the following factors influences you to walk into a store?						Total
		Brand name and logo	Kind of merchandise on display	Outer view of store	Promotional offering at entrance (sales/discounts)	Store exterior graphic and signage	Window display setting	
Good layout, moving/browsing and ambience encourage you to shop in a particular apparel store?	Agree Count	26	15	3	42	6	10	102
	%	57.8%	41.7%	27.3%	60.9%	54.5%	58.8%	54.0%
	Disagree Count	2	0	0	1	1	1	5
	%	4.4%	0.0%	0.0%	1.4%	9.1%	5.9%	2.6%
	Neutral Count	8	7	2	11	2	2	32
	%	17.8%	19.4%	18.2%	15.9%	18.2%	11.8%	16.9%
	Strongly agree Count	9	14	6	14	2	3	48
	%	20.0%	38.9%	54.5%	20.3%	18.2%	17.6%	25.4%
	Strongly disagree Count	0	0	0	1	0	1	2
	%	0.0%	0.0%	0.0%	1.4%	0.0%	5.9%	1.1%
Total	Count	45	36	11	69	11	17	189
	%	100.0%	100.0%	100.0%	100.0%	100.0%	100.0%	100.0%

Chi-Square Tests

	Value	df	Asymptotic Significance (2-sided)
Pearson Chi-Square	21.357[a]	20	.376
Likelihood Ratio	20.152	20	.448
N of Valid Cases	189		

a. 18 cells (60.0%) have expected count less than 5. The minimum excepted count is 12.

Interpretation: *We do not reject the null hypothesis since the p-value exceeds our chosen significance level (α = 0.05). Forum display does not significantly influence impulse buying behaviour (p > 0.05).*

Crosstab

Which of the following factors influences you to walk into a store?

		Brand name and logo	Kind of merchandise on display	Outer view of store	Promotional offering at entrance (sales/ discounts)	Store exterior graphic and signage	Window display setting	Total
In store Merchandise and mannequin display promotes impulsive buying?	Agree							
	Count	27	18	6	32	4	12	99
	%	60.0%	50.0%	54.5%	46.4%	36.4%	70.6%	52.4%
	Disagree							
	Count	2	1	1	6	1	0	11
	%	4.4%	5.6%	9.1%	5.8%	0.0%	0.0%	4.8%
	Neutral							
	Count	10	10	2	19	6	2	49
	%	22.2%	27.8%	18.2%	27.5%	54.5%	11.8%	25.9%
	Strongly agree							
	Count	6	6	2	12	1	3	30
	%	13.3%	16.7%	18.2%	17.4%	9.1%	17.6%	15.9%
	Strongly disagree							
	Count	0	0	0	2	0	0	2
	%	0.0%	0.0%	0.0%	2.9%	0.0%	0.0%	1.1%
Total	Count	45	36	11	69	11	17	189
	%	100.0%	100.0%	100.0%	100.0%	100.0%	100.0%	100.0%

Chi-Square Tests

	Value	df	Asymptotic Significance (2-sided)
Pearson Chi-Square	14.180[a]	20	.821
Likelihood Ratio	15.513	20	.746
N of Valid Cases	189		

a. 18 cells (60.0%) have expected count less than 5. The minimum expected count is 12.

Interpretation: We do not reject the null hypothesis since the p-value is more significant than our chosen significance level ($\alpha = 0.05$). Forum display does not significantly influence impulse buying behaviour ($p > 0.05$).
H0: Floor merchandising does not significantly influence impulse buying behaviour.

Crosstab

		Which of the following factors influences you to walk into a store?						Total
		Brand name and logo	Kind of merchandise on display	Outer view of store	Promotional offering at entrance (sales/ discounts)	Store exterior graphic and signage	Window display setting	
Do you feel price boards promote impulse buying?	Agree Count	22	9	6	21	3	5	66
	%	48.9%	25.0%	54.5%	30.4%	27.3%	29.4%	34.9%
	Disagree Count	2	1	1	6	1	0	11
	%	4.4%	2.8%	9.1%	8.7%	9.1%	0.0%	5.8%
	Neutral Count	1	2	2	8	1	3	17
	%	2.2%	5.6%	18.2%	11.6%	9.1%	17.6%	9.0%
	Strongly agree Count	0	0	0	1	0	0	1
	%	13.3%	13.9%	9.1%	18.8%	9.1%	0.0%	13.8%
	Strongly disagree Count	14	19	1	20	5	9	68
	%	31.1%	52.8%	9.1%	29.0%	45.5%	52.9%	36.0%
Total	Count	45	36	11	69	11	17	189
	%	100.0%	100.0%	100.0%	100.0%	100.0%	100.0%	100.0%

Chi-Square Tests

	Value	df	Asymptotic Significance (2-sided)
Pearson Chi-Square	27.929[a]	25	.311
Likelihood Ratio	32.067	25	.156
N of Valid Cases	189		

a. 25 cells (69.4%) have expected count less than 5. The minimum excepted count is 06

Interpretation: Since the p-value is greater than our chosen significance level ($\alpha = 0.05$), we do not reject the null hypothesis. Floor merchandising does not significantly influence impulse buying behaviour ($p > 0.05$).

Crosstab

Which of the following factors influences you to walk into a store?

			Brand name and logo	Kind of merchandise on display	Outer view of store	Promotional offering at entrance (sales/ discounts)	Store exterior graphic and signage	Window display setting	Total
Do you think video displayed in digital screens and walls help you to get information about new arrivals and promotions?	No	Count	2	2	2	4	0	0	10
		%	4.4%	5.6%	18.2%	5.8%	0.0%	0.0%	5.3%
	Undefined	Count	5	11	1	13	3	4	37
		%	11.1%	30.6%	9.1%	18.8%	27.3%	23.5%	19.6%
	Yes	Count	38	23	8	52	8	13	142
		%	84.4%	63.9%	72.7%	75.4%	72.7%	76.5%	75.1%
Total		Count	45	36	11	69	11	17	189
		%	100.0%	100.0%	100.0%	100.0%	100.0%	100.0%	100.0%

Chi-Square Tests

	Value	df	Asymptotic Significance (2-sided)
Pearson Chi-Square	11.151[a]	10	.346
Likelihood Ratio	11.386	10	.328
N of Valid Cases	189		

a. 9 cells (50.0%) have expected count less than 5. The minimum excepted count is 58.

Interpretation: Since the *p*-value is greater than our chosen significance level ($\alpha = 0.05$), we do not reject the null hypothesis. Floor merchandising does not significantly influence impulse buying behaviour ($p > 0.05$).

Crosstab

Which of the following factors influences you to walk into a store?

		Brand name and logo	Kind of merchandise on display	Outer view of store	Promotional offering at entrance (sales/ discounts)	Store exterior graphic and signage	Window display setting	Total
Do you get attracted to signs of special offers inside the store would make positive impact in your mind and tend to buy more products?	Agree Count	27	15	6	36	5	9	98
	%	60.0%	41.7%	54.5%	52.2%	45.5%	52.9%	51.9%
	Disagree Count	1	1	0	1	0	0	3
	%	2.2%	2.8%	0.0%	1.4%	0.0%	0.0%	1.6%
	Neutral Count	4	6	1	4	0	2	17
	%	8.9%	16.7%	9.1%	5.8%	0.0%	11.8%	9.0%
	Strongly agree Count	13	13	4	26	6	5	67
	%	28.9%	36.1%	36.4%	37.7%	54.5%	29.4%	35.4%
	Strongly disagree Count	0	1	0	2	0	1	4
	%	0.0%	2.8%	0.0%	2.9%	0.0%	5.9%	2.1%
Total	Count	45	36	11	69	11	17	189
	%	100.0%	100.0%	100.0%	100.0%	100.0%	100.0%	100.0%

Chi-Square Tests

	Value	df	Asymptotic Significance (2-sided)
Pearson Chi-Square	11.530a	20	.931
Likelihood Ratio	13.617	20	.849
N of Valid Cases	189		

a. 19 cells (63.3%) have expected count less than 5. The minimum excepted count is 17.

Interpretation: Since the *p*-value is greater than our chosen significance level ($\alpha = 0.05$), we do not reject the null hypothesis. Floor merchandising does not significantly influence impulse buying behaviour ($p > 0.05$).

H0: Brand name does not significantly influence impulse buying behaviour.

Crosstab

		Which of the following factors influences you to walk into a store?						Total
		Brand name and logo	Kind of merchandise on display	Outer view of store	Promotional offering at entrance (sales/ discounts)	Store exterior graphic and signage	Window display setting	
Do you feel brand display creates an urge to buy products?	No — Count	1	4	5	16	1	3	30
	%	2.2%	11.1%	45.5%	23.2%	9.1%	17.6%	15.9%
	No answer — Count	14	19	1	20	5	9	68
	%	31.1%	52.8%	9.1%	29.0%	45.5%	52.9%	36.0%
	Yes — Count	29	13	5	33	5	5	90
	%	64.4%	36.1%	45.5%	47.8%	45.5%	29.4%	47.6%
	Yes | No — Count	1	0	0	0	0	0	1
	%	2.2%	0.0%	0.0%	0.0%	0.0%	0.0%	0.5%
Total	Count	45	36	11	69	11	17	189
	%	100.0%	100.0%	100.0%	100.0%	100.0%	100.0%	100.0%

Chi-Square Tests			
	Value	df	Asymptotic Significance (2-sided)
Pearson Chi-Square	30.525[a]	15	.010
Likelihood Ratio	31.533	15	.007
N of Valid Cases	189		

a. 11 cells (45.8%) have expected count less than 5. The minimum excepted count is 06.

H0: Since the p-value is less than our chosen significance level ($\alpha = 0.05$), we do reject the null hypothesis. Brand name significantly influences impulse buying behaviour ($p < 0.05$).

5. FINDINGS AND CONCLUSION

This study was done to look at some external factors that affect consumers' impulsive buying patterns. The learning endeavoured to clarify the link between the various forms of visual merchandising and the customer's impulsive purchasing behaviour to investigate the relationship further. The prior learning was that visual merchandising could positively influence consumers' impulsive purchasing behaviour.

The findings demonstrated that brand name and logo, which are reportedly a component of visual marketing, had a substantial impact on consumers' impulsive purchasing behaviour. This study showed that consumers have a close relationship with brand names. When a company successfully retains a client base, customers feel compelled to purchase anytime he or she sees the brand name and logo, even though the transaction may be completely impulsive.

Visual merchandising, including window displays, forum displays, floor merchandising and shop brand names, operate as potent stimulants, prompting and inciting people to make impulse purchases as soon as they enter a store, according to all available evidence. This study makes a strong case for the significance of visual merchandising in deliberate impulse purchases.

As a result, there is a significant correlation between visual merchandising and impulse purchase behaviour regarding brand names and logos.

6. LIMITATION OF THE STUDY

The research included the following drawbacks:

- The sample was geographically restricted, and data were only gathered from Bangalore city.

- The mechanism was only suitable of quantitative approach. Because they were aware of what is happening and influence in advance, the poll asked respondents to provide answers based on their unforeseen purchasing experience.
- The results of the qualitative research for this study could vary.

REFERENCES

Bashar, A., and I. Ahmad (2012). "Visual merchandising and consumer impulse buying behavior: an empirical study of Delhi & NCR," *International Journal of Retail Management & Research* 2277–4750.

Bastow, S., D. Zetocha, and G. Passewitz (1991). *Visual Merchandising: A Guide for Small Retailers*. Lowa: University Publications.

Bayley, G., and C. Nancarrow (1998). "Impulse purchasing: a qualitative exploration of the phenomenon," *Qualitative market Research: An International Journal* 99–114.

Beatty, S.E., and M. E. Ferrell (1998). "Impulse buying: modelling its precursors," *Journal of Retailing* 169–191.

Dadfar, I. (2009). *Identification and Prioritization of the Effective Factors on Buying Decision of the Cars of the Iran Khodro*. Tehran.

Emotiv (2022). Retrieved from https://www.emotiv.com/glossary/consumer-psychology/.

Jondry Adrin Hetharie, S.A. (2019). "SOR (stimulus-organism-response) model application in observing the influence of impulsive buying on consumer's post-purchase regret," *International Journal of Scientific & Technology Research* 2829–2841.

Kollat, D. T., and R. P. Willet (1969). Is impulse purchasing really a useful concept for marketing decisions? *The Journal of Marketing* 79–83.

Maria, P., A. Susana, B. Vera., M. Fernando, M. Rui, and L. Jose (2010). "The effect of visual merchandising on fashion stores in shopping center," *5th International Textile, Clothing & Design Conference – Magic World of Textiles*, Dubrovnik, Croatia, pp. 3–6.

McGoldrick, P.J. (2002). *Retail Marketing*. Berkshire: McGraw-Hill Education.

Mehta, N.C. (2012). "Visual merchandising: impact on consumer behavior," *Global Business and Technology Association* 607–614.

Mills, K.P. (1995). *Applied Visual Merchandising*. Englewood Cliffs.

Muruganantham., & Bhakat, R. S. (2013). A Review of Impulse Buying Behavior. *International Journal of Marketing Studies*, 149-160.

NEWS (2021). Retrieved from The University of Tennessee knoxville: https://news.utk.edu/2021/03/03/covid-19-pandemic-prompts-shifts-in-retail-marketing-strategies/

Oakley, M. (1990). *Design Management: A Handbook of Issues and Methods*. Oxford: Oxford University Press.

Omar, O. (1999). *Retail Marketing*. London: Pitman Publishing.

Park, M., and S. Lennon (2009). "Brand name and promotion in online shopping contexts," *Journal of Fashion Marketing and Management* 149–160.

Sen (2000).

Solomon, M.R. (2006). *Consumer Behaviour : A European Perspective*. Harlow England New York: Financial Times/Prentice Hall.

Vinamra, J., S. Ashok, and N. Pardeep (2012). "Impact of visual merchandising on consumer behavior towars women's apparel," *International Journal of Research in Management* 106–117.

17. An Empirical Study on the Determinants of Green Consumption among Gen Z Customers

B Apoorva Vaishnavi[1], Kirti Aija[1], Abisekh Kumar JV[1], Anuska Sanyal[1], and Subhadarshini Khatua[2]

[1]MBA Candidate, Woxsen University, Hyderabad, Telangana, India
[2]Assistant Professor, Woxsen University, Hyderabad, Telangana, India

ABSTRACT: With better standards of living, rapid growth of population is leading towards more rapid and significant consumption of resources in domestic and industrial segments. Unsustainable patterns of consumption are adversely affecting the environment. Customers should be engaged in a mutual and active conversation to promote green consumption. There are several environmental concerns world is facing these days, such as water pollution, climate change, harmful waste disposal, and global warming.

These concerns have significantly affected consumer behavior and psychology, especially Gen Z. Young people are quickly affected by environmental concerns as they have strong curiosity and access to updated information. Gen Zs are supposed to lead responsible and civilized lifestyle. Hence, they are aware of their consumption behavior and the role it plays in environmental change. So, they are quickly replacing their existing products with more eco-friendly, green products as a great way to reduce harmful effects.

The aim of this study is to determine the factors affecting green consumption in Gen Z. Researchers have analyzed responses based on the available responses from the participants. The findings suggest that Gen Z consumers are completely aware of environmental concerns and green behavior of businesses. A majority of Gen Z customers are willing to buy green products as per the availability. Environmental concerns played a vital role in their green buying behavior. Due to this reason, they have moved towards more sustainable path in terms of buying behavior. However, some customers may skip greener products if they are overpriced.

KEYWORDS: Green consumption, buying behavior, Gen Z, young customers, environmental concerns, green products, Gen Z customers

Chapter 17 DOI- 10.4324/9781003397175-17

1. INTRODUCTION

Gen Z is a very smart generation in different aspects. They are supposed to be the most educated and tech-savvy generation as they do not remember the world without smart devices much (Parker, 2020). They were born and brought up in the world where digital technology has been more matured and they have learned using it very well. Some have even lived in the world without computers. Similarly, green marketing is a relatively new concept of marketing. It is based on profitability with sustainability. It has influenced almost every aspect of business, i.e., ranging from production to advertising and packaging to public relations.

Usually, green consumption is costly in the beginning but it is beneficial in the long term (Bhardwaj, 2021). People are getting more and more health conscious and interested to lead healthy living and save the environment at the same time. In order to attract more people towards green marketing, it is vital to know the factors influencing the green buying behavior. This study is also based on factors affecting green consumption in Gen Z. When society is deeply concerned on saving the environment, brands start reacting to the same to meet society's needs (Polonsky, 1994).

Some companies took no time in embracing green marketing concepts like recycling, using eco-friendly materials for production, etc. Consumers have been more aware of the environmental concerns like global warming, pollution, etc. these days. In order to meet the eco-friendly demand of customers, companies should understand the green consumption pattern of customers. The process of decision-making is very important for customers to know exactly what they need (Stankevich, 2017). Gen Z is a very important cohort because they are very keen to redefine the concept of green marketing. Gen Z is very analytical and quick to make decisions (Francis and Hoefel, 2018). They seriously want to change the world and solve the problem of climate change. They are cosmopolitan with a potential and desire to mobilize across the world (Tanner, 2020).

1.1. Background

Green marketing is a relatively recent concept of marketing as it relies on more eco-friendly behavior and choices. A lot of customers want to contribute towards nature and marketing should be aimed to meet consumer demand (Bhardwaj, 2021). In order to save the environment while making profit for business and give value to the customers, it is vital to know what will help in greener buying behavior in Gen Z. A customer is an individual who uses and purchases the products. Marketing strategies must consider green marketing for most businesses.

These days, a lot of consumers are concerned over pollution, increasing population, and climate change. Hence, companies should make the most of their green consumption pattern to save the environment. Considering the rise of global warming and climate change, previous five years from 2015 to 2019 were recorded warm by the "World Meteorological Organization (WMO)". The sea level across the world has been increased to 3.6mm year-on-year from 2005 to 2015. A lot of such changes have been made because of the rise of water level as it warms up (BBC News, 2020).

Consumers have never been so concerned to the environment as they are now and they believe that they can collectively make change to the commercial processes of brands (BBC News, 2020). Companies now need to focus on their green behavior and strive to adopt eco-friendly practices even out of their business like donating to charity as part of their Corporate Social Responsibility (CSR) initiatives. Companies also need to focus on reducing chemical waste and landfills. They have to significantly reduce the amount of waste by reusing the same. There is also a need to reduce food waste by at least 50% (Frigo, 2021).

Green marketing is used by the companies to improve their corporate image and to convey their green behavior towards the environment (Sherman, 2020). However, customers must be careful while choosing the products as some companies are involved in green washing. The term "green washing" refer to the push of green marketing when products are not actually organic. Customers are ready to pay premium price for green products but they do not know whether those products meet the needs of ecological sustainability. Communication plays a vital role in green buying behavior. Companies should show their concern for the environment, so that they can make way to new markets (Frigo, 2021).

2. LITERATURE REVIEWS

When it comes to buy fashion products like apparels, Gen Z are the most lucrative customers as they are more willing to move to greener processes than other generational cohorts. Arora and Manchanda (2022) explored the mediating role of "positive attitude towards green clothing with sustainable perceived value and intention to buy sustainable apparel" among Gen Z customers. They also explored the "moderated mediation model" with consumer knowledge on green apparel and materialistic values as moderators. They gathered data from 308 undergraduate students from Delhi and NCR region through a survey method. The findings suggested a "partial mediating role of positive attitude towards sustainable clothing and confirmed the moderating role of materialistic values and consumer knowledge on green apparel". In addition, consumer knowledge and materialistic values were found to be the key moderators. A significant relationship was observed among "positive attitude for sustainable clothing, green perceived value, and intention to opt for sustainable clothing".

In this day and age, researchers and policymakers have focused a lot on environmental and social responsibility of apparel industry. Getting more responsible demand from young customers is a way to deal with harmful effects of clothing sector. Vlastelica et al., (2023) conducted a study to test the determinants of socially and environmentally responsible behavior of young consumers for clothing in a developing country. Convenient sampling was used in the study to gather responses from 439 participants in Serbia through "structured online survey". Along with 2-step "structural equation modeling (SEM)", "exploratory factor analysis (EFA)" was conducted to test structural relations. Findings suggested green consumption behavior as the most important factor of "responsible apparel consumption" followed by receptivity to sustainable communication and conscious consumption. Findings play a vital role on the body of knowledge towards responsible consumption.

Buying behavior is widely affected by the gap between action and intention and skepticism among customers in context of green products. Priyadarsini *et al.*, (2022) conducted a study to determine several factors affecting millennials' buying behavior of green products. This study was supposed to be helpful in making "environmental consumer behavior" and "green marketing" theories. This study is based on "theory of planned behavior". It is aimed to understand consumers' behavior. It is observed that important green marketing factors like "Green Packaging and Branding (GPB)", "Environmental Concerns and Beliefs (ECB)" and "Green Product, Premium and Pricing (GPPP)" have a significant and positive impact on "Consumer Beliefs Towards the Environment (CBTE)". A positive impact on environmental beliefs is expected by improving the spending on branding and green packaging. Meanwhile, there is a negative impact of "Eco- Labelling (EL)" on "CBTE" due to skepticism among consumers.

Song *et al.*, (2020) analyzed the mediating role of "perceived consumer effectiveness (PCE)", "product attributes" and "Environmental consciousness" on "eco-label-informed" buying decision of Gen Z in China as per the "theory of planned behavior". The findings suggest that (1) "eco-label-informed purchase" has significant rise in product attributes and PCE, (2) both of these threads have intermediating relation between environmental awareness and eco-labeling positively, and (3) eventually make Gen Z to make buying decision. They attempted to have more detailed study on "green consumption" and explore theoretical relation between PCE, product attributes, and environmental concern among Gen Z customers in China.

Lavuri *et al.*, (2021) determine the factors playing a vital role on promoting green buying decision and compare millennials and Gen Z customers in context of green buying behavior. They collected data from 372 participants from both generations in three states in south India. They adopted snowball and purposive sampling techniques to select respondents. They analyzed the data with IBM SPSS 23 with Pearson Correlation, Factor Analysis, one sample *t*-test and Multiple Linear Regression. There was no significant association between subjective norms (SNs) and "Green Purchase Intention (GPI)". There is a significant impact of variables like "environmental concern (EC)", "media exposure (ME)", "Perceived Behavioral Control", "Environmental Knowledge

(EK)" and "Environmental Attitude (EA)". The buying intention of customer had a significant impact on their green buying behavior.

2.1. Research Gap

Environmental awareness is very important for green consumption of customers (Mostafa, 2009). This study is conducted to fill this knowledge gap. This topic is also very relevant to the modern world as a lot of people are concerned about climate change and green choices can make a lot of change in future. This study has been conducted on factors affecting green consumption of Gen Z customers as it is important to know how consumers generally make better choices for being eco-friendly.

2.2. Research Question

Considering the above arguments, here is the main research question to be investigated and answered:

- What are the factors affecting the green consumption of Gen Z consumers?

2.3. Research Objectives

This study is aimed to know the factors affecting green consumption and buying behavior of Gen Z customers and how consumers can choose more eco-friendly products. Hence, the objective of this study is:

- To determine the green buying behavior of Gen Z customers
- To know the factors affecting green consumption in Gen Z

3. RESEARCH METHODOLOGY

The "stimulus–organism–response (SOR)" model is widely used in scientific study to determine the buying behavior of consumers in different contexts. Originally, the "stimulus" part had environmental stimuli in the previous SOR model. Those stimuli were tested basically with external factors affecting buying behavior. However, several internal factors (like neurological, biochemical and physiological aspects of people) and external factors (like government and corporate actions) have affected buying behavior (Jacoby, 53). The extended SOR model was considered to develop the research framework for this study to determine the phenomenon of green consumption (Chan *et al.*, 2017; Peng and Kim, 2014).

3.1. Theoretical Framework

On the basis of the SOR model, the theoretical framework of this study is indicated in Figure 1.

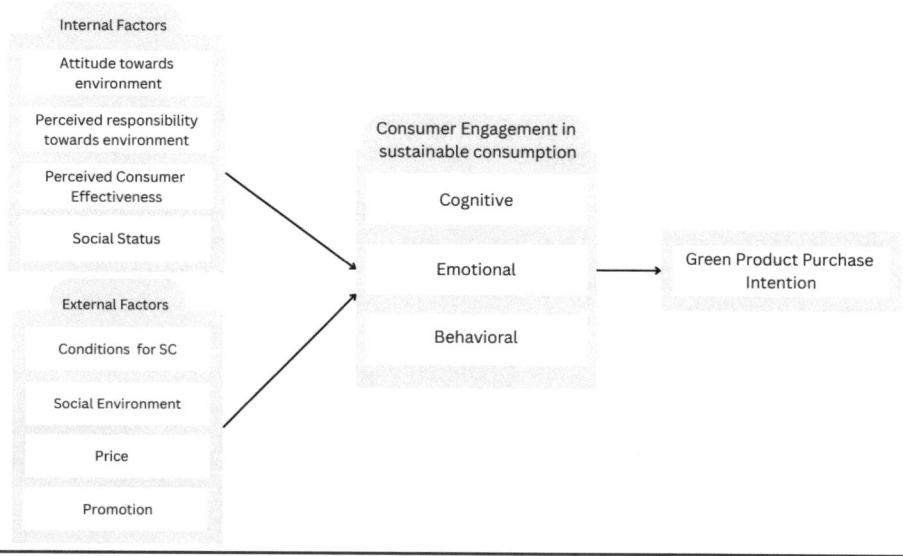

Figure 1. Research model as per the SOR Model.

Existing studies on the factors affecting green buying behavior and determinants of engagement have a lot of evidences to prove that both external and internal factors are vital predictors for sustainable consumption and consumer engagement. Though there was a lack of studies to determine the consumer engagement in green consumption of Gen Z, here is the hypothesis that is loosely based on above arguments:

H1 – Internal factors (like "perceived responsibility", "environmental attitude" and "perceived behavioral efficiency") would positively impact "consumer engagement in sustainable consumption"

H2 – External factors (like "social environment", "conditions for SC" and "promotion of SC") would positively impact "consumer engagement in sustainable consumption"

There is evidence on recent studies related to consumer engagement that it improves customer satisfaction, loyalty, connect, empowerment, trust, emotional bond, and commitment (18,22). Hence, third hypothesis assumes that it can improve green consumption too –

H3 – "Consumer engagement" has positive impact on green consumption in Gen Z

3.2. Research Design

A quantitative research design has been adopted for this study to gather data from 122 respondents. A self-structured questionnaire was prepared on the basis of "7-Point Likert Scale (from 1 = Completely Disagree to 7 = Completely Agree) to measure green consumption of Gen Z. The sample population included Gen

Z customers who have made green buying decision quite recently. Research data were analyzed using SPSS 22.0 software and Excel spreadsheet.

4. STUDY ANALYSIS

4.1. Demographic Variables

In this study, majority of participants have completed 12th. When it comes to highest level of education to be achieved, out of 112 participants, 47 (42%) have completed 12th, 41 (36.6%) have completed graduation, and 24 (21.4%) have completed 10th (Table 1).

Table 1. Education level.

What is the highest level of education you have completed?		Frequency	Percent	Valid percent	Cumulative percent
Valid	10th	24	21.4	21.4	21.4
	12th	47	42.0	42.0	63.4
	Graduation	41	36.6	36.6	100.0
	Total	112	100.0	100.0	

Out of the 112 participants, 63 (56.9%) participants are unemployed, while 49 (43.8%) participants are employed (Table 2).

Table 2. Employment status.

Are you employed?		Frequency	Percent	Valid percent	Cumulative percent
Valid	No	63	56.3	56.3	56.3
	Yes	49	43.8	43.8	100.0
	Total	112	100.0	100.0	

When asked about the support for family, 36 (32%) participants said that their parents and siblings earn to support their family, 19 (17%) participants voted for siblings and similar percentage of participants voted for mother, father, and both parents (Table 3).

Table 3. Family support.

Who earns income to support your family?		Frequency	Percent	Valid percent	Cumulative percent
Valid	Both the parents	19	17.0	17.0	17.0
	Father	19	17.0	17.0	33.9
	Mother	19	17.0	17.0	50.9
	Parents and siblings	36	32.1	32.1	83.0
	Siblings	19	17.0	17.0	100.0
	Total	112	100.0	100.0	

In this study, 38 (33.9%) participants said that they have 3 people currently living in their household, including themselves. Similarly, 28 (25%) participants have 4 family members, 28 participants have 5 family members, 9 (8%) participants have 6 family members and 9 participants have 7 family members in their household (Table 4).

Table 4. Total family members.

How many people are currently living in your household, including yourself	Frequency	Percent	Valid percent	Cumulative percent
Valid 3	38	33.9	33.9	33.9
4	28	25.0	25.0	58.9
5	28	25.0	25.0	83.9
6	9	8.0	8.0	92.0
7	9	8.0	8.0	100.0
Total	112	100.0	100.0	

When asked about the residence, 69 (61.6%) participants live with family, 16 (14.3%) participants have no permanent address, 10 (8.9%) participants said that it is rented for money by themselves, 13 (11.6%) participants said that their residence is owned or being bought by them, 3 participants said that it is occupied without payment or rent and 1 participant lives with friend (Table 5).

Table 5. Residence.

		Frequency	Percent	Valid percent	Cumulative percent
Valid	I have no permanent residence	16	14.3	14.3	14.3
	I live with family	69	61.6	61.6	75.9
	I live with friends	1	.9	.9	76.8
	It is occupied without payment or money or rent	3	2.7	2.7	79.5
	It is owned or being bought by you	13	11.6	11.6	91.1
	It is rented for money by you	10	8.9	8.9	100.0
	Total	112	100.0	100.0	

4.2. Internal Factors

In this study, there are three internal factors considered to impact consumer engagement, namely "perceived responsibility, environmental attitude and perceived behavioral efficiency". In order to find out the first hypothesis, one sample t-test was performed. In this test, the value of significance ($p < 0.05$), which shows significant and positive impact of those internal factors on "consumer engagement in sustainable consumption. Hence, H1 is approved (Table 6).

Table 6. One-sample test on internal factors.

	Test value = 0					
	t	df	Sig. (2-tailed)	Mean Difference	95% confidence interval of the difference	
					Lower	Upper
Environmental attitude	40.677	111	.000	50.16071	47.7171	52.6043
Perceived responsibility	34.026	111	.000	21.67857	20.4161	22.9411
Perceived behavioral efficiency	41.961	111	.000	20.40179	19.4383	21.3652

4.3. External Factors

Like internal factors, three external factors are also chosen that are known to impact consumer engagement, i.e., "social environment, conditions for sustainable consumption (SC) and promotion of SC". One sample t-test was performed to find out the second hypothesis. Here the value of significance for all three factors are 0.000 ($p < 0.05$). Again, null hypothesis is rejected and H1 is accepted, i.e. "External factors (like social environment, conditions for SC and promotion of SC) would positively impact consumer engagement in sustainable consumption" (Table 7).

Table 7. One-sample test on external factors.

	Test Value = 0					
	t	df	Sig. (2-tailed)	Mean difference	95% Confidence interval of the difference	
					Lower	Upper
Conditions for SC	40.213	111	.000	21.04464	20.0076	22.0817
Social environment	38.475	111	.000	29.73214	28.2009	31.2634
Promotion of SC	41.894	111	.000	29.08036	27.7049	30.4559

4.4. Consumer Engagement

There are two groups of factors affecting "consumer engagement in sustainable consumption (CESC)" and each group consists of three factors to determine CESC. The internal and external groups of factors are discussed above. Each factor was measured by asking for opinions based on 7-Point Likert Scale. There is a high level of reliability in Cronbach's alpha coefficients of all the 6 factors (Table 8).

Table 8. Reliability statistics of factors.

Cronbach's Alpha	Cronbach's alpha based on standardized items	No. ofitems
.965	.987	6

Green product purchasing is represented as "green consumption" of Gen Z. This aspect was selected as one of the common outcomes of consumer engagement. Seven-point Likert scale was used to measure the individual items.

Table 9. Descriptive statistics of factors.

Factors	Mean	SD
Internal factors		
Perceived responsibility	21.679	6.75
Environmental attitude	50.161	13.05
Perceived behavioral efficiency	20.40	5.16
External factors		
Conditions for SC	21.04	5.57
Promotion of SC	29.05	7.24
Social environment	29.73	8.19
CESC		
Emotional	53.63	4.66
Cognitive	52.78	5.32
Behavioral	47.67	4.72

Considering the descriptive statistics of factors in Table 9, it is found that overall mean scores for internal scores were higher, which show that participants have stronger internal concern for sustainability of products. In consumer engagement or CESC, the mean scores showed more emotional engagement of participants than behavioral engagement. Hence, the green buying behavior of Gen Z customers show similar willingness to opt for greener option irrespective of environment status. Hence, H3 is proved that "CESC has a positive impact on green consumption in Gen Z".

5. RESULTS

This study attempted to explore factors affecting "consumer engagement in sustainable consumption (CESC)" of Gen Z consumers. This study attempts to combine the construct of customer engagement with green consumption. Engagement is considered to be the strategic and important tool to promote consumer behavior. Though consumer engagement is specific to the context, CESC consists of area which lacks in empirical and conceptual background. Meanwhile, new research path is observed with external and internal factors in the domain of sustainability. Hence, this study plays a vital role in developing theories related to the field of green consumption and consumer behavior.

Our research framework consists of the construct of consumer engagement, its determinants, and consequence of buying green product on the "stimulus–organism–response" model, which is applied significantly in studies related to the consumer behavior (Koklić, 2019; Peng and Kim, 2014; Chan et al., 2017). This study is also based on the SOR model to determine CESC and its confirmation could be a vital outcome. This study has identified 3 internal and 3 external factors, namely "environmental attitude, perceived responsibility, perceived behavioral efficiency, conditions for SC, social environment and the promotion of SC," which were found to have a positive and significant impact on CESC. There is a stronger impact of internal factors on CESC as compared to external ones.

Generally, the findings suggest that Gen Z customers who are more concerned on the environment and green buying are more aware of their roles towards the planet and understand that one person's actions can make change and they are more willing to purchase green products. However, there is a research gap in this domain, which does not help in sensible comparisons with results from same kind of study. There are several studies based on green consumption, but they lack in the construct of "consumer engagement", although they observed positive attitude towards sustainability concerns (Wu et al., 2016; Geng et al., 2017; Leonidou et al., 2010). In addition, perceived effectiveness and responsibility positively affected consumer behavior in previous studies (Wang et al., 2014; Geng et al., 2017).

This study has made a significant contribution to research academia in context of green consumption by focusing on "consumer engagement in sustainable consumption (CESC)". In social sciences, engagement is based on several concepts. In terms of consumer engagement, studies usually reported positive engagement in perceived value, customer loyalty, and behavior (Brodie et al., 2011; Hollebeek, 2011). In green consumption, a process-oriented approach defines sustainable behavior of customers as part of three stages – "purchase, use, and disposal of services and products" (Geiger et al., 2018; Kim et al., 2012). All in all, the findings suggest that consumer engagement has a positive influence on green consumption in Gen Z. Consumer engagement can help businesses to achieve desirable outcomes in consumers' buying behavior.

Green consumption is well regarded to play a vital role in conservation of environment. Preferring green products to lead healthy lifestyle has become a trend these days, which ultimately mitigates burden on the environment. Green consumption is vital for businesses to understand modern trend and shift the attention of customers towards eco-friendly and green products. It is also important to study the concept of green marketing and know the way customers are attracted towards eco-friendly solutions.

The situation of climate change is getting worse every year and it is the right time to find the ways to control it. Green consumption is one of the important ways to improve this situation. It reduces the use of non-renewable sources, while meeting customers' needs. People are getting more and more inclined towards environmental concerns and looking forward to change their behavior to preserve nature. Companies have a lot of stakeholders to serve like investors, consumers, and employees. Hence, companies need to be eco-friendly. Modern consumers prefer companies which are conscious for the environment and strive to promote green consumption. Hence, they start several online and advertising campaigns to promote their sustainable behavior (Choudhary and Gokarn, 2013).

Green consumption is aimed to reuse products, reduce waste, change processes, increase the use of herbal products, and promote eco-friendly communication. These days, many people are focused on controlling waste. Several efforts are made to control water waste and amount of waste into landfill. Hence, packaging of the products has become biodegradable. Products must be modified to control impact on environment, for example, ingredients must be sourced only from natural materials which are safe for pets and humans. Companies promoting eco-friendly products would definitely want to make profit at the end of the day.

Companies attract customers who are willing to pay more for green products to protect the nature and reduce carbon footprint. Companies are more encouraged to carefully use and save resources like electricity and water. Using alternative sources of energy and renewable processes are also the part of green marketing, apart from taking only what is needed from the nature.

Communication is a key factor in green marketing, so that customers can understand the benefits of products and commit to the environment (Kiran, 2012). Companies are widely making their marketing strategies as per the increasing demand for green products and services to offer eco-friendly products to their customers. Companies should develop effective and legal green marketing and product strategy to avoid green washing and improve profitability the right way (Sheth and Partvatiyar, 1995). In the process of greenwashing, companies claim to be eco-friendly even when they are not. It is an unsupported claim to misguide customers to believe that their products are eco-friendly and they support green behavior. (Kenton, 2020).

A lot of companies may be engaged in this practice as their services and products are not green labeled. Companies need to make eco-friendly products from scratch to avoid greenwashing the customers. Green Wrap from Fuji Xerox is a classic example of green marketing. Green Wrap is committed to secure the environment in the long term by developing waste-free products efficiently. Around 60% of pulp for Green Wrap comes straight from milk packaging and other recycled waste (Bhat, 1993).

The company needs to promote green performance and support the importance of sustainability. This way, if a company supports unsustainable practices like poor condition of employees, they cannot claim to be sustainable. It affects their credibility. Green positioning is clearly visible in The Body Shop. They do not support animal logs. Instead, they promote fair trade, protect the environment and human rights (Dibb and Simkin, 1991). The company should focus on saving vital resources for the customers. For example, an automaker may focus on fuel efficiency while promoting new vehicles.

Only the product should not be eco-friendly. Packaging is the first part of the product observed by the customer. Packaging should be eco-friendly too. Rodrigue *et al.*, (2001) emphasized on the value of green logistics. Impact on environment has been second to level of thinking. Green logistics should be used locally and internationally. If the packaging is not eco- friendly, consumers will not trust green values of the company too. So, companies should look at every stage of life cycle of the product and its disposal. In this day and age, food waste is a primary concern and only a small part of food item goes for recycling. There are still services to help sell food waste at lower rate and control waste (Veni *et al.*, 2017). There are also old products available at discounted prices in Finland. A lot of people can buy food products as they get old and cheaper, resulting in reducing food waste. Gen Z is the first generation that has grown up in the age of technology. Gen Z is already aware of digital world and they cannot imagine a world without internet and smartphones. Social media has always been there and they know how to make the most of technology in various areas (Dimock, 2019).

6. CONCLUSION

This study has contributed to find the determinants of green consumption among Gen Z customers. There has been a lack of research on consumer engagement in green buying behavior so far, especially through emotional, cognitive, and behavioral aspects of consumers. This innovative study covers both internal and external factors of CESC. The findings suggested the need to consider the mediating role of engagement when it comes to determine green buying behavior of Gen Z. The SOR model is applied in this study to determine CESC. In an internal state of consumer, it is argued that engagement can be developed with both internal and external factors. Consumers are highly engaged with internal factors like "perceived responsibility, environmental attitude and perceived behavioral efficiency".

Recent studies found positive attitude of customers towards eco-friendly product promotion and applying various engaging methods to promote green consumption. Any kind of engagement approaches can be helpful to social organizations. One-way communication related to any damage due to unsustainable consumption and overconsumption is not much efficient as a lot of engaging ways are available to promote sustainable consumption, such as using and disposing of resources, services, and goods, purchasing, and other practices which encourage public acknowledgement.

There are also suggestions on the basis of results of this study. Social organizations should come up with steps to promote positive behavior towards global environmental concerns and reinforce positive evaluation of sustainable behavior. In addition, companies should promote accessibility to green products regarding convenience and take other measures to improve more green consumption. Social ad campaigns should be promoted to encourage individual responsibility and actions to protect the environment. Finally, whole society should be addressed for positive change in performance and perception.

6.1. Limitations of the Study

Along with contributions, this study also has its limitations, which are still positive as they open further research path. This study is geographically limited with small sample population. Considering cultural contexts, the results of further studies might vary. In addition, this research is based only on Gen Z, i.e., younger people. So, it is recommended to cover other generation cohorts in future studies. This study topic is novel in green marketing field. This study has identified six major factors categorized in two groups which influence CESC. There are chances that there are other factors, which could be important to determine engagement of consumers. Further studies can consider other variables and factors that are not considered in this study. There are different forms of expression in sustainable consumption, which reflect in possible directions for future studies.

6.2. Future Scope and Suggestions

Since this study is focused only on Gen Z customers, its results may not be generalized to other generation cohorts as their taste, style, preference, attitude, habits, and thinking may be different. It has also been suggested for carrying out a comparative analysis of various generations in future studies. This study has also not focused on religious and cultural beliefs, which are vital demographic variables that may affect green behavior and consumption. Hence, future studies can investigate whether religious and cultural beliefs affect green buying behavior of consumers.

Sociodemographic aspects of Gen Z like income, family status, education, etc. might also affect their consumption pattern, which are also not explored in this study. In addition, consumers' choices might also vary in various categories of products. So, it is suggested to do studies on customers' choices on different types of eco-friendly products and services.

REFERENCES

Arora, N., and Manchanda, P. (2022). Green perceived value and intention to purchase sustainable apparel among Gen Z: The moderated mediation of attitudes. *Journal of Global Fashion Marketing, 13*(2), 168-185.

BBC News. (2020). What is climate change? A really simple guide. Retrieved 2021-01- 12 from https://www.bbc.com/news/science-environment-24021772.

Bhardwaj, A. (2021). A Complete Guide on Green Marketing, Its Importance & Benefits. Retrieved from https://startuptalky.com/green-marketing/.

Bhat, V. N. (1993). Green marketing begins with green design. *Journal of Business & Industrial Marketing*, 8(4), 26-31.

Brodie, R. J., Hollebeek, L. D., Jurić, B., and Ilić, A. (2011). Customer engagement: Conceptual domain, fundamental propositions, and implications for research. *Journal of service research*, 14(3), 252-271.

Chan, T. K., Cheung, C. M., and Lee, Z. W. (2017). The state of online impulse-buying research: A literature analysis. *Information & Management*, 54(2), 204-217.

Choudhary, A., and Gokarn, S. (2013). Green Marketing: A means for sustainable development. *Journal of Arts, Science & Commerce*, 4(3), 3.

Dibb, S., and Simkin, L. (1991). Targeting, segments and positioning. *International Journal of Retail & Distribution Management*.

Dimock, M. (2019). Defining generations: Where Millennials end and Generation Z begins. *Pew Research Center*, 17(1), 1-7.

Frigo, C. (2021). Green marketing: What it is and why companies can't do without it. Retrieved from https://www.doxee.com/blog/marketing/green-marketing-for-companies/.

Geiger, S. M., Fischer, D., and Schrader, U. (2018). Measuring what matters in sustainable consumption: An integrative framework for the selection of relevant behaviors. *Sustainable development*, 26(1), 18-33.

Geng, D., Liu, J., and Zhu, Q. (2017). Motivating sustainable consumption among Chinese adolescents: An empirical examination. *Journal of Cleaner Production*, 141, 315-322.

Hollebeek, L. (2011). Exploring customer brand engagement: definition and themes. *Journal of strategic Marketing*, 19(7), 555-573.

Kenton, W. (2020). Greenwashing. Retrieved from https://www.investopedia.com/terms/g/greenwashing.asp.

Kim, S. Y., Yeo, J., Sohn, S. H., Rha, J. Y., Choi, S., Choi, A. Y., and Shin, S. (2012). Toward a composite measure of green consumption: an exploratory study using a Korean sample. *Journal of family and economic issues*, 33(2), 199-214.

Kiran, K. U. (2012). Opportunity and Challenges of Green Marketing with special references to Pune. *International Journal of Management and Social Sciences Research*, 1(1), 18-24.

Koklić, M. K. (2019). Effect of Specialty Store Environment on Consumer's Emotional States: The Moderating Role of Price Consciousness. *MARKET/TRŽIŠTE*, 31(1), 7-22.

Leonidou, L. C., Leonidou, C. N., and Kvasova, O. (2010). Antecedents and outcomes of consumer environmentally friendly attitudes and behaviour. *Journal of Marketing Management*, 26(13-14), 1319-1344.

Mostafa, M. M. (2009). Shades of green: A psychographic segmentation of the green consumer in Kuwait using self-organizing maps. *Expert systems with Applications*, 36(8), 11030-11038.

Peng, C., and Kim, Y. G. (2014). Application of the stimuli-organism-response (SOR) framework to online shopping behavior. *Journal of Internet Commerce*, 13(3-4), 159-176.

Priyadarsini, M. K., PraveenKumar, T., Lakshmi, B. A., Jyotsna, S. A., and Swetha, A. (2022). Do Millennial Exhibit Environmentally Responsive Consumption Behaviors—A Study on Determinants of Green Purchase Decision?. In *International Conference on Chemical, Bio and Environmental Engineering* (pp. 771-784). Springer, Cham.

Robert, D., and John, R. (1982). Store atmosphere: an environmental psychology approach. *Journal of retailing*, *58*(1), 34-57.

Rodrigue, J. P., Slack, B., and Comtois, C. (2001, July). The paradoxes of green logistics. In *World Conference on Transport Research (WCTR). Seoul.*

Sherman, F. (2020). What Are the Benefits of Green Marketing? Retrieved from https://smallbusiness.chron.com/benefits-green-marketing-68744.html.

Sheth, J., and Parvatiyar, A. (1995). Ecological imperatives and the role of marketing. *Environmental marketing: Strategies, practice, theory, and research*, 3-20.

Song, Y., Qin, Z., and Qin, Z. (2020). Green marketing to Gen Z consumers in China: Examining the mediating factors of an eco-label–informed purchase. *Sage Open*, *10*(4), 2158244020963573. Lavuri, R., Jusuf, E., and Gunardi, A. (2021). Green sustainability: factors fostering and behavioural difference between Millennial and Gen Z: mediating role of green purchase intention. *Ekonomia i* Środowisko.

Veni, D. K., Kannan, P., Edison, T. N. J. I., and Senthilkumar, A. (2017). Biochar from green waste for phosphate removal with subsequent disposal. *Waste Management*, *68*, 752-759.

Vlastelica, T., Kostić-Stanković, M., Rajić, T., Krstić, J., and Obradović, T. (2023). Determinants of Young Adult Consumers' Environmentally and Socially Responsible Apparel Consumption. *Sustainability*, *15*(2), 1057.

Wang, P., Liu, Q., and Qi, Y. (2014). Factors influencing sustainable consumption behaviors: a survey of the rural residents in China. *Journal of Cleaner Production*, *63*, 152-165.

Wu, C. S., Zhou, X. X., and Song, M. (2016). Sustainable consumer behavior in China: An empirical analysis from the Midwest regions. *Journal of Cleaner Production*, *134*, 147-165.

18. The Role of Digital Marketing for Business Growth

Pasupulatei Kundan[1], Badrinath Reddy[2], K Bharath[3], Subhadarshini Khatua[4]

[1]MBA Student, Woxsen University, Hyderabad

[2]MBA Student, Woxsen University, Hyderabad

[3]MBA Student, Woxsen University, Hyderabad

[4]Assistant Professor, Woxsen University, Hyderabad

ABSTRACT: This study aims to examine the effects that digital advertising has on contemporary companies. A variety of digital marketing strategies and their possible impact on business success are analyzed. The article will examine the pros and cons of digital advertising and the obstacles that companies experience when trying to implement it. Additionally, it will compare and contrast online and traditional marketing efforts to determine the efficacy of digital marketing strategies. There will also be an examination of how digital marketing influences such things as customer retention, enthusiasm, and advocacy and how it influences brand awareness. The research will also investigate the efficacy of digital marketing initiatives by examining the tools and platforms utilized to deploy them.

Quantitative method is used in this study. In-depth interviews with digital marketing experts and surveys of companies using digital marketing will be used as qualitative methodologies. Data from online channels like websites, blogs, and social media will be analyzed using quantitative approaches. Information for the study will come from various places, such as online databases, specialized journals, and official reports from the relevant authorities. Statistical packages like SPSS will be used to examine the data gathered during the study.

Overall, this study will comprehensively evaluate digital marketing's effects on organizations and provide helpful recommendations for making the most of these techniques.

KEYWORDS: Digital marketing, growth, technology, traditional marketing, digital transformation, online service

Chapter 14 DOI- 10.4324/9781003397175-18

1. INTRODUCTION

The practice of promoting a company, product, or service through the use of digital technology, mostly through the use of the internet, but also incorporating mobile phones, display advertising, and any other type of digital medium is referred to as digital marketing. (Barone, 2022). Systems that are based on the internet and have the ability to develop, accelerate, and transmit product value from the producer to the final customer through the use of digital networks are known as digital marketing channels. It encompasses a wide variety of endeavours including search engine optimization (SEO), content marketing, email marketing, social media marketing, pay-per-click (PPC) advertising, and even more (Alexander, 2022).

Digital marketing makes use of a wide variety of technology to assist firms in connecting with the consumers they are trying to reach. Analytics allows digital marketers to monitor the success of their campaigns and make adjustments to expand their audience and stimulate more interaction with their content. The field of digital marketing is undergoing rapid development, and in order for organizations to continue being effective and competitive, they need to stay up with these changes. The ever-evolving nature of the digital landscape necessitates that businesses be able to adjust their plans accordingly and remain current with the most recent trends and technologies (Chaffey, 2018).

Digital marketing is a strong tool for businesses that want to reach their target customers in a cost-effective and scalable way in today's digital age. Digital marketing may help businesses reach their customers in a number of different ways. Businesses have the ability to engage with their target audience, enhance the number of leads they create, and boost their sales by using the power of digital platforms. Customers who are geographically distributed or who are difficult to target using traditional marketing approaches can easily be reached with digital marketing, which is another significant benefit of this form of marketing. Traditional marketing channels, such as television, radio, and print media, may not be able to bring in all of a company's potential clients, but digital marketing makes it possible for businesses to do so (Scott, 2020).

The marketing strategy of a successful organization must always include digital marketing as an integral component. Reaching their target clients, generating leads, and increasing sales are all goals that may be accomplished by businesses with the correct strategy combination. When it comes to achieving success in today's digital environment, digital marketing is taking on an increasingly critical role for organizations. Businesses are able to boost their sales, create leads, and reach their target clients by utilizing the power of digital platforms (Gillis, 2022).

The industry of digital marketing is one that is always undergoing change, and in order for organizations to maintain their level of success and competitiveness, they need to stay up with these developments. The ever-evolving nature of the digital landscape necessitates that businesses be able to adjust their plans accordingly and remain current with the most recent trends and technologies. In general, digital marketing is an effective tool that companies may utilize to communicate with and engage the customers they are trying to reach. An efficient digital marketing plan may be created for a company such that it can promote both growth and success if the appropriate combination of strategies is used (Roger, 2021).

2. LITERATURE REVIEW

The primary duty of a digital marketer is to oversee marketing efforts that promote a brand and its goods. They contribute significantly to increasing brand recognition in addition to bringing in traffic, prospects, and customers. Applying digital technologies that include digital platforms (Web, email, databases, as well as mobile/wireless and digital TV) to advertising initiatives that seek to achieve financially viable client acquisition and retainment (in underneath a multichannel purchasing procedure and buyer lifecycle) by enhancing our knowledge of our clients (of their profiles, behavior patterns, value, as well as loyalty drivers), after which attempting to deliver fully integrated targeted advertising and online services that can be used to increase customer loyalty. Chaffey's formulation, which emphasises the notion that the business model should drive electronic marketing rather than technology, reflects the association marketing approach (Dave Chaffey, 2015).

Furthermore, It has been highlighted that digital marketing has taken the role of tedious advertising and marketing techniques in the present era. Furthermore, it is so powerful that it has the ability to significantly increase the government efficiency while simultaneously aiding in economic revival (Mort and Drennan, 2002). Businesses in Singapore have looked into the usefulness of digital marketing technologies as methods for getting outcomes. More importantly, the growth of digital marketing is the result of the accelerating rate of technical advancement and changing market dynamics.

Additionally, Chaston and Mangles (2003) analyzed and worked on the impact of marketing strategy on the adoption of the Internet by small manufacturing businesses in the United Kingdom. They applied a quantitative methodology to investigate whether companies that have adopted a connection marketing approach as compared to a transactional marketing orientation will employ the Internet in business-to-business marketplaces in a different way. The research was carried out by mailing questionnaires to a sample of 298 UK small businesses, which were defined as producers of either mechanical or electronic elements, with a central objective on business-to-business selling, somewhere around 10 and 50 employees, and just not local branches of British or multinational corporations. It was determined that there was insufficient evidence to support the assertion that organizations that prioritize relationships view online markets differently than their competitors that are more transactionally focused.

In his post, a number of elements were mentioned as having an effect on organic rankings, namely information, keywords, internal links, tags, as well as the page rank of a website. The responsibility of a business is to enhance the pertinent components and strive for the highest placement on the result page (Nelson, Todd and Wixom, 2005).

Additionally, it has been asserted that word-of-mouth is connected to enlisting new members and increasing visitors to websites, webpages, or online gatherings—both of which foster awareness within the marketing communication context communication. Facebook, the most popular social networking site, has opened up new marketing possibilities and provided advertisers with exposure to masses of individuals who are curious to learn more about their products and services. It is necessary for the business to set up effective communications approaches to attract the customers and improve their experience with a particular product or service in order for this to be successful (Mangold and Faulds, 2009).

Since social networking sites like Facebook store user information on all of their users, they are thought to be superior to traditional forms of advertising in that they ensure that marketing messages are sent to a retailer's particular target market. Retailers may improve the customer experience with their brand by utilizing the data stored on social media platforms. For merchants, social media platforms are a terrific platform for creating an experience (Curran, Graham and Temple, 2011).

In their research, they discovered that businesses may increase brand awareness by being creative when interacting with customers on social media platforms. As more consumers utilize social media (including Twitter, Facebook, MySpace, and LinkedIn) and depend on them for information when making purchases, promotion through these platforms has become increasingly important (Shankar et al., 2011).

Additionally, it has been determined that marketing professionals must have a complete understanding of the goals and initiatives for online social marketing, as well as how to put these initiatives into practice using performance assessment indicators. Because social media is widely available to young people and is increasingly being used by them, market dynamics are altering globally. Together in company's marketing communication plan, effective synergy methodologies must be implemented (Hanna, Rohm and Crittenden, 2011).

According to this research, digital marketing, a hybrid of traditional and internet-based advertising that is less expensive and faster to reach the buyer directly, is the best strategy for firms to advertise locally or globally. Therefore, both marketing strategies can help traders and marketers do business when they are contrasted. Each has advantages and disadvantages. Traditional marketing enables customers to experience and interact with the actual goods or services, but the effect is constrained. In contrast to digital advertising, which will go beyond the borders and provide products and services to the demography of internet users. Additionally, it would be quicker, more efficient, and practical to use the internet for marketing. The advantages of digital marketing typically outweigh those of traditional marketing (Salehi *et al.*, 2012).

He discusses digital marketing in his article, which is an intriguing subject for marketing scholars in particular. It is a novel way to introduce a service or merchandise to a specific global market. This article offers a creative approach to digital advertising in online trade, showing why marketers need this advancement for their industry. Additionally, it gives marketing managers more time to devote to tasks with a greater value, such developing marketing plans for the expansion of the business (Heng and Yazdanifard, 2013).

It was determined that although the phrase "digital marketing" is relatively new, its impact is enormous, frightening, likewise difficult. Despite the fact that digital marketing is one of the most important components of the promotion mix, businesses struggle to understand and implement it. As a result, they are looking for a coherent strategy to start and include it. Today's social media platforms, like Facebook, Google Plus, Twitter, and others, have been successful in influencing the target audience's opinions and worldview. A sizeable, reliable clientele, reliable data, and real-time consumer experience feedback were all used in this digital marketing campaign. Digital marketing, in general, refers to the application of modern technology to marketing initiatives with the goal of better satisfying the needs and preferences of consumers in order to increase customer knowledge (Chaffey and Smith, 2013).

According to their article, blogs have been an effective tool for digital marketing, particularly for goods where customers can read evaluations and leave comments about their own experiences. This has led to an increase in sales revenue. Online evaluations have demonstrated to be an extremely successful part of marketing plans for businesses (Shankar *et al.*, 2011).

Online services are being used regularly by more and more people every day. The fact that the world has changed, though, is something that traditional marketers find difficult to accept. When new chances are provided to them, they are sluggish to seize them. B2B-enhanced supply chain processes require careful real-world business processes are taken into consideration, automated systems are adjusted to reflect business practices, and content and technology are integrated with key information systems (Stuart *et al.*, 2014).

With the continually evolving needs of individuals around the world, they have added new elements to their item. It has presented the world with brand-new ideas that have a modernized slant while also opening up a vast array of options and facilities. India is home to the third-highest number of internet users in the world, which is also causing the country's popularity of online shopping to grow quickly. Online shopping has a large future because it has the potential to be very popular in India and, of course, everywhere else in the world. It reduces the amount of time wasted when a person goes shopping in a store and encounters various traffic jams, automobile problems, and other hurdles along the road. Online shopping would be the greatest choice for the average person, who seldom has time to unwind on the weekends, as he would not need to make travel plans in order to purchase products that are readily available online and provide the highest level of satisfaction. Many small and medium-sized businesses are finding the growing popularity of online shopping to be a blessing in disguise, as it presents them with excellent opportunities to grow and prosper their operations. Additionally, it enables them to partner with leading Indian online portals to advertise their services and display their goods. The phrase "a seek of a new world where all your searches end at one location" can be used to describe online buying (Tanna, Raval and Raval, 2014).

The article claims that global online activity is still increasing quickly and that digital advertising is becoming a more important element of competitive gain including both B2C as well as B2B marketing. A lot of emphasis has been paid to the great advantages of digital marketing, but much less has been paid to the actual challenges that organizations are facing as they go digital. In this paper, we highlight these issues based on the results of a survey conducted among 777 marketing professionals selected as convenience samples from across the globe. The results demonstrate that bridging talent gaps, changing organizational design, and implementing meaningful KPIs are where firms across industries have the greatest opportunity for growth (Leeflang *et al.*, 2014).

They assert in their article that marketers face new challenges and opportunities in the digital age. Digital marketing is the term used to describe how advertisers promote their products or services using electronic media. Attracting clients and giving them the chance to interact with the business online are the primary aims of digital marketing. In this essay, the importance of digital marketing for both organizations and consumers is emphasized. We examine the effect that digital marketing has on the companies' revenues. Additionally, the differences between offline and online marketing are covered in this essay. In this study, the many forms of digital marketing, their effectiveness, and how they impact a sales volume have all been covered (Yasmin, Tasneem and Fatema, 2015). The differences between offline and online marketing are also covered in this essay. The various forms of digital marketing, their effectiveness, and how they impact a company's business have all been explored in this study. The researched sample, made up of 150 businesses and 50 executives, aims to show the effectiveness of digital marketing. The gathered data has been analyzed using a variety of statistical techniques and tools.

Curran, Graham, and Temple (2011) describe in their paper how the world has evolved towards a digital environment. For today's businesses, having a website is crucial, as is using the internet to engage with customers. Even if you are addressing a huge particular demographic, there are some classic marketing strategies that are effective. However, using digital marketing is essential if you want to remain pertinent in the contemporary world. Because digital marketing is believed to be more participatory, quantifiable, and targeted than traditional marketing, the terms "digital marketing" as well as "Internet marketing" can be used interchangeably, despite the fact that they apply to completely distinct strategies. Among the Internet marketing techniques used are link building, SEO, and search engine marketing (SEM). It also encompasses offline retailers that sell digital content, such as e-books, optical discs, games, call-backs, cellphone ringtones, short message service (SMS), multimodal messaging service (MMS), as well as SMS. In the twenty-first century, digital marketing has emerged as a new type of advertising. Various types of digital marketing strategies, including SEO, SEM, SMM, and PPC, are covered in this study paper. The potential of digital marketing is discussed in this essay, as well as some of the concerns involved. It additionally gives business owners digital marketing guidance.

The purpose of this study is to better understand the origins of digital marketing, its advantages, the changes that it has undergone, and the contrasts between digital and traditional marketing. Recent improvements in information technology and the growth of broadband internet access have sped up accessibility to shopping websites. As a result of such modifications, businesses have to adapt to the digital world. Therefore, just as communication laws vary, so do the domain and definition of marketing. The development of IT technologies has led to the digitalization of traditional marketing methods. Businesses that use technology well can easily communicate with customers while providing goods or services. Similar to traditional marketing, developing strong client relationships and recognizing and satisfying their needs and demands are crucial components of digital marketing. Businesses that use digital marketing distinguish out through two-way communication, but those who decide not to compete eventually lose ground to their rivals. Utilizing social networking and search engines to successfully reach your target audience is the main advantage of digital marketing (Das and Lall, 2016).

In their study, the authors propose creating and describing a research framework in digital advertising that emphasises the steps in both the marketing process and the development of marketing strategies where the use of digital technologies has had and will continue to have a big impact. We evaluate the published research in the relatively broad digital marketing arena using the framework to arrange innovations and existing studies all around components and touch points of the framework. In this section, we discuss the changing problems at the touch sites and their environs as well as the related research questions. In order to investigate the challenges from the standpoint of the firm, we finally integrate these highlighted topics and establish a study agenda for future studies in digital marketing (Kannan and Li, 2017). Examining the most effective digital marketing methods was the aim of this study work. This study uncovered a number of digital marketing tactics that can be applied by enterprises, organizations, educational institutions, non-profits, and other professional fields. The material made it abundantly evident that a number of classic marketing principles should be combined with technology. According to Author, the strategy shift from focusing on the consumer to the customer rather than the product is the key distinction between traditional marketing and digital marketing (Piñeiro-Otero and Martínez-Rolán, 2016).

This inquiry will highlight the importance of computerized media showcasing in the information era by highlighting the necessary and supplemental data acquired. The term "advanced displaying" refers to marketing strategies where organizations may track the performance of a campaign over time, including what is being seen, how frequently, and to what extent, as well as information about how deals are changing hands and associated processes. Reviews show that people choose advertisements as a more effective method of product promotion, which is evident from the data collected. Before purchasing a product, Indian customers check for user reviews and more information about the product, including details about its quality and price. The impact of advancements on Indian customers' ability to make a purchase is significant. Online shopping is frequently done by consumers in India. They are encouraged to shop for clothing and electronics online. Indian youth and young Indians are currently examples of people that watch television programmes online. The main motivation may be uninteresting, but it allows them to watch programmes they had previously skipped for a variety of reasons. People embrace online news sources since they don't have to wait around for continuous daily publications, similar to what is happening with the daily newspaper (Srikar and Hussain, 2008).

3. RESEARCH METHODOLOGY

This study will use a quantitative approach as its research technique. As part of this methodology, there will be a survey administered to digital marketing experts and business owners, and the data for this survey will be gathered from secondary sources. This will contain questions about the impact that digital marketing has had on business, such as how it has helped promote customer engagement, sales, and loyalty among customers. The survey will also include questions regarding the manner in which digital marketing strategies have been implemented, the manner in which these strategies have been reviewed, and the manner in which these tactics have impacted the overall success of organizations (Ancillai, 2019).

In addition, the collecting of secondary data from a wide variety of sources is going to be a part of the research approach that will be used for this study. These sources consist of academic publications and books, reports from various industries, and other resources based on the internet such as websites and forums. The evaluation of the effect that digital marketing has on the overall performance of a company is the reason for collecting these sources. As a result of this, it will be essential to take into consideration the dependability and validity of such sources in order to guarantee that the data gathered is precise and pertinent to the aims of the research (Redjeki and Affandi, 2021). In addition, the data that was gathered will be examined making use of quantitative and qualitative techniques such as descriptive statistics, content analysis, and thematic analysis. This will allow us to draw more relevant inferences from the data as well as gain a better understanding of the impact that digital marketing has on the performance of businesses (López García *et al.*, 2019).

In conclusion, the findings of the research will be presented in a manner that is both clear and succinct, opening the door for additional conversation and investigation into the impact that digital marketing has on the success of businesses. This will provide a comprehensive understanding of the subject matter as well as aid to inform further research in the field (Mogaji, Soetan, and Kieu, 2020).

4. CONCLUSION

As a result, it is clear from the explanation above that digital marketing has a big impact on many business outcomes. Due to a dearth of empirical and research-based data, evaluating the relationship between digital marketing and business success is challenging. The purpose of the current study is to determine how digital marketing affects business performance. One of the earliest studies to establish the connection between them and analyze the moderating impact of environmental factors, such as a dynamic industry and a competitive pressure to adopt new technologies. Furthermore, it has been noticed that the adoption of online marketing apps is influenced by the exterior competitive pressure that influences corporate performance. However, environmental factors like tough competition and a competitive economy had minimal impact on how well organizations did. The study's findings were proved to be relevant in analyzing the influence of digital marketing on performance. Furthermore, it did not seem like the environmental factor had a moderating impact. It is advised that businesses use the most recent technological advancements for marketing efforts to contact clients efficiently and effectively in order to prosper in a highly competitive environment utilizing secondary research and analysis.

4.1. Limitations of digital marketing for businesses

High Competition—As digital marketing can be so competitive, companies need to develop marketing strategies that allow them to differentiate themselves from their competitors. (Mehralian and Khazaee, 2022) Difficult to measure success–Measuring the success of digital marketing strategies and keeping track of return on investment may be challenging. (Ritz, Wolf and McQuitty, 2019) Technical difficulties—Digital marketing demands a significant amount of technical expertise and comprehension, both of which can be challenging to acquire and put into practice (Wang, 2020). Time-consuming—The development and management of digital marketing initiatives demand major investments of both time and resources on the part of enterprises (Saura, Palacios-Marqués and Ribeiro-Soriano, 2021). Cost- Digital marketing can be pricey because it requires firms to make investments in software, tools, and services in order to successfully operate the marketing campaigns (Mkwizu, 2019).

4.2. Research background

Research on the effects of digital marketing on businesses is important because it enables businesses to understand how their digital marketing efforts are impacting their business goals, such as increased sales, brand awareness, and customer engagement. This research is also important because it enables businesses to understand how their digital marketing efforts are impacting their business (Bomen, 2021). Businesses are able to make more educated judgments about their marketing strategy and change their marketing methods in order to maximize their return on investment if they have a solid grasp of the impact that their digital marketing efforts have had. In addition, research on the effects that digital marketing has on firms can help define industry trends, which in turn enables businesses to maintain their competitive edge in a digital landscape that is constantly evolving (Techfunnel, 2018).

Research on the effects that digital marketing has on businesses is vital for a number of reasons, the most important of which is that it enables firms to understand their own digital marketing efforts and get the most out of their investments (Behera *et al.*, 2019).

Businesses today have a tremendous resource at their disposal in the form of digital marketing, which enables them to connect with their target audiences in an efficient and cost-effective manner. It can assist companies in increasing sales, building relationships with customers, creating brand recognition, and monitoring how customers behave (David, 2022). The term "digital marketing" refers to a broad spectrum of advertising methods, including SEO, social media, email marketing, content marketing, native advertising, display advertising, mobile marketing, and more. Increased sales, decreased expenses, enhanced customer satisfaction and loyalty, and enhanced brand reputation are some of the business benefits that may be attributed to the use of digital marketing strategies. Digital marketing can also assist companies in better targeting their audiences, tracking the behavior of their customers, and providing improved service to those customers. The use of digital marketing can also give organizations useful insights into the tastes and habits of their customers, so enabling them to better tailor their marketing campaigns (Abhimbola, 2018).

The study of how organizations are impacted by digital marketing is still in its infancy and is undergoing rapid development as we speak. Research has been carried out to determine the levels of customer happiness, customer loyalty, and customer involvement in response to digital marketing efforts. A comparison of the efficacy of traditional marketing methods with that of digital marketing strategies has also been the subject of research. In addition, studies have been carried out to investigate the effect that digital marketing has on the total customer experience as well as the percentage of customers that choose to remain loyal to a brand (Dastane, 2020).

In general, digital marketing is a potent instrument that can assist firms in more efficiently and cheaply reaching out to their target audiences (customers). Ongoing research is being conducted in an effort to better understand the complete impact that digital marketing has on firms and to design methods that are more effective (Omar and Atteya, 2020).

4.3. Research Findings

According to the carried out and described descriptive research, five characteristics can be used to identify successful digital marketing campaigns in order to classify digital marketing for the growth of a business:

- Successful online business activities are thought to depend on a website's high level of quality, which helps in reaching customers outside of the local area. Additionally, the audience can access it via digital means (Veleva and Tsvetanova, 2020).
- Strong social media presence allows businesses to communicate with customers directly through Facebook, Twitter, and Instagram applications, allowing them to build long-lasting relationships that are more reliable, better, and sustainable (Hofacker et al., 2020).
- For determining online ranking at search engines that illustrates online popularity, SEO is regarded as a crucial and vital component of digital marketing (Gangeshwer, 2013).
- Email marketing is still a successful method of reaching out to customers, but for best results, it must be done with careful forethought. Because users are interested in the company, it has been found that emails to them may result in quicker responses. Establishing a long-term relationship with clients through personalized offers and efficient communication motivates them to stick around and make additional purchases (Chester and Montgomery, 2017).
- Mobile-friendly apps have also been discovered to have an impact on attracting clients, as was mentioned earlier. Most users look for their selected things using mobile phone apps to evaluate online shops. To provide simple and rapid access to clients, businesses must create effective and user-friendly mobile applications. (Giantari *et al.*, 2022)

4.4. Relevance of digital transformation programmes

34%: We have no plans to run a digital transformation programme 31%: We are planning to introduce a programme within the next 12 months

25%: We have just started a digital transformation programme (within the last 2 years) 10%: We have had a digital transformation process in place for >2 years

Number of respondents: 329 (Chaffey, 2020).

Research studies have demonstrated the effectiveness of digital marketing elements such as internet ads, email campaigns, advertising on social media text messaging marketing, affiliate marketing, SEO, and PPC advertising. There are other advantages of using digital media for marketing, which have been looked at in previous studies. These advantages and benefits include (i) up-to-date services or products relevant data (ii) increased client engagement (iii), clear availability of information (iv), ease of comparison with rivals or service providers (v), 24/7 accessibility of online shopping (vi), the capacity to share the content with peers or other relevant individuals (vii), selling prices, and (viii) the capability to end up making an immediate purchase through electronic devices and systems provided by businesses (Sodikin, 2020). Numerous studies have looked at the effect of marketing strategies on how well businesses do in terms of consumer reviews of their products or services. Hotels, online services, and any other enterprises that connect with customers or do business utilizing digital devices and internet technologies have all been found to benefit from customer product or service reviews (Nuseir and Aljumah, 2020).

It has been investigated how digital marketing affects business outcomes. It is well known that the usage of internet-based techniques and systems is a big and crucial factor that benefits a range of organizations. A research of the hotel industry and hotel management was conducted to determine the effect of digital marketing on a hotel's performance. The study found that employing knowledge technology systems and equipment for marketing goals can significantly improve performance because it has been shown to be effective in luring customers due to a quick strategy and the delivery to customers of the necessary information. Researchers found that by making sensible investments and using online feedback to determine the attitude, opinion, and level of customer satisfaction, managers may control business operations in response to feedback (Mehralian and Khazaee, 2022). According to studies, managers are more likely to improve the quality of their work and have an impact on how customers perceive them if they take feedback into account, appreciate it, and adapt their marketing plan. It has also been claimed that social media and hotel websites have an impact on visitor behavior. According to the study, customers' perceptions of websites and social media are shaped by their utility, enjoyment, social interaction, and satisfaction, all of which affect consumers' desire to make purchases (Pandey, Nayal, and Rathore, 2020).

REFERENCES

Abhimbola (2018). Digital Marketing Is Bound to Make an Impact on Your Business. Learn Why! [online] Mauco Enterprises. Available at: https://mauconline.net/5-reasons-you-should-be-talking-about-digital-marketing/.

Adil, M. (2022). 10+ Advantages And Disadvantages Of Online Digital Marketing» Adil Blogger. [online] adilblogger.com. Available at: https://adilblogger.com/advantages-disadvantages-online-digital-marketing/.

Alexander, L. (2022). What Is Digital Marketing? [online] Hubspot. Available at: https://blog.hubspot.com/marketing/what-is-digital-marketing.

Ancillai, C. (2019). Advancing social media driven sales research: Establishing conceptual foundations for B-to-B social selling. Industrial Marketing Management, 82(5), pp.293–308. doi:https://doi.org/10.1016/j.indmarman.2019.01.002.

Banicek, N. (2019) *How to choose the right PPC channel for your business, How to Choose the Right PPC Channel for Your Business*. Available at: https://clutch.co/agencies/ppc/resource/how-choose-right-ppc-channel-business (Accessed: February 15, 2023).

Barone, A. (2022). Digital Marketing. [online] Investopedia. Available at: https://www.investopedia.com/terms/d/digital-marketing.asp.

Behera, R.K., Gunasekaran, A., Gupta, S., Kamboj, S. and Bala, P.K. (2019). Personalized Digital Marketing Recommender Engine. *Journal of Retailing and Consumer Services*, 53, p.101799. doi:https://doi.org/10.1016/j.jretconser.2019.03.026.

Bomen (2021). Impact of Digital Marketing on Business And Its Growth. [online] TFOT. Available at: https://thefutureofthings.com/16083-the-impact-of-digital-marketing-on-businesses-its-growth/.

Bouchrika, I. (2022). How to Write Research Methodology: Overview, Tips, and Techniques «Guide 2 Research. [online] research.com. Available at: https://research.com/research/how-to-write-research-methodology.

Busca, L. and Bertrandias, L. (2020). A Framework for Digital Marketing Research: Investigating the Four Cultural Eras of Digital Marketing. *Journal of Interactive Marketing*, 49(1), pp.1–19.

Chaffey, D. (2018). What is Digital Marketing? A visual summary - Smart Insights. [online] Smart Insights. Available at: https://www.smartinsights.com/digital-marketing-strategy/what-is-digital-marketing/.

Chaffey, D. (2020) Digital transformation strategy to grow your business | Smart Insights, *Smart Insights*, [online] Available from: https://www.smartinsights.com/digital-marketing-strategy/what-is-digital-transformation-strategy-and-how-does-it-fuel-business-growth/.

Chaffey, D. and Smith, P. (2013) *Emarketing Excellence*, Routledge.

Chaston, I. and Mangles, T. (2003) Relationship marketing in online business-to-business markets, *European Journal of Marketing*, 37(5/6), pp. 753–773.

Chester, J. and Montgomery, K. (2017) The Role of Digital Marketing in Political Campaigns, *INTERNET POLICY REVIEW Journal on Internet Regulation*, 6(4), [online] Available from: https://policyreview.info/node/773/pdf.

Curran, K., Graham, S. and Temple, C. (2011) Advertising on Facebook, *ResearchGate*, unknown, [online] Available from: https://www.researchgate.net/publication/268289406_Advertising_on_Facebook.

Das, S. K. and Lall, D. G. S. (2016) Traditional marketing VS digital marketing: An analysis, *International Journal of Commerce and Management Research*, **2**(8), pp. 05-11, [online] Available from: http://www.managejournal.com/archives/2016/vol2/issue8/2-7-24.

Dastane, O. (2020). Impact of Digital Marketing on Online Purchase Intention: Mediation Effect of Customer Relationship Management. *Journal of Asian Business Strategy*, 10(1), pp.142–158. doi:https://doi.org/10.18488/journal.1006.2020.101.142.158.

Dave Chaffey (2015) *Digital business and e-commerce management: strategy, implementation and practice*, 6th ed, Harlow, Pearson.

David (2022). How Digital Marketing will Impact Businesses in 2019. [online] Talentedge. Available at: https://talentedge.com/blog/digital-marketing-will-impact-businesses/.

Dunakhe, K. and Panse, C. (2021). Impact of digital marketing – a bibliometric review. *International Journal of Innovation Science*, ahead-of-print(ahead-of-print). doi:https://doi.org/10.1108/ijis-11-2020-0263.

Durmaz, Y. and Efendioglu, I. H. (2016) Travel from Traditional Marketing to Digital Marketing, *Global Journal of Management and Business Research*, **16**(2), pp. 35–40.

Faruk, M., Rahman, M. and Hasan, S. (2021). How Digital Marketing Evolved over time: a Bibliometric Analysis on Scopus Database. Heliyon, [online] 7(12), p.e08603. doi:https://doi.org/10.1016/j.heliyon.2021.e08603.

Fred (2022). Importance of Digital Marketing for the Success of a Business. [online] www.amuratech.com. Available at: https://www.amuratech.com/blog/importance-of-digital-marketing.

Gangeshwer, D. K. (2013) E-Commerce or Internet Marketing: A Business Review from Indian Context, *International Journal of u- and e-Service, Science and Technology*, **6**(6), pp. 187–194.

Giantari, I., Yasa, N., Suprasto, H. and Rahmayanti, P. (2022) The role of digital marketing in mediating the effect of the COVID-19 pandemic and the intensity of competition on business performance, *International Journal of Data and Network Science*, **6**(1), pp. 217–232, [online] Available from: http://m.growingscience.com/beta/ijds/5154-the-role-of-digital-marketing-in-mediating-the-effect-of-the-covid-19-pandemic-and-the-intensity-of-competition-on-business-performance.html.

Gillis (2022). What is digital marketing? Everything you need to know | Definition from TechTarget. [online] SearchCustomerExperience. Available at: https://www.techtarget.com/searchcustomerexperience/definition/digital-marketing.

Gregor (2022). Importance of Digital Marketing for the Success of a Business. [online] Edureka. Available at: https://www.edureka.co/blog/importance-of-digital-marketing-for-the-success-of-a-business/ [Accessed 12 Feb. 2023].

Hanna, R., Rohm, A. and Crittenden, V. L. (2011) We're all connected: The power of the social media ecosystem, *Business Horizons*, **54**(3), pp. 265–273.

Heng, C. and Yazdanifard, R. (2013) Generation Gap; Is There any Solid Solution? From Human Relation Point of View, *International Journal of Economy, Management and Social Sciences*, **2**(10), [online] Available from: https://citeseerx.ist.psu.edu/document?repid=rep1&type=pdf&doi=4b3ca59ac28e7efa1b39825fb6e8e371913b6e6d (Accessed 18 November 2022).

Hofacker, C., Golgeci, I., Pillai, K. G. and Gligor, D. M. (2020) Digital marketing and business-to-business relationships: a close look at the interface and a roadmap for the future, *European Journal of Marketing*, **54**(6), pp. 1161–1179, [online] Available from: https://www.emerald.com/insight/content/doi/10.1108/EJM-04-2020-0247/full/html.

Kannan, P. K. and Li, H. A. (2017) Digital marketing: a framework, Review and Research Agenda, *International Journal of Research in Marketing*, **34**(1), pp. 22–45, [online] Available from: https://www.sciencedirect.com/science/article/pii/ S0167811616301550.

Kaur (2021). What Is Research Methodology and Why Is it Important? [online] Indeed Career Guide. Available at: https://www.indeed.com/career-advice/career-development/ research-methodology.

Kim, J., Kang, S. and Lee, K.H. (2019). Evolution of digital marketing communication: Bibliometric analysis and network visualization from key articles. *Journal of Business Research*, 130(3). doi:https://doi.org/10.1016/j.jbusres.2019.09.043.

Krishen, A.S., Dwivedi, Y.K., Bindu, N. and Kumar, K.S. (2021). A broad overview of interactive digital marketing: A bibliometric network analysis. *Journal of Business Research*, 131(0148-2963), pp.183–195. doi:https://doi.org/10.1016/j. jbusres.2021.03.061.

Leeflang, P. S. H., Verhoef, P. C., Dahlström, P. and Freundt, T. (2014) Challenges and solutions for marketing in a digital era, *European Management Journal*, **32**(1), pp. 1–12, [online] Available from: https://econpapers.repec.org/article/eeeeurman/ v_3a32_3ay_3a2014_3ai_3a1_3ap_3a1-12.htm.

López García, J.J., Lizcano, D., Ramos, C.M. and Matos, N. (2019). Digital Marketing Actions That Achieve a Better Attraction and Loyalty of Users: An Analytical Study. Future Internet, [online] 11(6), p.130. doi:https://doi.org/10.3390/fi11060130.

Mangold, W. G. and Faulds, D. J. (2009) Social media: the New Hybrid Element of the Promotion Mix, *Business Horizons*, **52**(4), pp. 357–365, [online] Available from: https://www.sciencedirect.com/science/article/pii/S0007681309000329.

Mehralian, M. M. and Khazaee, P. (2022) Investigating the Interrelationships between Digital Marketing and Marketing Intelligence and Their Effect on Business Strategy, *papers.ssrn.com*, Rochester, NY, [online] Available from: https://papers.ssrn. com/sol3/papers.cfm?abstract_id=4190278.

Mehralian, M.M. and Khazaee, P. (2022). Effect of Digital Marketing on the Business Performance of MSMEs during the COVID-19 Pandemic: The Mediating Role of Customer Relationship Management. [online] papers.ssrn.com. Available at: https:// ssrn.com/abstract=4195985.

Menon (2021). The Benefits of Digital Marketing. [online] Lucid Advertising. Available at: https://www.lucidadvertising.com/blog/benefits-digital-marketing/.

Mkwizu, K.H. (2019). Digital marketing and tourism: opportunities for Africa. International Hospitality Review, ahead-of-print(ahead-of-print). doi:https://doi. org/10.1108/ihr-09-2019-0015.

Mogaji, E., Soetan, T.O. and Kieu, T.A. (2020). The implications of artificial intelligence on the digital marketing of financial services to vulnerable customers. *Australasian Marketing Journal (AMJ)*, 29(3). doi:https://doi.org/10.1016/j.ausmj.2020.05.003.

Mort, G. S. and Drennan, J. (2002) Mobile digital technology: Emerging issue for marketing, *Journal of Database Marketing & Customer Strategy Management*, **10**(1), pp. 9–23, [online] Available from: https://link.springer.com/article/10.1057%2Fpalgrave. jdm.3240090.

NELSON, R. R., TODD, P. A. and WIXOM, B. H. (2005) Antecedents of Information and System Quality: An Empirical Examination Within the Context of Data Warehousing, *Journal of Management Information Systems*, **21**(4), pp. 199–235.

Nirmal (2015) What works best in digital marketing today?, *www.nirmal.com.au*, [online] Available from: https://www.nirmal.com.au/what-works-best-in-digital-marketing-today/ (Accessed 15 February 2023).

NIU (2019). Advantages and disadvantages of digital marketing. [online] nibusinessinfo. co.uk. Available at: https://www.nibusinessinfo.co.uk/content/advantages-and-disadvantages-digital-marketing.

Nuseir, M. T. and Aljumah, A. I. (2020) (PDF) The Role of Digital Marketing in Business Performance with the Moderating Effect of Environment Factors among SMEs of UAE, *ResearchGate*, [online] Available from: https://www.researchgate.net/publication/339789814_The_Role_of_Digital_Marketing_in_Business_Performance_with_the_Moderating_Effect_of_Environment_Factors_among_SMEs_of_UAE.

Omar, A.M. and Atteya, N. (2020). The Impact of Digital Marketing on Consumer Buying Decision Process in the Egyptian Market. *International Journal of Business and Management*, 15(7), p.120. doi:https://doi.org/10.5539/ijbm.v15n7p120.

Pandey, N., Nayal, P. and Rathore, A. S. (2020) Digital marketing for B2B organizations: structured literature review and future research directions, *Journal of Business & Industrial Marketing*, **ahead-of-print**(ahead-of-print).

Patel (2022). What is Digital Marketing and Its benefits? [online] Engaio Digital. Available at: https://engaiodigital.com/digital-marketing/.

Piñeiro-Otero, T. and Martínez-Rolán, X. (2016) Understanding Digital Marketing—Basics and Actions, *Management and Industrial Engineering*, pp. 37–74, [online] Available from: https://www.researchgate.net/publication/312190728_Understanding_Digital_Marketing-Basics_and_Actions.

Prasanna (2022). Advantages And Disadvantages Of Digital Marketing | What is Digital Marketing?, 8+ Digital Marketing Advantages and Disadvantages. [online] A Plus Topper. Available at: https://www.aplustopper.com/advantages-and-disadvantages-of-digital-marketing/.

Redjeki, F. and Affandi, A. (2021). Utilization of Digital Marketing for MSME Players as Value Creation for Customers during the COVID-19 Pandemic. *International Journal of Science and Society*, [online] 3(1), pp.40–55. doi:https://doi.org/10.54783/ijsoc.v3i1.264.

Ritz, W., Wolf, M. and McQuitty, S. (2019). Digital marketing adoption and success for small businesses. *Journal of Research in Interactive Marketing*, [online] 13(2), pp.179–203. doi:https://doi.org/10.1108/jrim-04-2018-0062.

Roger (2021). What is Digital Marketing? Definition & Examples | Wrike Guide. [online] www.wrike.com. Available at: https://www.wrike.com/digital-marketing-guide/what-is-digital-marketing/.

Salehi, M., Mirzaei, H., Aghaei, M. and Abyari, M. (2012) Dissimilarity of E-marketing VS traditional marketing, *International Journal of Academic Research in Business and Social Sciences*, 2(1), p. 510, [online] Available from: https://axware.nl/sites/default/files/fulltext/548.pdf.

Saura, J.R., Palacios-Marqués, D. and Ribeiro-Soriano, D. (2021). Digital marketing in SMEs via data-driven strategies: Reviewing the current state of research. *Journal of Small Business Management*, 3(5), pp.1–36. doi:https://doi.org/10.1080/00472778.2021.1955127.

Scott (2020). 6 Types of Digital Marketing: When and How To Use Them? [online] Simplilearn.com. Available at: https://www.simplilearn.com/types-of-digital-marketing-article.

Shankar, V., Inman, J. J., Mantrala, M., Kelley, E. and Rizley, R. (2011) Innovations in Shopper Marketing: Current Insights and Future Research Issues, *Journal of Retailing*, **87**, pp. S29–S42.

Simplil (2021). 8 Benefits of Digital Marketing: All You Should Know. [online] Simplilearn. com. Available at: https://www.simplilearn.com/digital-marketing-benefits-article.

Sodikin, M. (2020) Competitive Advantages of Sharia Banks: Role of Ihsan Behavior and Digital Marketing in New Normal, *Journal of Digital Marketing and Halal Industry*, **2**(1), p. 1.

Srikar, K. and Hussain, A. (2008) Digital Marketing Impact of Digital Marketing in a SME, *International Journal of Advances in Engineering and Management*, **9001**, p. 816, [online] Available from: https://ijaem.net/issue_dcp/Digital%20Martketing%20 Impact%20of%20Digital%20Marketing%20in%20a%20Sme.pdf (Accessed 15 February 2023).

Stuart, J. A., Maddalena, L. A., Merilovich, M. and Robb, E. L. (2014) A midlife crisis for the mitochondrial free radical theory of aging, *Longevity & Healthspan*, **3**(1), [online] Available from: https://longevityandhealthspan.biomedcentral.com/ articles/10.1186/2046-2395-3-4.

Tanna, D., Raval, Z. and Raval, D. (2014) Social Media and Public Awareness, *https:// www.researchgate.net/publication/265163948_Social_Media_and_Public_Awareness*, [online] Available from: https://www.researchgate.net/publication/265163948_Social_ Media_and_Public_Awareness (Accessed 5 February 2023).

Techfunnel (2018). Impact of Digital Marketing on Business. [online] Techfunnel. Available at: https://www.techfunnel.com/martech/impact-digital-marketing-business/.

Todd (2020). Importance of Digital Marketing for businesses in 2021. [online] GreatLearning Blog: Free Resources what Matters to shape your Career! Available at: https://www.mygreatlearning.com/blog/importance-of-digital-marketing-for-businesses/.

Varadarajan, R., Welden, R.B., Arunachalam, S., Haenlein, M. and Gupta, S. (2021). Digital Product Innovations for the Greater Good and Digital Marketing Innovations in Communications and Channels: Evolution, Emerging Issues, and Future Research Directions. *International Journal of Research in Marketing*, 6(2). doi:https://doi. org/10.1016/j.ijresmar.2021.09.002.

Veleva, S. S. and Tsvetanova, A. I. (2020) Characteristics of the digital marketing advantages and disadvantages, *IOP Conference Series: Materials Science and Engineering*, **940**(1), p. 012065.

Vieira, V.A. (2019). In pursuit of an effective B2B digital marketing strategy in an emerging market. *Journal of the Academy of Marketing Science*, 47(6), pp.1085–1108. doi:https://doi.org/10.1007/s11747-019-00687-1.

Wang, F. (2020). Digital marketing capabilities in international firms: a relational perspective. International Marketing Review, ahead-of-print(ahead-of-print). doi:https://doi.org/10.1108/imr-04-2018-0128.

Wang, J., Duncan, D., Shi, Z. and Zhang, B. (2013) WEB-based GEne SeT AnaLysis Toolkit (WebGestalt): update 2013, *Nucleic Acids Research*, **41**(W1), pp. W77–W83.

Yasmin, A., Tasneem, S. and Fatema, K. (2015) Effectiveness of Digital Marketing in the Challenging Age: An Empirical Study, *The International Journal of Management Science and Business Administration*, 1(5), pp. 69–80, [online] Available from: https://researchleap.com/effectiveness-of-digital-marketing-in-the-challenging-age-an-empirical-study/.

19. Employees' psychological ownership and competitive advantage

V. Sudeepa Sree, P. V. S. L. Anusha, and K. Sushmitha Reddy

School of Business, Woxsen University, Hyderabad, Telangana, India

ABSTRACT: In the globalized work environment, human capital in terms of expertise and skills of employees are equally available in the labor market. Similarly, corporations are equally endowed with adequate financial and physical resources. Given the fact that similar human resources and corporate resources are equally available, creating a competitive advantage for a particular company is increasingly becoming difficult. We have attempted to understand and explore the role of psychological ownership in fostering competitive advantage in the long run both for employees' well-being and organizational productivity. We have reviewed pertinent literature on psychological ownership, its antecedents, and the moderators that influence workplace outcomes and augment competitive advantage. Finally, managerial implications are also indicated.

KEYWORDS: Psychological ownership (PO), employee commitment, well-being, competitive advantage, attrition

1. INTRODUCTION

Commitment plays an important role in workplace productivity. Earlier people were not quitting their jobs frequently as there were not many options and there was an underlying attitude that the entity of work is synonymous to social identity. However, with the rapid and dynamic economic development, there is an increase in the notion of competitiveness in work environments from both the point of view of employers and employees. Given this context, the working population is increasingly looking for higher benefits and the same is the case of the employers in terms of extracting the best from the employees. This competitive spree has led to an unprecedented level of a disengaged workforce and psychological ownership towards the work and the organization has become a farfetched reality. Apart from the basic benefits that are derived from work, other attitudinal factors like extra-role behaviors, organizational citizenship behavior, and commitment towards the work play a crucial role in augmenting individual and organizational productivity. In this paper, we have attempted to understand and explore the role of psychological ownership (PO) construct in

Chapter 19 DOI- 10.4324/9781003397175-19

fostering competitive advantage for the companies in the long run. It is very crucial for any business or organization to have good workplace commitment from their employees to ensure harmony and a smooth flow of efforts within the organization. Committed employees tend to work better and show a sense of loyalty to their work and organization. This helps to improve the overall productivity of the company. Employee motivation plays a major role in improving the level of commitment among employees. Comparing the past corporate environment to the current corporate scenario, there is a huge difference in the overall profits of the companies and businesses. In the past years (early 2000's) people used to love their jobs and there was also less competition in the market. But later on, with the rise of technological advancements, the scope of the job market increased and better opportunities have arisen. So, eventually, people started leaving their current jobs to find better work that is feasible to their constraints. So, now to solve this current situation of job crisis and commitment, it is worth exploring the role of psychological ownership to build a better work environment so that employees can be retained by the company.

2. PSYCHOLOGICAL OWNERSHIP AND COMPETITIVE ADVANTAGE

Three possible pathways for PO improvement are suggested by Pierce et al. (2001, 2003) in their pioneering work. They refer to these pathways as the "courses" to mental proprietorship. Through exercising "experienced command over the goal of ownership, personal knowledge of the objective of possession, and venture of the self into the objective of possession," they guarantee that PO can be established. According to organization-based PO variables and occupation-based PO factors, authoritative conduct researchers have ranked predecessors of PO based on experiences from these lines of grant.

2.1. Organization-Based Psychological Ownership

According to researchers, hierarchical PO can be developed using a number of techniques, such as strong initiative styles, participatory dynamic methodologies, stock proprietorship plans, benefit-sharing plans, representative strengthening, and the advancement of suitable workplaces. Avey et al. (2009)'s study demonstrated a strong correlation between organization-based PO and ground-breaking initiatives. Also, studies have shown that an increase in organization-based PO is strongly associated with responsible behavior, fulfilling work, and authoritative citizenship ways of behaving (OCBs). In the opposite direction, PO and turnover expectations were related (Bernhard and O'Driscoll, 2011). Conditional authority approaches combined with moral reflections are unquestionably related to authoritative-based PO. These initiative techniques actually have an impact on positive work behaviors (Avey et al., 2012; Bernhard and O'Driscoll, 2011). Contrary to conventional wisdom, decentralized administrative methods result in the dismantling of hierarchically structured PO (Bernhard and O'Driscoll, 2011). Another study led by Zhu et al. (2013) demonstrated that the compassionate authority style supports the development of hierarchically structured PO. Studies have also revealed a strong correlation between organization-based PO and workers' navigation support (Chi and Han, 2008; Han et al., 2010; Liu et al., 2012). These findings support Penetrate et al.'s (2001, 2003) claim that the degree of participatory direction facilitates representatives' "accomplished control" over the goal of proprietorship and, as a result, aids in the development of association-based PO. In contrast, Chi and Han's (2008) study focuses on the specifics of how a participatory dynamic's ability to grow hierarchical-based PO levels while enhancing procedural fairness. Chi and Han (2008) found that benefit-imparting plans to higher levels of association-based PO had positive results. Giving representatives independence is also strongly linked to organization-based PO (Henssen et al., 2014; Mayhew et al., 2007).

According to O'Driscoll et al. (2006), a positive and disorganized workplace has a favorable influence on the company. Also, those representatives who experience work control report increased organization PO, according to research (Peng and Puncture, 2015).

2.2. Job-Based Psychological Ownership

The exploration of the origins of job-based PO has received relatively little study. Compared to organization-based PO, fewer studies have been done on the causes of employees' feelings about job-based PO. According to research on job-based PO, factors that support its growth include autonomy, effective leadership techniques, and mental and spiritual maturity. For example, the study reported by Mayhew et al. (2007) indicates that job-based PO is positively correlated with greater autonomy. Leadership philosophies and practices are equally important in job-based PO. According to research by Bernhard and O'Driscoll (2011), job-based PO is favorably correlated with both transformational and transactional leadership. Additionally, the connection between job-based PO and workers' spiritual and emotional intelligence is favorable (Kaur et al., 2013).

3. BENEFITS OF PSYCHOLOGICAL OWNERSHIP

Numerous findings have demonstrated the positive impacts of PO; however, the majority of these studies only analyze individuals as a whole. Many research show a favorable relationship between PO and individual attitudes such as organizational commitment, emotional commitment, job satisfaction, organization-based self-esteem, work engagement, and desire to remain (e.g., Avey et al., 2012; Bernhard & O'Driscoll, 2011; Han et al., 2010; Hou et al., 2009; Knapp et al., 2014; Liu et al., 2012; Mayhew et al., 2007; Pan et al., 2014; Peng & Pierce, 2015; Ramos et al., 2014; Sieger et al., 2011; Zhu et al., 2013).

In addition to the desirable work attitudes, PO studies have discovered favorable relationships between helpful behavior, OCBs, voice behavior, and stewardship behavior (e.g., Bernhard & O'Driscoll, 2011; Park et al., 2013; Ramos et al., 2014; Van Dyne & Pierce, 2004; Zhu et al., 2013). PO was found to have a positive association with knowledge-sharing behavior (Han et al., 2010). Moreover, PO was found to have a pessimistic association with burnout (Kaur et al., 2013).

Figure 1 shows a list of factors that serve as antecedents of PO and the moderating factors that affect PO and workout results based on the discussion above.

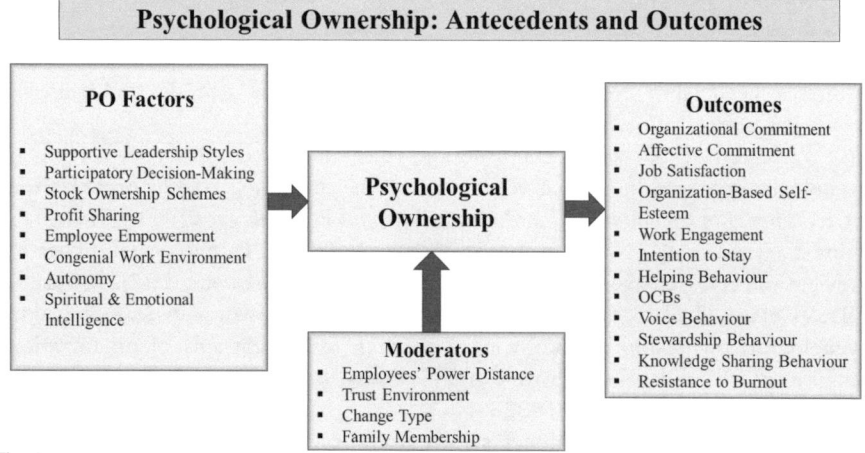

Figure 1. *Source: Dawkins et al. (2017).*

3.2. Role of Psychological Ownership in Creating Competitive Advantage

Together with well-known management scholars, famous industry executives such as Carly Fiorina of Hewlett-Packard Co, Bill Gates of Microsoft, and Andy Grove of Intel Corp, all agree that positive human resources are one of the most significant indicators of any organization. Positive psychological resources, in addition to functional abilities, are critical (Luthans & Youssef-Morgan, 2017; Luthans et al., 2015). In an increasingly flat society, attaining a competitive edge only through physical resources such as technology and financial advantages is becoming increasingly challenging (Kraaijenbrink et al., 2010; Luthans and Stajkovic, 1999). This tendency has occurred mostly because of easy access to technology and financial resources. Because psychological resources are not readily replicated, an organization's intangible resources (psychological ownership) might play a key influence in its growth and sustainability. Psychological ownership, as demonstrated in the preceding sections, has a variety of good outcomes in terms of organizational commitment, psychological attachment, work satisfaction, organization-based self-esteem, job involvement, and desire to remain. Positive work attitudes and habits will assist an organization's long-term sustainability and competitive edge.

REFERENCES

Avey, J. B., Avolio, B. J., Crossley, C. R., & Luthans, F. (2009). Psychological ownership: Theoretical extensions, measurement, and relation to work outcomes. *Journal of Organizational Behavior, 30*, 173–191.

Avey, J. B., Wernsing, T. S., & Palanski, M. E. (2012). Exploring the process of ethical leadership: The mediating role of employee voice and psychological ownership. *Journal of Business Ethics, 107*, 21–34.

Bernhard, F., & O'Driscoll, M. P. (2011). Psychological ownership in small family-owned businesses: Leadership style and nonfamily employees' work attitudes and behaviors. *Group and Organization Management, 36*, 345–384.

Chi, N. W., & Han, T. S. (2008). Exploring the linkages between formal ownership and psychological ownership for the organization: The mediating role of organizational justice. *Journal of Occupational and Organizational Psychology, 81*, 691–711.

Dawkins, S., Tian, A. W., Newman, A., & Martin, A. (2017). Psychological ownership: A review and research agenda. *Journal of Organizational Behavior, 38*(2), 163–183.

Han, T. S., Chiang, H. H., & Chang, A. (2010). Employee participation in decision making, psychological ownership and knowledge sharing: Mediating role of organizational commitment in Taiwanese high-tech organizations. *International Journal of Human Resource Management, 21*, 2218–2233.

Henssen, B., Voordeckers, W., Lambrechts, F., & Koiranen, M. (2014). The CEO autonomy stewardship behavior relationship in family firms: The mediating role of psychological ownership. *Journal of Family Business Strategy, 5*, 312–322.

Higgins, E. T. (1997). Beyond pleasure and pain. *American Psychologist, 52*, 1280–1300.

Higgins, E. T. (1998). Promotion and prevention: Regulatory focus as a motivational principle. In Zanna, M. P. (Ed.), *Advances in experimental social psychology* (Vol. 30, pp. 1–46). New York: Academic Press.

Hou, S. T., Hsu, M. Y., & Wu, S. H. (2009). Psychological ownership and franchise growth: An empirical study of a Taiwanese taxi franchise. *International Journal of Entrepreneurial Behavior & Research, 15*, 415–435.

Kaur, D., Sambasivan, M., & Kumar, N. (2013). Effect of spiritual intelligence, emotional intelligence, psychological ownership and burnout on caring behavior of nurses: A cross-sectional study. *Journal of Clinical Nursing, 22*, 3192–3202.

Knapp, J. R., Smith, B. R., & Sprinkle, T. A. (2014). Clarifying the relational ties of organizational belonging: Understanding the roles of perceived insider status, psychological ownership, and organizational identification. *Journal of Leadership & Organizational Studies, 21*, 273–285.

Kraaijenbrink, J., Spender, J. C., & Groen, A. J. (2010). The resource-based view: A review and assessment of its critiques. *Journal of Management, 36*(1), 349–372.

Liu, J., Wang, H., Hui, C., & Lee, C. (2012). Psychological ownership: How having control matters. *Journal of Management Studies, 49*, 869–895.

Luthans, F., & Stajkovic, A. D. (1999). Reinforce for performance: The need to go beyond pay and even rewards. *The Academy of Management Executive, 13*(2), 49–57.

Luthans, F., & Youssef-Morgan, C. M. (2017). Psychological capital: An evidence-based positive approach. *Annual Review of Organizational Psychology and Organizational Behavior, 4,* 339–366.

Luthans, F., Youssef-Morgan, C. M., & Avolio, B. J. (2015). *Psychological capital and beyond.* Oxford, New York, NY: Oxford University Press.

Mayhew, M. G., Ashkanasy, N. M., Bramble, T., & Gardner, J. (2007). Study of the antecedents and consequences of psychological ownership in organizational settings. *Journal of Social Psychology, 147,* 477–500.

O'Driscoll, M. P., Pierce, J. L., & Coghlan, A. M. (2006). The psychology of ownership: Work environment structure, organizational commitment, and citizenship behaviors. *Group & Organization Management, 31,* 388–416.

Pan, X. F., Qin, Q. W., & Gao, F. (2014). Psychological ownership, organization-based self-esteem and positive organizational behaviors. *Chinese Management Studies, 8,* 127–148.

Park, C. H., Song, J. H., Yoon, S. W., & Kim, J. (2013). A missing link: Psychological ownership as a mediator between transformational leadership and organizational citizenship behavior. *Human Resource Development International, 16,* 558–574.

Peng, H., & Pierce, J. L. (2015). Job- and organization-based psychological ownership: Relationship and outcomes. *Journal of Managerial Psychology, 30,* 151–168.

Pierce, J. L., Kostova, T., & Dirks, K. T. (2001). Toward a theory of psychological ownership in organizations. *Academy of Management Review, 26,* 298–310.

Pierce, J. L., Kostova, T., & Dirks, K. T. (2003). The state of psychological ownership: Integrating and extending a century of research. *Review of General Psychology, 7,* 84–107.

Ramos, H. M., Man, T. W. Y., Mustafa, M., & Ng, Z. Z. (2014). Psychological ownership in small family firms: Family and non-family employees' work attitudes and behaviors. *Journal of Family Business Strategy, 5,* 300–311.

Sieger, P., Bernhard, F., & Frey, U. (2011). Affective commitment and job satisfaction among non-family employees: Investigating the roles of justice perceptions and psychological ownership. *Journal of Family Business Strategy, 2,* 78–89.

Zhu, H., Chen, C. C., Li, X. C., & Zhou, Y. H. (2013). From personal relationship to psychological ownership: The importance of manager–owner relationship closeness in family businesses. *Management and Organization Review, 9,* 295–318.

20. Association of social media usage with social connectedness, social cognition, and mental health among the youth: Past, present, and future

Dr. Sulu Priya B.[1] and Dr. P.M. Ramesh Kumar[2]

[1]Associate Professor & Head, P.G. Department of Social Work, Dwaraka Doss Goverdhan Doss Vaishnav College (Autonomous), India.

[2]Assistant Professor, Department of Commerce, Dwaraka Doss Goverdhan Doss Vaishnav College (Autonomous), India.

Email- sulupriyamsw@gmail.com, Email – rameshkumarpm@dgvaishnavcollege.edu.in

ABSTRACT: Youth is one of the most energetic and challenging stages in one's life as it is a culmination of his past experiences and foundation for his future life. Sense of belongingness is innate in human nature, from the time immemorial. Social connectedness refers to the sense of belongingness while interacting with others and the resultant satisfaction. Social cognition refers to the innumerable ways in which individuals collect and use information to explain and predict their own behavior. Social media includes the web based virtual networks that enables sharing of ideas and information faster. There are 4.5 billion social media users around the world. This paper aims to assimilate the association between usage of social media on the social connectedness, social cognition, and the mental health of the youthful population of our country. Youth refers to individuals between the age group 15–29 years as per national youth policy 2014. With the advent of technology, growing internet accessibility, especially after the pandemic induced lockdown; there is a drastic change in the life styles of people. This research paper aims to understand and analyze the views of the youth and the determinants of social media addiction with a supported primary survey and quantitative data. The level of influence that social media has on the social cognition and social connectedness are significantly affecting the mental health. While social media has enhanced social connectedness virtually; the social cognition and mental health of people have taken a topsy turvy ride. The paper attempts make an analysis of the past situation before the internet age, take us through the present, and looks at the future exhorting the need to get awakened to use social media wisely.

Chapter 20 DOI- 10.4324/9781003397175-20

KEYWORDS: Social media, social connectedness, social cognition, youth, mental health, pandemic, techno stress, memory

1. A PRELUDE

Social media is so inextricable interwoven in our life, right from waking up till one hit bed. The frequent usage of social media especially during the formative years of life among the youth and adolescents invites significant attention (Ahn, 2011). The paper focuses on major aspects of impact of social media usage on social connected ness, social cognition, and mental health among the youth. It looks at how the youth distance themselves from near and dear ones emotionally and physically at the same time maintaining thousands of connections in social media handles. This paradoxical situation creates a feeling in them that they are well connected and start doing things to be displayed in the social platform (Acquier et al., 2017; Valenzuela et al., 2019). Life events whether irrespective of being sorrowful or joyful, become a matter of public display. After a point of time life revolves around YouTube, Instagram, Facebook, Twitter, and other social media arena. Every aspect of life becomes public and feeling of accomplishment is connected to the number of likes and subscribers from unknown parts and people of the world (Krogh, 2022). A dip in the number of likes and subscription is enough to push them to acute depression and chronic mental health issues (Qualman, 2012). This platform which was intended to support the knowledge transfer and economy has become a deterrent factor in distancing people from real life and situations to reel life. This paper looks into the youth who get addicted to social media that substitutes their best friend and dearest kith and kin in the name of paradoxical social connectedness and alienate them from the experiences of real world. This research paper adds to the literature on youth and their psychosocial wellbeing. It has implications for researchers working with the educational, developmental, and social psychology. This paper focuses on the following questions: (1) What are the potential implications of social media usage among youth?; (2) What are the major research gaps?; and (3) What are the possible ways by which youth can be made aware of the potential dangers of procrastination due to the increased social media usage in their formative years?

2. REVIEW OF LITERATURE

2.1. The Advent of the Internet Age: Past

It is estimated that social media had its beginning in May, 1844 through the evidence of electronic dots in a telegraph machine. The periods of 1980's and 1990's saw the advent of online media through emails, bulletin board messaging, and chatting. In 1999, live journal publishing site was launched leading to the origin of Pyra Labs which is a publishing platform for bloggers. It was subsequently acquired by Google in 2003. Basic online networking had started in 2001 by Friendster. LinkedIn launched in 2002 as a job search and networking platform, has more than 675 million users by 2020. Google + was launched in 2012.

Though the social media sites have established the minimum age to use its platform as 13 years of age, a survey by National Commission for Protection of Child Rights (NCPCR) had found that 38% of the 10-year-old in India have Facebook account and over 24% had Instagram accounts, reports ET Now Digital (2021).

2.2. Social Media Usage – The Present

A study conducted during June and July 2020, in all parts of India on the history of internet use and duration of use on daily base depicted that the social media is extensively used for social connectedness, socializing, and for recreation. While those in 18–25 years age bracket preferred Instagram, those belonging to 25–40 years and those above 40 years, preferred Facebook as the social media platform (Bohra, 2022). Youth are the major beneficiaries of social media usage due to easy access, parental approval, and compatibility of usage under the pretext of knowledge transfer. Chauhan and Yachu (2022) had reported through Indian express that Facebook and Instagram has 97.2 million and 69 million users from individuals aged 18–24 years, pointing towards the extent of its reach.

2.3. Social Connectedness, Social Media, and Youth

United Nations (2013) has defined "youth" as a person between the ages of leaving compulsory education, and finding their first job, i.e., between the ages 15 and 24 years. The UNICEF-GALLUP Study (2021) points that young people use social media to "stay informed."

Meta and Gallup's Study on Social Connections (2022) confirms that social connections are essential in the mental well-being of people. People spanning in seven countries: Brazil, Egypt, India, France, Indonesia, Malaysia, Mexico, and the United States had confirmed that governments and organizations are focusing on understanding about the impact of people's connections. Penn State (2019) had found that people who had "felt-love" and "connection" in their daily life had reported higher levels of psychological well-being and optimism.

It was observed that grabbing of user's attention by social media decreases the need and desire to sieve irrelevant information (Ophir et al., 2009). Shorter attention span has been observed among students who switch between social media and education (Rosen et al., 2013). Higher incidence of attention related errors was reported among students who use smartphones with less attention on its consequences (Marty-Dugas et al., 2018).

Feeling of loneliness and disconnectedness adds to depression, reports a survey by Cigna Health Insurance company among 2,00,000 Americans. The disconnectedness among the highly connected youth population pointed that hour of social media usage instead of real-life connection augments feeling of loneliness and inadequacy. The edited and curated reels of people paint a comparison in the mind of the viewer's adding to the feeling of depression (Amatenstein, 2019).

Passal et al. (2016), had done a quantitative analysis on the available literature on social anxiety and internet use based on the time spent online, experience of comfort of being online, and problematic internet usage. It has confirmed a highly positive correlation. Age was the moderating variable for internet use and social anxiety relationship. Hawes et al. (2020) had made a path breaking observation in their study among 763 adolescent population. Preoccupation with body image related content in the social media, frequency of visits to social media sites and the amount of time spent in these sites had displayed more sensitivity towards their appearance. It was observed to have an association with anxiety and sensitivity towards rejection based on appearance, leading to depression.

O'Day and Heimberg (2021) had done a multiple data base research on the literature published up to 2020 in the realm of social media usage. The relationship between social anxiety and loneliness was confirmed to be associated with problematic social media usage. Loneliness was established to be a risk factor for online engagement. Feder et al. (2020) had reported frequent social media usage to increased symptoms of psychopathology.

The study conducted in Cape Town, among 250 youth subjects, has proved the association between use of social media, disturbances in sleep schedules and depression. A higher incidence of sleep disturbance has been reported among high frequency social media users with a positive effect on depression (De Doncker and McLean, 2022).

2.4. Linkages between Use of Social Platforms and Cognition

Though, link between use of social platforms and cognition has been widely researched (Sherman et al., 2016) its effects on cognition has not been documented much (Lara and Bokoch, 2021).

Scott (2011) had found out that there are two significant cognitive resources woven with our cognitive abilities which are "working memory" and "attention." The addictive tendency to check notifications, while at work or in school is increasingly disrupting the normal functioning and performance (Brown et al. 2019; Uncapher et al., 2016). The tendency to get carried away in each technological distraction takes a longer duration to come back to the work mode, more often leading to compromise on the work quality. Catastrophic consequences have been reported among motor vehicle drivers due to these frequent technological disturbances. Hence the connectedness between social media and social cognition is to be explored to enhance the quality of our daily living. Impact of the environment leading to frequency of use and distractibility also needs to be documented (Wiradhany and Nieuwenstein, 2017).

Lara and Bokoch (2021) were unable to emphasize the connect between usage of social platforms, functioning of memory, and ability to filter information. The effect on the mood of the individuals has been explored. But, the effect of social media on cognition is not yet established. Those engaged in media multitasking were observed to have a dip in the performance of working memory (Uncapher et al., 2016).

2.5. Social Media and Mental Health

Karim et al. (2020) had analyzed the literature on effect of social media usage on the mental health and has confirmed a general association between social media usage and mental health issues. In addition to creating a pressure to create an ideal image/life that others want to see, prolonged use of social media and the quest for social sanction leads to depression, anxiety, and stress.

Adasme and Cataldo (2022), had reported about the effect of techno stress created by a techno mediated study on performance among 189 students and 172 educators and had proven that young people are more affected negatively than their educators due to techno overload. Techno mediated study had become popularized since Covid-19 lockdown.

Based on the Crisis and Emergency Risk Communication Model and Health Belief Model, Zhong et al. (2021), had explored the interconnection between mental health and usage of social media platforms among inhabitants of Wuhan soon after the identification of Corona virus. A health behavior change was predicted related to secondary trauma and depression. Change was not detected between health-related behavior and mental health. The first outbreak of the virus was announced to the world through social media and it was helpful to people to gain informational, emotional, and peer support from all over the globe. However, the uncontrolled use of social media, had led to mental health issues. A break in using the content in social platforms has been observed to promote well-being during the pandemic, which is crucial in mitigating the harm inflicted by the pandemic on the mental health of the people.

Derks et al. (2021) in their quantitative diary study had examined the impact of smart phone use at work to employee well-being leading to end of the day exhaustion. Interruptive smart phone usage during work costs heavily at work especially when they try to integrate work and family to match the demands of work and home, simultaneously. Kels et al. (2020) had verified through a systematic review that the hours used in browsing social platforms and subsequent addiction is directly correlated with depression, anxiety, and distress. This finding is confirmed by Nereim et al. (2020) in an exploratory study on social media and emotional wellbeing of adolescents. The observation was that passive social media usage like reading blogs and posts are more strongly linked to depression compared to the active role of making the posts.

2.6. Need and Significance of the Study

The crucial effect of online platforms on social cognition, social connectedness, as well as mental wellbeing warrants the attention of academia, industry (future workforce) policy makers, and government as well as private organizations. The understanding that profuse social media usage can deteriorate mental health leading to anxiety, depression, and loneliness is a major eye opener. The undesirable digital cues are raised by sharing of pictures in social media platform is potentially hazardous. Behavior of people engrossed in problematic social media usage has become synonymous with that of addiction affected people. It has been observed that the addictive behavior originating from social media usage is synonymous with established features of drug addictions especially among the youth. A quantitative approach to understand the social media addiction behavior of youth is need of the hour and addressed in the present study.

3. METHODOLOGY

The current study had an empirical and descriptive research design, and a survey approach was used to collect primary responses from Tamil Nadu youth living in Greater Corporation of Chennai. To identify respondents for the study, the survey method and non-probability judgment sampling approach were used. To collect primary data from youth in the study area, a structured questionnaire with seven elements has been finalized. After removing extreme values and replies that were inappropriate for the study, a sample size of 520 was evaluated. Cronbach's alpha reliability co-efficient was used to evaluate the consistency and reliability of the structured questionnaire, and the value of 0.903 indicates that the scales are consistent and highly trustworthy.

4. RESULTS AND DISCUSSION

Data analysis was performed on primary data acquired from youth living in the study region using SPSS Version 27.0. To comprehend the survey's findings, statistical procedures such as simple percentage analysis, descriptive statistics, and simple linear regression analysis were used. The simple percentage analysis has been applied to understand the demographic profile of the participants in the primary survey. The results are presented in the Table 1.

Gender: The majority of the respondents are females (56.9%) followed by males (43.1%).

Educational Qualification: The sizeable portion of the respondents are post-graduates (39.2%) followed by graduates (32.3%), professionals (10.0%), school educated (8.5%), diploma holders (6.9%), and others (3.1%).

Table 1. Demographic background.

Demographic profile	Frequency	Percentage
Gender		
Male	224	43.1
Female	296	56.9
Educational qualification		
School level	44	8.5
Under graduate	168	32.3
Post graduate	204	39.2
Professional	52	10.0
Diploma	36	6.9
Others	16	3.1
Occupational status		
Self-employed	36	6.9
Professional	28	5.4
Salaried	216	41.5
Students	240	46.2
Monthly family income (in Rs.)		
<Rs. 15,000	144	27.7
Between Rs. 15,000 to Rs. 30,000	204	39.2
Between Rs. 30,000 to Rs. 45,000	88	16.9
Above Rs. 45,000	84	16.2
Nature of living		
Urban	376	72.3
Semi urban	144	27.7
Social media usage per day (in hours)		
Less than 3 hours	460	88.5
3–4 hours	36	6.9
Above 4 hours	24	4.6

Occupational Status: The maximum number of respondents is students (46.2%) followed by salaried employees (6.9%), self-employed (5.4%), and professionals (5.4%).

Monthly Family Income (Rs): Around 39.2% of the participants belong in the income group between Rs. 15,000 and Rs. 30,000 followed by a 27.7% of the subjects belonging to less than Rs. 15,000/-monthly income (27.7%), between Rs. 30,000 and Rs. 45,000 (16.9%), and above Rs. 45,000 (16.2%).

Nature of Living: The majority of the respondents are hailing from urban places (72.3%) followed by semi-urban places (27.7%).

Social Media Usage Per Day (In Hrs): The majority of the youth participated in the survey opined that they use less than 3 hours (88.5%) per day followed by between 3 and 4 hours (6.9%) and more than 4 hours per day (4.6%).

Table 2 results of the Kolmogorov–Smirnova and Shapiro–Wilk tests of normality demonstrate that the data in Table 2 has a normal distribution since the mean values are higher than the standard deviation and the standard deviation is less than one-third of the mean for factors such as Emotional Competence, Information and Education Seeking, Entertainment, Conscientiousness, and Cognitive Communication. Furthermore, "Cronbach's alpha reliability co-efficient were applied to measure the reliability and consistency of the data and the results show that the values for variables like Emotional Competence, Information and Education Seeking, Entertainment, Conscientiousness, Cognitive Communication, and Social Media Addiction are greater than 0.700 (≥0.700), demonstrating the consistency and reliability of the data collected from the respondents.

4.1. Multiple Regression Analysis on Significant Predictors of Social Media Use Among the Youth

Multiple regression analysis of Ordinary Least Square (OLS) model has been applied to explore the influence of Cognitive Communication, Information and Education Seeking, Entertainment, Emotional Competence, and Conscientiousness dimensions among the youth involved in the study. The results are presented in the Table 3.

Table 3 shows that influence on independent factors of Cognitive Communication, Information and Education Seeking, Entertainment, Emotional Competence, and Conscientiousness factors on social media addiction to use among youth. The result of the linear regression analysis reveals that OLS model has a goodness of fit for multiple regression analysis and the linear combination of Cognitive Communication Factor, Information and Education Seeking Factor, and Entertainment Factor on Social Media Addiction to Use Total Score {F-Value = 371.360; P-Value = < 0.001}. Cognitive Communication Factor, Information and Education Seeking Factor, and Entertainment Factor have significant and Positive Influence on Social Media Addiction to Use among youth participated in the study whereas, Emotional Competence Factor and Conscientiousness Factor do not have any significant influence on social media addiction to use. Therefore, higher is Cognitive Communication, Information and Education Seeking, and Entertainment perception among youth; higher is the social media addiction to use.

Table 2. Descriptive statistics of social media addiction to use and its predictors.

Variables	Descriptive statistics					Tests of normality		Reliability statistics		
	Mean (SD) Statistic	Std. error	Variance Statistic	Skewness Statistic (Std. error = 0.107)	Kurtosis Statistic (Std. error = 0.214)	Kolmogorov–Smirnov[a] Statistic	Shapiro–Wilk Statistic	Corrected item–total correlation	Squared multiple correlation	Cronbach's alpha
Emotional competence	2.762 (0.894)	0.039	0.799	−0.475	−0.810	0.136^{*}	0.946^{*}	0.665	0.520	0.806
Information and education seeking	3.315 (1.000)	0.044	1.001	−0.658	−0.412	0.150^{*}	0.934^{*}	0.815	0.700	0.716
Entertainment	3.410 (1.029)	0.045	1.060	−0.904	0.104	0.168^{*}	0.903^{*}	0.864	0.771	0.854
Conscientiousness	3.210 (1.037)	0.045	1.075	−0.559	−0.379	0.123^{*}	0.944^{*}	0.839	0.747	0.833
Cognitive communication	3.540 (1.033)	0.045	1.067	−1.094	0.499	0.192^{*}	0.877^{*}	0.840	0.757	0.875
Social media addiction	3.590 (1.099)	0.048	1.208	−1.052	0.334	0.213^{*}	0.867^{*}	0.785	0.686	0.901

[a]Lilliefors significance correction.

Table 3. Antecedents of social media addiction to use.

Model	Variables	Mean (SD)	Model summary						Coefficients		Collinearity statistics	
			R	R square	Adj. R square	Std. error of estimate	F-Value (Sig.)	Durbin–Watson	Standardized coefficients Beta	t-Value (Sig.)	Tolerance	VIF
Dependent variable			0.827	0.683	0.682	0.620	371.360 (0.000)*	1.816	***	***		
Social media addiction		3.590 (1.099)										
Independent variables												
3	(Constant)								***	2.478 (0.014)*	***	
	Cognitive communication	3.540 (1.033)							0.536	13.193 (0.000)*	0.371	2.693
	Information and education seeking	3.315 (1.000)							0.185	4.745 (0.000)*	0.405	2.471
	Entertainment	3.410 (1.029)							0.175	3.834 (0.008)*	0.293	3.415
	Emotional competence	2.762 (0.894)							−0.056	−1.583 (0.114)	0.488	2.048
	Conscientiousness	3.210 (1.037)							−0.060	−1.224 (0.221)	0.257	3.889

aDependent variable: social media addiction.

5. LIMITATIONS AND FUTURE RESEARCH

More and more qualitative research and vertical cohort studies are required to know the influence that social media has on the mental health as it is an emerging field (Karim et al., 2020). Not only the adverse effects but also the extent of damage among diverse population is yet to be determined. More research is required to identify social media usage on the dimensions of frequency and motivation to use (O'Day and Heimberg, 2021). Impact of the environment on the frequency of social media usage also needs to be explored. Effect of technologies to augment learning has to be explored as increased reliance on technology for education exists today. Covid-19 induced lockdown is the reason for prevalent social media, smart phone usage and technical aids for education. There is an urgent need to find out whether the online tools are indeed causing benefits compared to the destruction proven on depression, loneliness, detachment, and the impact on the social connectedness.

The present study is limited to Greater Chennai Corporation as the geographical area selected for the study. The perceptions are subject to change due to changes in the socio-economic changes in the perception of the respondents. The limitations associated with non-probability judgment sampling are also applicable for the present study.

Majority of the studies highlight the potential harm posed by excessive digital indulgence on the health of youth. There are some observations which point towards slight positive impact on depression and loneliness. Deeper understanding on the changes in the cognitive aspects and social media usage needs to be done. Further studies are required to establish a deeper understanding of the impact of the environment on this phenomenon. Social media use and its impact on mood have been thoroughly studied but social media and its effect on cognition needs more focus. The effect of social media on one's attention span or working memory is to be recorded as it can have an impact on one's memory interfering with his productivity.

6. FUTURE OF CYBER SPACE

The future of cyber space is bright but the impact it creates on every stakeholder is worth understanding. The lack of awareness on privacy and safety measures while handling social media content, make majority of the population to be victims of cybercrime and cyber bullying. The alarming rate of social media addiction among the youth has an impact on their behavior, relationships, productivity, and social life.

7. CONCLUSION

This research paper which provides a snap shot on the social media usage and youth, lays the foundation to understand future trends related to mental health issues and social media usage. It provides the peer review literature on social cognition, connectedness, and mental health among the youth due to internet usage. This gives insights to Social Workers, Psychologists, and Behavioral psychologists to handle today's youth. Adhering to empirical evidences on social media usage during most part of the productive life makes the youth vulnerable to relationship issues, depression and loneliness. It affects their overall productivity in life. Hence, it is imperative to educate the youth about safety and privacy settings while accessing social media. Regular capacity building workshops on safe internet practices are to be imparted in educational institutions. Laws related to the age and content while youngsters access social media platforms, need to be made stringent. Safe social media practice, awareness on cybercrime, ability to take charge of their life has to be imparted to the youth through their parents, teachers, and peers. We cannot deny the fact that social media has its own benefits. But the individual need to be trained to evaluate his priority, analyze the usage of time and take charge of his life. The youth should be sensitized to use the media wisely and widely to accentuate knowledge and professional gains rather than the entertainment purposes.

REFERENCES

Acquier, A., Daudigeos, T., & Pinkse, J. (2017). Promises and paradoxes of the sharing economy: An organizing framework. *Technological Forecasting and Social Change, 125*, 1–10.

Ahn, J. (2011). The effect of social network sites on adolescents' social and academic development: Current theories and controversies. *Journal of the American Society for Information Science and Technology, 62*(8), 1435–1445.

Allen, K. A., Ryan, T., Gray, D. L., McInerney, D. M., & Waters, L. (2014). Social media use and social connectedness in adolescents: The positives and the potential pitfalls. *The Educational and Developmental Psychologist, 31*(1), 18–31.

Amatenstein, S. (2019). Not so social media: How social media increases loneliness. *Psycom – Mental Health and Wellbeing.* Retrieved from: https://www.psycom.net/how-social-media-increases-loneliness.

Berryman, C., Ferguson, C. J., & Negy, C. (2018). Social media use and mental health among young adults. *Psychiatric Quarterly, 89*(2), 307–314.

Bohra, A. S. (2022). Patterns of internet usage among youths in India. *Social Media Matters.* Retrieved from: https://www.socialmediamatters.in/patterns-of-internet-usage-among-youths-in-india.

Bravo-Adasme, N., & Cataldo, A. (2022). Understanding techno-distress and its influence on educational communities: A two-wave study with multiple data samples. *Technology in Society, 70, 102045.*

Brown, S. G., Tenbrink, A. P., & LaMarre, G. (2019). Performance while distracted: The effect of cognitive styles and working memory. *Personality and Individual Differences, 138*, 380–384.

Chauhan, S., & Yachu, S. (2022). Mental health in India: Impact of social media on young Indians. *The Indian Express.* Retrieved from: https://indianexpress.com/article/

lifestyle/health/mental-health-in-india-impact-of-social-media-on-young-indians-facebook-instagram-youtube-twitter-7778499/.

De Doncker, K., & McLean, N. (2022). Social media, sleep difficulties and depressive symptoms: A case study of South African youth in Cape Town. *Technology in Society*, 70, 102038.

Derks, D., Bakker, A. B., & Gorgievski, M. (2021). Private smartphone use during worktime: A diary study on the unexplored costs of integrating the work and family domains. *Computers in Human Behavior*, 114, 106530.

Feder, K. A., Riehm, K. E., & Mojtabai, R. (2020). Is there an association between social media use and mental health? The timing of confounding measurement matters – Reply. *JAMA Psychiatry*, 77(4), 438–438.

Hawes, T., Zimmer-Gembeck, M. J., & Campbell, S. M. (2020). Unique associations of social media use and online appearance preoccupation with depression, anxiety, and appearance rejection sensitivity. *Body Image*, 33, 66–76.

Jolly, E., Tamir, D. I., Burum, B., & Mitchell, J. P. (2019). Wanting without enjoying: The social value of sharing experiences. *PLoS One*, 14(4), e0215318.

Karim, F., Oyewande, A. A., Abdalla, L. F., Ehsanullah, R. C., & Khan, S. (2020). Social media use and its connection to mental health: A systematic review. Cureus, 12(6), e8627. https://doi.org/10.7759/cureus.8627.

Krogh, S. C. (2022). 'You can't do anything right': How adolescents experience and navigate the achievement imperative on social media. *Young*, 31(1), 5–21. https://doi.org/10.1177/11033088221111224.

Lara, R. S., & Bokoch, R. (2021). Cognitive functioning and social media: Has technology changed us? *Acta Psychologica*, 221, 103429.

Marty-Dugas, J., Ralph, B. C., Oakman, J. M., & Smilek, D. (2018). The relation between smartphone use and everyday inattention. *Psychology of Consciousness: Theory, Research, and Practice*, 5(1), 46.

Nereim, C. D., Bickham, D. S., & Rich, M. O. (2020). Social media and adolescent mental health: Who you are and what you do matter. *Journal of Adolescent Health*, 66(2), S118–S119.

O'Day, E. B., & Heimberg, R. G. (2021). Social media use, social anxiety, and loneliness: A systematic review. *Computers in Human Behavior Reports*, 3, 100070.

O'Reilly, M., Dogra, N., Whiteman, N., Hughes, J., Eruyar, S., & Reilly, P. (2018). Is social media bad for mental health and wellbeing? Exploring the perspectives of adolescents. *Clinical Child Psychology and Psychiatry*, 23(4), 601–613.

Ophir, E., Nass, C., & Wagner, A. D. (2009). Cognitive control in media multitaskers. *Proceedings of the National Academy of Sciences*, 106(37), 15583–15587.

Penn State. (2019, November 25). Feeling loved in everyday life linked with improved well-being. *ScienceDaily*. Retrieved November 13, 2022 from: https://www.sciencedaily.com/releases/2019/11/191125121005.htm.

Prizant-Passal, S., Shechner, T., & Aderka, I. M. (2016). Social anxiety and internet use – A meta-analysis: What do we know? *What are we missing? Computers in Human Behavior*, 62, 221–229.

Qualman, E. (2012). *Socialnomics: How social media transforms the way we live and do business*. John Wiley & Sons.

Rasmussen, E. E., Punyanunt-Carter, N., LaFreniere, J. R., Norman, M. S., & Kimball, T. G. (2020). The serially mediated relationship between emerging adults' social media use and mental well-being. *Computers in Human Behavior*, 102, 206–213.

Rosen, L. D., Carrier, L. M., & Cheever, N. A. (2013). Facebook and texting made me do it: Media-induced task-switching while studying. *Computers in Human Behavior*, 29(3), 948–958.

Scott, J. G. (2011). Attention/concentration: The distractible patient. *In The little black book of neuropsychology* (pp. 149–158). Boston, MA: Springer.

Sherman, L. E., Payton, A. A., Hernandez, L. M., Greenfield, P. M., & Dapretto, M. (2016). The power of the like in adolescence: Effects of peer influence on neural and behavioral responses to social media. *Psychological Science*, 27(7), 1027–1035.

Uncapher, M. R., K Thieu, M., & Wagner, A. D. (2016). Media multitasking and memory: Differences in working memory and long-term memory. *Psychonomic Bulletin & Review*, 23(2), 483–490.

Valenzuela, S., Halpern, D., Katz, J. E., & Miranda, J. P. (2019). The paradox of participation versus misinformation: Social media, political engagement, and the spread of misinformation. *Digital Journalism*, 7(6), 802–823.

Wiradhany, W., & Nieuwenstein, M. R. (2017). Cognitive control in media multitaskers: Two replication studies and a meta-analysis. *Attention, Perception, & Psychophysics*, 79(8), 2620–2641.

Zaghouani, W. (2018). A large-scale social media corpus for the detection of youth depression (project note). *Procedia Computer Science*, 142, 347–351.

Zhong, B., Huang, Y., & Liu, Q. (2021). Mental health toll from the coronavirus: Social media usage reveals Wuhan residents' depression and secondary trauma in the COVID-19 outbreak. *Computers in Human Behavior*, 114, 106524.

21. Impact of Augmented Reality in Marketing on Retail Industry in the Context of India

Devapujala Kodhanda Revan[1], Nirlipta Rath[1], Suraj Karar[1], and
Subhadarshini Khatua[2]
[1]Woxsen University
[2]Assistant Professor, Woxsen University

ABSTRACT:

BACKGROUND—Augmented reality (AR) has been a part of daily living thanks to the evolution of Industry 4.0 (4th industrial revolution). It has great potential to redefine the way people shop and consume products and services. Irrespective of significant growth in the retail industry, there is less research on the "impact of this technology on consumers in the retail industry".

OBJECTIVE—Considering the above arguments, this study is aimed to understand the impact of the use of AR apps on marketing and sales of retail industry and to suggest how retail businesses can improve customer interaction with the use of AR.

METHODOLOGY—In order to fulfill the above objectives, this study is "based on primary data collected from an online survey" with a self-structured questionnaire. Total 119 responses were collected through a random sampling method. The responses were analyzed using IBM SPSS v22 software.

RESULTS—This study has key theoretical and practical implications for retail managers. This study has elaborated the concept of AR in the retail industry which accounts for perception of customers in the use of AR. The findings would help retail managers to leverage AR apps as part of their marketing strategy to make long-term relationships with customers.

KEYWORDS: AR, augmented reality, marketing strategy, retail industry, Industry 4.0, AR apps

Chapter 21 DOI- 10.4324/9781003397175-21

1. INTRODUCTION

AR or "augmented reality" is one of the fastest growing technologies widely adopted in the retail sector (Lavoye *et al.*, 2021; Kumar, 2022). It refers to a medium to make virtual content life-like in the field of vision of a user, ranging from highly functional (assisted reality) purposes to highly realistic (mixed reality) experiences where virtual elements cannot be differentiated from real elements (Rauschnabel *et al.*, 2022). Since 1960 when Ivan Sutherland announced a prototype of the world's first head-mounted display, AR has come a long way in the past decade (Rauschnabel, 2018; Caboni and Hagberg, 2019).

A lot of tech giants like Microsoft, Google, Amazon, Apple, etc. are using AR apps to engage customers (Kumar and Srivastava, 2022). In retail markets, the AR industry has been proposed to reach over $7.9 billion across the world by the end of 2023 (Markets and Markets, 2019). According to Shopify, AR products had 94% more conversion as compared to products without AR features (Kumar *et al.*, 2023). There has been a rise in studies on applications of AR technologies in business activities over the past decade (Yim *et al.*, 2017; Wedel *et al.*, 2020). There are also some studies which cover various aspects of AR marketing like consumer response and media features, consumer experience, consumer acceptance, and impact of AR on consumer behavior (Javornik, 2016, Huang *et al.*, 2019, McLean and Wilson, 2019, Kumar and Srivastava, 2022; Kumar, 2022).

1.1. Background

The never-ending tussle between makers and retailers for attracting consumers has created a phenomenon called "retail power pendulum" because each party wants control over the effort to increase profits and reduce costs. In this day and age, information, digital technologies, and processing have redefined the retail sector and AR is no exception AR makes the use of smartphones, headsets, or tablets to put virtual representation of products on the real world (Jayanti and Singh, 2010). This virtual overlay provides partial (like Google Glass), minimal (i.e., using smartphones), and total (i.e., by using Oculus Rift) experience to the user in AR. This way, shoppers can augment the real world with virtual information and make almost perfect decisions (Heller *et al.*, 2016).

Mobile devices and advancement in technology has blurred the line between digital and real world over the past decade. Hence, the traditional way to do business has been shifted to e-commerce, mobile commerce and online commerce. This development has been made with the significant growth of 4G and 5G technologies. Applications can easily augment any environment with digital data in real time. A lot of researchers have discussed the technicalities of AR and its effects on customers along with usability of the system (Nee *et al.*, 2012; Azuma *et al.*, 2001; Carmigniani *et al.*, 2011).

Till date, there has been relatively less research conducted on the potential benefits of AR for retail businesses and customers (Van Esch *et al.*, 2016; Scholz and Duffy, 2018). In marketing, AR can improve customer loyalty and customer satisfaction, positive word of mouth, and repeat purchases, while helping makers and retailers to enhance the profitability. Bulearca and Tamarjan (2010) have recently observed that AR devices have been helpful to consumers, especially by aligning their buying behavior with shopping needs. Consumers have a lot of details before and during the process. Their uptake seems to be slow when using AR in retail settings. Hence, more studies are needed to know the impact of AR on marketing and sales in the retail industry to improve consumer relationships (Chylinski *et al.* 2014).

2. LITERATURE REVIEWS

The existing body of literature discusses various attributes of AR like augmentation, interactivity, novelty, physical control, vividness, etc. (Lavoye *et al.*, 2021; Kumar, 2021). Augmentation and interactivity are two major characteristics of AR (Javornik, 2016; Yim *et al.*, 2017; Park and Yoo, 2020). There are two perceptions to study interactivity—(1) customer perception and (2) technical attributes. The technical attribute means capability of a technological system to enable people to interact and be engaged more smoothly with content (Yim *et al.*, 2017). There are certain aspects to be used for explaining interactivity like mapping, range, and speed (Steuer *et al.*, 1995).

From the point of view of customer perception, experience with the media is subjective. The technical aspect is widely used for explaining interactivity in the context of AR. It was observed a significant positive impact of interactivity on utilitarian and hedonic values. As discussed in research, the impact of interactivity on utilitarian and hedonic motivation in the context of using AR Qin *et al.*, (2021). It is explored the importance of AR on customer engagement and observed utilitarian and hedonic values of interactivity on the users, which ultimately caused brand inspiration and engagement (Nikhashemi *et al.*, (2021); Kumar and Srivastava, 2022).

Augmentation refers to the ability to overlap the real world with a virtual backdrop (Javornik, 2016). Augmentation is basically categorized into three types—product augmentation, environment augmentation, and body augmentation. AR influences affective response and hedonic value of customers (Watson *et al.*, 2018). In this research article assumed "perceived augmentation" as a silent attribute of media of AR and observed that there is a significant influence of "perceived augmentation" on the flow of experience like enjoyment and fun Javornik (2016). In addition, various studies have confirmed that "perceived augmentation" has a huge impact on utilitarian and hedonic values as a media attribute (Javornik, 2016; Hilken *et al.*, 2017; Kumar and Srivastava, 2022).

According to the "theory of planned behavior", the intentions can predict the true behavior of consumers very well (Azjen, 1980). In addition, values are important factors of preference behavior and consumer choice (Gutman, 1982). AR has been well regarded to affect consumer behavior significantly (Wedel *et al.*, 2020; Wang *et al.*, 2021; Hilken *et al.*, 2017; Whang *et al.*, 2021; Moriuchi *et al.*, 2021). The outcomes of AR are of three types—"behavioral intentions", "attitudinal outcomes" and "decision making assistance". Since "attitudinal outcomes and decision-making assistance" have ultimately caused "behavioral outcomes", the effect of AR attributes like augmentation and interactivity would be made on behavioral intent with utilitarian and hedonic values (Kumar, 2022).

AR is truly an immersive technology to create highly fun and engaging experiences, causing hedonic value (Javornik, 2016; Rauschnabel, 2021). The AR has created the hedonic value to affect behavioral intentions. Cultural influence also plays a vital role among customers (Hofstede *et al.*, 2010)." Culture influences beliefs, values, and users' perception significantly (Eisend, 2019). According to the "means-end chain theory", values are the key drivers of actions and behaviors of the user and culture affects behavioral intentions (Gutman, 1982).

2.1. Research Gap

The existing literature provides various attributes of AR like augmentation, interactivity, novelty, control, and environments to the customers. However, there is a lack of studies discussing the "impact of AR" on marketing in the retail sector, especially in the context of India. Hence, this study fills this knowledge gap and opens further research paths for researchers and marketers to understand the benefits of AR for their businesses to attract more customers.

2.2. Research Questions

R1: How AR can impact marketing and sales of a retail business?

R2: How can retail organizations improve customer interaction with the use of AR?

2.3. Research Objective

Considering the above arguments, this study is aimed to understand the impact of the use of AR apps on marketing and sales of retail industry and to suggest how retail businesses can improve customer interaction with the use of AR.

2.3.1. Hypotheses

H1: There is a significant impact of AR on marketing and sales of retail business.
H2: Customer interaction is significantly improved with the use of AR in retail organization.

2.3.2. Research Methodology

In order to fulfil the above objectives, this study is "based on primary data collected from 119 participants" through random sampling. Self-structured questionnaire was prepared for this study and shared through Google Forms. Data were gathered and analyzed using IBM SPSS v22 software. "One sample *t*-test was conducted" to understand the impact of AR on the retail industry. Participants of the study are regular customers of various businesses who are using AR apps of those businesses. Those participants were selected to understand their perception towards AR apps and interaction.

3. DATA ANALYSIS

3.1. Demographics

In this study, 33 (28%) participants are 40 to 50 years old, 46 (39%) participants are 31 to 40 years old, 22 (19%) participants are 26 to 30 years old, and 18 (15%) participants are 18 to 25 years old (Table 1).

Table 1. Age group.

		Frequency	Percent	Valid percent	Cumulative percent
Valid	18 to 25 years	18	15.1	15.1	15.1
	26 to 30 years	22	18.5	18.5	33.6
	31 to 40 years	46	38.7	38.7	72.3
	40 to 50 years	33	27.7	27.7	100.0
	Total	119	100.0	100.0	

Out of 119 participants, 53 (45%) participants have done post-graduation, 43 (36%) participants are graduates, and 23 (19%) participants have done high schooling (Table 2).

Table 2. Academic qualifications.

		Frequency	Percent	Valid percent	Cumulative percent
Valid	graduate	43	36.1	36.1	36.1
	High school	23	19.3	19.3	55.5
	Post-graduate	53	44.5	44.5	100.0
	Total	119	100.0	100.0	

In this study, 78 (66%) participants are private employees, 18 (15%) participants are students, 14 (12%) participants are self-employed, and 9 (8%) participants are entrepreneurs (Table 3).

Table 3. Occupation.

		Frequency	Percent	Valid percent	Cumulative percent
Valid	entrepreneur	9	7.6	7.6	7.6
	Private Employee	78	65.5	65.5	73.1
	Self-employed	14	11.8	11.8	84.9
	Student	18	15.1	15.1	100.0
	Total	119	100.0	100.0	

In this study, 56 (47%) participants have been using AR apps for their regular shopping, while 21 (18%) participants have recently started using AR apps. In addition, 33 (28%) participants have been using AR apps for 3 years and 8 (7%) participants have been using such apps for 2 years (Table 4).

Table 4. Usage of AR apps.

How long have you been using the AR app of your regular business?		Frequency	Percent	Valid percent	Cumulative percent
Valid	1 year	56	47.1	47.1	47.1
	2 years	8	6.7	6.7	53.8
	3 years	33	27.7	27.7	81.5
	More than 5 years	1	.8	.8	82.4
	Recently started using	21	17.6	17.6	100.0
	Total	119	100.0	100.0	

3.2. Customer Interaction with AR apps

There are 92% participants in this study, who use AR apps every time they shop for any product (Table 5).

Table 5. I use AR every time when I shop for any product.

		Frequency	Percent	Valid percent	Cumulative percent
Valid	Neutral	10	8.4	8.4	8.4
	Agree	45	37.8	37.8	46.2
	Strongly Agree	64	53.8	53.8	100.0
	Total	119	100.0	100.0	

In this study, 52% participants agree that the AR app does not always meet their expectations, while 35% participants disagree (Table 6).

Table 6. The AR app does not always meet my expectations.

		Frequency	Percent	Valid percent	Cumulative percent
Valid	Strongly Disagree	18	15.1	15.1	15.1
	Disagree	24	20.2	20.2	35.3
	Neutral	15	12.6	12.6	47.9
	Agree	37	31.1	31.1	79.0
	Strongly Agree	25	21.0	21.0	100.0
	Total	119	100.0	100.0	

There are 60% participants who get positive experience overall when making a buying decision using AR apps, while 40% participants do not think so (Table 7).

Table 7. Overall, I get positive experience when making buying decision using AR.

		Frequency	Percent	Valid percent	Cumulative percent
Valid	Disagree	16	13.4	13.4	13.4
	Neutral	32	26.9	26.9	40.3
	Agree	32	26.9	26.9	67.2
	Strongly Agree	39	32.8	32.8	100.0
	Total	119	100.0	100.0	

There are 76% participants who find using AR apps fun and engaging in this study (Table 8).

Table 8. Using AR app is fun and engaging.

		Frequency	Percent	Valid percent	Cumulative percent
Valid	Neutral	29	24.4	24.4	24.4
	Agree	48	40.3	40.3	64.7
	Strongly Agree	42	35.3	35.3	100.0
	Total	119	100.0	100.0	

In this study, all the participants feel like an object is already present in their room when they use AR apps (Table 9).

Table 9. I feel like object is already here in my room when I use this app.

		Frequency	Percent	Valid percent	Cumulative percent
Valid	Agree	44	37.0	37.0	37.0
	Strongly Agree	75	63.0	63.0	100.0
	Total	119	100.0	100.0	

3.3. Impact of AR on retail marketing

According to 77% participants, AR apps improve sales for retail businesses (Table 10).

Table 10. AR app enhances sales for retail businesses.

		Frequency	Percent	Valid percent	Cumulative percent
Valid	Neutral	27	22.7	22.7	22.7
	Agree	60	50.4	50.4	73.1
	Strongly Agree	32	26.9	26.9	100.0
	Total	119	100.0	100.0	

Table 11. You prefer retail business with AR app than those without AR apps.

		Frequency	Percent	Valid percent	Cumulative percent
Valid	Strongly Disagree	30	25.2	25.2	25.2
	Disagree	22	18.5	18.5	43.7
	Agree	40	33.6	33.6	77.3
	Strongly agree	27	22.7	22.7	100.0
	Total	119	100.0	100.0	

In this study, there are 63% participants who make better decisions with experiential shopping experience using AR apps, while 22% participants do not think so (Table 12).

Table 12. Experiential shopping helps make better decisions.

		Frequency	Percent	Valid percent	Cumulative percent
Valid	Strongly Disagree	15	12.6	12.6	12.6
	Disagree	11	9.2	9.2	21.8
	Neutral	18	15.1	15.1	37.0
	Agree	45	37.8	37.8	74.8
	Strongly agree	30	25.2	25.2	100.0
	Total	119	100.0	100.0	

There are 92% participants in this study, who always find the exact match of the product they are looking for while using AR apps (Table 13).

Table 13. I always find the exact match of the product while using AR.

		Frequency	Percent	Valid percent	Cumulative percent
Valid	Neutral	9	7.6	7.6	7.6
	Agree	62	52.1	52.1	59.7
	Strongly Agree	48	40.3	40.3	100.0
	Total	119	100.0	100.0	

According to 87% participants, AR has a great potential to boost the retail industry in terms of marketing and sales (Table 14).

Table 14. AR has a great potential in retail industry.

		Frequency	Percent	Valid percent	Cumulative percent
Valid	Neutral	16	13.4	13.4	13.4
	Agree	50	42.0	42.0	55.5
	Strongly Agree	53	44.5	44.5	100.0
	Total	119	100.0	100.0	

According to 85% participants, AR convinces customers more when shopping a product or looking for a product than anything else (Table 15).

Table 15. AR convinces customers more when looking for a product.

		Frequency	Percent	Valid percent	Cumulative percent
Valid	Neutral	18	15.1	15.1	15.1
	Agree	72	60.5	60.5	75.6
	Strongly Agree	29	24.4	24.4	100.0
	Total	119	100.0	100.0	

According to 58% participants, AR helps improve customer satisfaction, while 22% participants do not think so (Table 16).

Table 16. AR helps in increasing the customer satisfaction.

		Frequency	Percent	Valid percent	Cumulative percent
Valid	Strongly Disagree	15	12.6	12.6	12.6
	Disagree	11	9.2	9.2	21.8
	Neutral	24	20.2	20.2	42.0
	Agree	31	26.1	26.1	68.1
	Strongly agree	38	31.9	31.9	100.0
	Total	119	100.0	100.0	

H1: There is a significant impact of AR on the marketing and sales of retail business

In order to solve the above hypothesis, one-sample t-test is performed on 6 variables of AR on retail marketing (Table 17). With a confidence interval of 95%, the value of Sig. (2-tailed) was 0.000 for all variables. Hence, H1 is approved, i.e., there is a significant impact of AR on the marketing and sales of retail business.

Table 17. One-sample test on retail marketing.

	Test value = 0				95% Confidence interval of the difference	
	t	df	Sig. (2-tailed)	Mean difference	Lower	Upper
AR app enhances sales for retail businesses	62.469	118	.000	4.042	3.91	4.17
You prefer retail business with AR app than those without AR apps	21.622	118	.000	3.101	2.82	3.38

Experiential shopping helps make better decisions	29.527	118	.000	3.538	3.30	3.78
I always find the exact match of the product while using AR	77.121	118	.000	4.328	4.22	4.44
AR has a great potential in retail industry	67.370	118	.000	4.311	4.18	4.44
AR convinces customers more when looking for a product	71.515	118	.000	4.092	3.98	4.21
AR helps in increasing customer satisfaction	28.568	118	.000	3.555	3.31	3.80

H2: Customer interaction is significantly improved with the use of AR in a retail organization

There are five variables of customer interaction in this study. This hypothesis is also tested with one-sample t-test on customer interaction (Table 18). At 95% confidence interval, the value of Sig. (2-tailed) is 0.000 for all the variables ($p < 0.05$). Hence, H2 is approved, i.e., customer interaction is significantly improved with the use of AR in a retail organization.

Table 18. One-sample test on customer interaction.

	Test value = 0					
					95% Confidence interval of the difference	
	t	df	Sig. (2-tailed)	Mean difference	Lower	Upper
I use AR every time when I shop for any product	75.017	118	.000	4.454	4.34	4.57
The AR app does not always meet my expectations	25.387	118	.000	3.227	2.98	3.48
Overall, I get positive experience when making buying decision using AR	39.427	118	.000	3.790	3.60	3.98

	Test value = 0					
					95% Confidence interval of the difference	
	t	*df*	Sig. (2-tailed)	Mean difference	Lower	Upper
Using AR app is fun and engaging	58.376	118	.000	4.109	3.97	4.25
I feel like object is already here in my room when I use this app	104.192	118	.000	4.630	4.54	4.72

4. RESULTS

The earlier section of data analysis yielded various important and interesting insights. It is observed that the majority of participants found AR apps fun and interesting and they are more willing to shop using AR apps. Similarly, most of the participants also believe that AR apps enhance sales and marketing for retail businesses. Hence, it is the right time for retail businesses to upgrade their operations with such latest technologies. The use of AR also significantly improves customer interaction as it has a lot of features. Most customers also feel that objects in the real world that they want to purchase.

AR provides indirect or direct insight of real surroundings of the customers, which are improved or altered by digital data like sound, video, GPS data or graphics. Customers can use life-like spaces using AR technology and even their faces and bodies to visualize objects like furniture, clothing, etc. Hence, AR technology combines digital and real worlds. Several organizations have started using AR technology in their apps to enhance customer experience, especially in the retail industry. For example, Timberland, an outdoor footwear, clothing, and accessories marketplace brought virtual fitting rooms powered by "Microsoft Kinect". This technology tracks facial features, gait, and other physical qualities to project trousers, shoes, jackets, and sweaters on the customers, so that they can check out various outfits virtually. Their new look can be sent to their inbox and also added to the photo gallery on the fan page of Timberland. Customers can share their look on social media with an email link.

5. CONCLUSION

Even though the recent boom in AR may not be as transformational as the introduction of e-commerce, retail businesses cannot afford to miss this opportunity and let go of this potential technological shift and shoppers' need. AR is a very promising concept in the retail industry and a lot of big organizations are already leading with this technology. Considering various benefits of AR, online businesses should understand its benefits and plan for investment. Considering the response of participants in this study, they can rest assured to get positive results in the long term by attracting more customers.

It is a great time for online and offline retailers in India to boost their interaction with customers using AR. Lenskart, an Indian online prescription glasses multinational chain, is already using AR to enable customers to test various lenses on their faces in real time using 3D Try-on feature. In addition, an online fashion giant, Myntra, is also working on their AR feature for customers to use their device camera to make buying decisions. Ecommerce market is already witnessing a significant boom in India. Adopting AR can really be a game-changer for retailers in terms of marketing.

REFERENCES

Azuma, R., Baillot, Y., Behringer, R., Feiner, S., Julier, S., and MacIntyre, B. (2001). Recent advances in augmented reality. *IEEE computer graphics and applications*, 21(6), 34-47.

Bulearca, M., and Tamarjan, D. (2010). Augmented reality: A sustainable marketing tool. *Global business and management research: An international journal*, 2(2), 237-252.

Caboni, F., and Hagberg, J. (2019). Augmented reality in retailing: a review of features, applications and value. *International Journal of Retail & Distribution Management*.

Carmigniani, J., Furht, B., Anisetti, M., Ceravolo, P., Damiani, E., and Ivkovic, M. (2011). Augmented reality technologies, systems and applications. *Multimedia tools and applications*, 51, 341-377.

Chyinski, M., De Ruyter, K., Sinha, A., and Northey, G. (2014). Augmented retail reality: situated cognition for healthy food choices. In *ANZMAC 2014 Conference*. ANZMAC (Australian and New Zealand Marketing Academy).

Eisend, M. (2019). Explaining digital piracy: A meta-analysis. *Information Systems Research*, 30(2), 636-664.

Gutman, J. (1982). A means-end chain model based on consumer categorization processes. *Journal of marketing*, 46(2), 60-72.

Heller, J., Chylinski, M., Northey, G., de Ruyter, K., Sinha, A., van Esch, P., ... and Mahr, D. (2016). Topology of augmented commerce: The state and directions for future research. In *Proceedings: Australian and New Zealand Marketing Academy Conference (ANZMAC) 2016: Marketing in a Post-Disciplinary Era* (pp. 88-95).

Hilken, T., de Ruyter, K., Chylinski, M., Mahr, D., and Keeling, D. I. (2017). Augmenting the eye of the beholder: exploring the strategic potential of augmented reality to enhance online service experiences. *Journal of the Academy of Marketing Science*, 45, 884-905.

Hofstede, G., Hofstede, G. J., and Minkov, M. (2005). *Cultures and organizations: Software of the mind* (Vol. 2). New York: McGraw-Hill.

Huang, T. L., and Liao, S. (2015). A model of acceptance of augmented-reality interactive technology: the moderating role of cognitive innovativeness. *Electronic Commerce Research*, 15, 269-295.

Huang, T. L., Mathews, S., and Chou, C. Y. (2019). Enhancing online rapport experience via augmented reality. *Journal of Services Marketing*.

Javornik, A. (2016). Augmented reality: Research agenda for studying the impact of its media characteristics on consumer behaviour. *Journal of Retailing and Consumer Services, 30*, 252-261.

Jayanti, R. K., and Singh, J. (2010). Pragmatic learning theory: an inquiry-action framework for distributed consumer learning in online communities. *Journal of Consumer Research, 36*(6), 1058-1081.

Kumar, H. (2022). Augmented reality in online retailing: a systematic review and research agenda. *International Journal of Retail & Distribution Management, 50*(4), 537-559.

Kumar, H., and Srivastava, R. (2022). Exploring the role of augmented reality in online impulse behaviour. *International Journal of Retail & Distribution Management*, (ahead-of-print).

Kumar, H., and Srivastava, R. (2022). Exploring the role of augmented reality in online impulse behaviour. *International Journal of Retail & Distribution Management*, (ahead-of-print).

Kumar, H., Gupta, P., and Chauhan, S. (2023). Meta-analysis of augmented reality marketing. *Marketing Intelligence & Planning, 41*(1), 110-123.

Lavoye, V., Mero, J., and Tarkiainen, A. (2021). Consumer behavior with augmented reality in retail: a review and research agenda. *The International Review of Retail, Distribution and Consumer Research, 31*(3), 299-329.

Markets and Markets (2019). Press Release. Available at: https://www.marketsandmarkets.com/PressReleases/augmented-reality-retail.asp.

McLean, G., and Wilson, A. (2019). Shopping in the digital world: Examining customer engagement through augmented reality mobile applications. *Computers in Human Behavior, 101*, 210-224.

Moriuchi, E., Landers, V. M., Colton, D., and Hair, N. (2021). Engagement with chatbots versus augmented reality interactive technology in e-commerce. *Journal of Strategic Marketing, 29*(5), 375-389.

Nee, A. Y., Ong, S. K., Chryssolouris, G., and Mourtzis, D. (2012). Augmented reality applications in design and manufacturing. *CIRP annals, 61*(2), 657-679.

Nikhashemi, S. R., Knight, H. H., Nusair, K., and Liat, C. B. (2021). Augmented reality in smart retailing: A (n)(A) Symmetric Approach to continuous intention to use retail brands' mobile AR apps. *Journal of Retailing and Consumer Services, 60*, 102464.

Park, M., and Yoo, J. (2020). Effects of perceived interactivity of augmented reality on consumer responses: A mental imagery perspective. *Journal of Retailing and Consumer Services, 52*, 101912.

Qin, H., Peak, D. A., and Prybutok, V. (2021). A virtual market in your pocket: How does mobile augmented reality (MAR) influence consumer decision making?. *Journal of Retailing and Consumer Services, 58*, 102337.

Rauschnabel, P. A. (2018). Virtually enhancing the real world with holograms: An exploration of expected gratifications of using augmented reality smart glasses. *Psychology & Marketing, 35*(8), 557-572.

Rauschnabel, P. A., Felix, R., Hinsch, C., Shahab, H., and Alt, F. (2022). What is XR? Towards a framework for augmented and virtual reality. *Computers in Human Behavior, 133*, 107289.

Scholz, J., and Duffy, K. (2018). We ARe at home: How augmented reality reshapes mobile marketing and consumer-brand relationships. *Journal of Retailing and Consumer Services*, *44*, 11-23.

Steuer, J., Biocca, F., and Levy, M. R. (1995). Defining virtual reality: Dimensions determining telepresence. *Communication in the age of virtual reality*, *33*, 37-39.

Tan, Y. C., Chandukala, S. R., and Reddy, S. K. (2022). Augmented reality in retail and its impact on sales. *Journal of Marketing*, *86*(1), 48-66.

van Esch, P., Northey, G., Chylinski, M., Heller, J., De Ruyter, K., Sinha, A., and Hilken, T. (2016). Augmented reality: Consumer saviour or disruptive agent in the retail power pendulum?. In *ANZMAC 2016 Conference*. ANZMAC.

Watson, A., Alexander, B., and Salavati, L. (2018). The impact of experiential augmented reality applications on fashion purchase intention. *International Journal of Retail & Distribution Management*, *48*(5), 433-451.

Wedel, M., Bigné, E., and Zhang, J. (2020). Virtual and augmented reality: Advancing research in consumer marketing. *International Journal of Research in Marketing*, *37*(3), 443-465.

Whang, J. B., Song, J. H., Choi, B., and Lee, J. H. (2021). The effect of Augmented Reality on purchase intention of beauty products: The roles of consumers' control. *Journal of Business Research*, *133*, 275-284.

Yim, M. Y. C., Chu, S. C., and Sauer, P. L. (2017). Is augmented reality technology an effective tool for e-commerce? An interactivity and vividness perspective. *Journal of interactive marketing*, *39*(1), 89-103.

22. A Gender-Based Comparative Study on Selfitis, Narcissism and Attachment Patterns among Young Adults Cognitive Effects of Social Media Usage on Youth

Ms. Aishee Ghosh

Masters Student Department: Psychology Institutional, Amity University, Kolkata

Correspondence details: Phone number: 9836262109, Email Address: aisheeghosh0101@gmail.com

ABSTRACT

BACKGROUND: The study aims to understand the impact of gender on the concepts of selfitis, narcissism and attachment patterns and also the relation between these concepts. Selfitis can be defined as a cacoethes for clicking photos of oneself and uploading on social media sites, which gives a boost to self-esteem and improves the sense of intimacy. Narcissism refers to a paramount degree of self-involvement by having an inordinate interest over one's own needs and physical appearance that often, makes the individual disregard the needs of those around them. The distinct way in which an individual connects and associates with other people is known as their attachment style. The four types of attachment styles of adults are Secure, Fearful, Preoccupied and Dismissive.

METHOD: The study was conducted through a Google form where all the questions were plotted along with specific instructions for the convenience of the subjects. Selfitis Behaviour Scale, Narcissistic Personality Inventory-40 and Attachment Styles Questionnaire were used to collect data for the respective dimensions. The data were collected from 300 participants, i.e., 150 males and 150 females, aged between 18 and 27 years. Snowball sampling method was used. The data were analysed with the help of descriptive statistics, MANOVA, ANOVA and correlation.

RESULTS: The results indicated that gender affects narcissism and the preoccupied and fearful attachment style. Correlation has been established between selfitis and narcissism, between selfitis and fearful style of attachment and preoccupied style of attachment and between narcissism and preoccupied style of attachment.

CONCLUSION: The study findings are helpful in understanding the reason behind the social media addiction, prevalent among the youth.

KEYWORDS: Selfitis, narcissism, attachment styles, gender, social media addiction

1. INTRODUCTION

The present study empirically explores the concepts of selfitis, narcissism and attachment patterns. The study assumes that there is an impact of gender on these concepts. These are the domains in which studies are still very few and the results are therefore tentative.

1.1. Selfitis

The concept of selfitis came up in 2014 when the media started focusing on the obsession with taking selfies.

According to Sorokowski *et al.* (2015), selfie involves taking a picture of oneself (with or without other people) with a camera placed at one arm distance or through a mirror's reflection. These pictures are often shared on social media sites leading to an obsession of clicking selfies, now termed as 'selfitis'.

The urge of constantly taking and posting selfies can be in order to gain validation by impressing others, which can be represented by self-presentation theory (Ma *et al.*, 2017). The trend of taking selfies has become more popular as while posting selfies via social media domains, one often goes viral (Frosh, 2015; Hess, 2015; Moon *et al.*, 2016; Rettberg, 2014; Roberts and Koliska, 2017).

Balakrishnan and Griffiths formulated the Selfitis Behaviour Scale in 2017. In lieu of that, they had identified environmental enhancement, social competition, attention seeking, mood modification, self-confidence, and subjective conformity to be influencing the initiation of selfitis.

Selfie can be categorised as a disorder as it often creates a privacy risk, can cause an addiction, can hamper interpersonal relationships and puts an emphasis on physical appearance. Selfie addiction paves the way for several other associated problems like low self-confidence, skin damage and plastic surgery to change facial features. It can even be causes for suicide and accidents (Safna, 2017). Dokur *et al.* (2018) suggested that the willingness to make the selfies more likeable, youngsters are engaging in risk taking behaviour, which often leads to fatal accidents.

The extreme use of technology, thus, often leads to digital diseases (Polat, 2017), like Internet addiction, social media addiction, online gaming addiction, cyberchondria, nomophobia and technoference (Balakrishnan and Griffiths,

2018). Selfitis can escalate to the form of such disorders if proper awareness in not present in the society.

1.2. Narcissism

The term 'narcissism' was proposed by Havelock Ellis in 1898. Narcissism refers to a paramount degree of self-involvement that makes a person disregard the needs of those around them.

Narcissistic traits at an extreme level would lead to narcissistic personality disorder, which is maladaptive. Narcissism can be initiated by antagonistic or hostile interpersonal styles, social isolation, or termination of social relationships (Semenyna, 2018).

According to Back *et al.* (2013), individuals having narcissistic traits choose certain expedients to gain social validation. Using assertive social styles to gain admiration is strongly associated with grandiose narcissism. Vulnerable narcissists employ rivalry by derogating and devaluing others and are often part of confrontational social practices (Back *et al.*, 2013; Foster *et al.*, 2016; Grijalva *et al.*, 2015).

Self-sufficiency, superiority, vanity, authority, entitlement, exhibitionism, and exploitativeness are the seven components included in Narcissistic Personality Inventory-40 (Raskin and Terry, 1988).

1.3. Attachment Patterns

John Bowlby, a British Psychoanalyst (1907–1990), postulated the Attachment Theory. Attachment styles refer to the distinct way in which an individual connects and associates with other people.

According to Bowlby, individual differences occur due to differential internal working models of relation. The attachment styles become resistant to changes with time. These internal models develop in childhood and become a guide for future choices in relationships and behaviour towards others. The individual differences in attachment patterns have been elaborately studied by Ainsworth *et al.* (1978). According to Ainsworth (1979), there are three styles of attachment, i.e., secure, insecure avoidant and insecure ambivalent/resistant. The early interactions with the mother forms a basis for developing attachment styles as stated by McLeod (2018).

Bartholomew and Horowitz (1991) developed a new model stating four types of attachment styles of adults, i.e., secure style of attachment, fearful style of attachment, preoccupied style of attachment and dismissive style of attachment. This model originated from Bowlby's (1973) two -dimensional model (positive vs. negative) of self and others.

Secure style: individuals having secure style of attachment have a positive self-concept. They usually are not apprehensive about others and their intentions. They are able to be in intimate, secure and loving relationships without any trials and tribulations.

Fearful style: people with this style have a tendency to doubt themselves as well as others. They are afraid of personal relationships due to the fear of getting hurt or deceived. They have trepidation about abandonment.

Preoccupied style: preoccupied individuals solicit acceptance and validation from others, to feel satisfied with themselves as hold a negative perception of themselves. Such people are willing to maintain close relationships with others.

Dismissive style: individuals who have a dismissing style of attachment focus more on themselves and are often not intrigued by the idea of making personal bonds with others.

Anxiety over desertion and avoidance of connection are the two aspects of attachment insecurity that has been highlighted by current attachment theories. It can be stated that individuals having caregivers who were attentive and responsive to their needs, tend to have low anxiety and avoidance issues, thus developing a balanced approach to support seeking and emotional regulation and form the secure style of attachment. On the other hand, individuals having caregivers who were inconsistently available and occasionally responsive, tend to develop attachment anxiety.

Individuals often put greater efforts to seek protection and intimacy, are hypersensitive to rejection, have obsessive thoughts regarding one's own flaws and trying to figure imminent relationship dangers which highlights the hyper-activation of the attachment system. Non- responsive caregivers give rise to avoidant attachment styles. These people tend to have attachment deactivation, which is characterised by avoiding intimate relationships, denial of requirement of attachment in life and suppressing their vulnerability from others (Mikulincer and Shaver, 2003).

1.4. Narcissism and Gender

Previous literature suggests that there is the existence of gender differences in Narcissism. According to Gutmann (1965), males are more individualistic and females are more interpersonal in nature. Being less individualistic and more interpersonal suggests less narcissism. Males focus more on self-assertion and self-expansion, whereas women focus on the sense of being with other individuals (Bakan, 1966).

Carlson did a study in 1970 that stated that males and females have differences in experience and representation of self.

Narcissism positively is related to self-absorption and self-esteem (Raskin, 1980). Females succumb to social pressure and hence, have considerably lower self-esteem when compared with the males. This could be a reason for females being more interpersonal and less narcissistic.

Schwalbe and Staples carried out a study in 1991 on self-esteem, self-absorption and social comparison. Men and women have different patterns of self-esteem as they experience different parameters for self-evaluation and self-enhancement.

Women tend to suffer from depression more than their male counterparts. Sex differences have been found to be linked with feelings of shame and guilt related to specific situations and self- image. In context of narcissism, men appear to be more prone to it than women (Wright *et al.*, 1989).

The existing literature on narcissism focuses more on men and due to this overrepresentation, theoretical understanding of gender differences has not yet been explored properly.

1.5. Narcissism and Attachment Patterns

It can be said that narcissism is developed in males due to certain forms of family structure. (Philipson, 1985). According to Reis *et al.* (2021), individuals having pathological narcissistic vulnerability, have higher emotional reactivity, face insecurity in attachment and are extremely sensitive to rejection.

1.6. Attachment Patterns and Gender

Previous literature has suggested that there is a significant impact of gender on attachment styles. According to Gutmann (1965), women are more interpersonal in relation and less individualistic. It can be stated that females experience themselves to be more intrinsically related to others when compared to males (Carlson, 1970; Gutmann, 1965). Men tend to show self-preoccupation as they have a preference for gaining mastery, indifferent aloofness and dismissing others. Females are unconditional in attachment and have more concern towards social affiliation making them more vulnerable in relationships, which makes them open to interpersonal relationships. Females are often seen to be more fearful and less secure, preoccupied and dismissive in their attachment styles (Bakan, 1966; Carlson, 1970). Females are at times found to be less dismissive in their attachment patterns when compared to their male counterparts (Downing, 2008). According to Schwalbe and Stapples (1991), men are more concerned regarding power and performance whereas women are more concerned regarding social and relational appraisals. Bilsker and Marcia (1991) suggested that women are higher in subjectivism.

1.7. Selfitis and Narcissism

In order to establish self-importance (Murray, 2015) and individual identity, people engage in a self-oriented action of selfie taking (Ehlin, 2014).

Buffardi and Campbell (2008) pointed out that the behaviour of taking selfies can be associated with narcissism. As, selfie taking and narcissism are often seen to be complementing each other (Halpern *et al.*, 2016), but people using a selfie-stick to capture selfies are often perceived to be moderately inconsiderate and narcissistic as well as comparatively less socially attractive (Bevan, 2017).

Selfie-taking behaviour is been affected by loneliness, need for attention and egocentrism (Charoensukmongkol, 2016). McCain *et al.* (2016) stated that social attractiveness boosts the behaviour of selfie taking, which further caters to narcissism.

1.8. Selfitis and Gender

A significant amount of studies have not been done to analyse and state whether there exists any effect of gender on selfitis. According to Fox and Rooney (2015), a relation could not be established between the selfie posting behaviour of women with narcissism, but the male population tend to display an opposite finding.

India has the highest number of Facebook users as it is well known social network for posting photos (Rao, 2020). According to Chauhan and Yachu (2022), the youth of the country displays ever-increasing dependency on social media with 97.2 million of them using Facebook and 69 million of them spending huge amount of time on Instagram. Hence, this age group was chosen for this study, as they are most prone to social media usage.

The **OBJECTIVES** of the present research study could be summarised as:

- To analyse the impact of gender on the concepts of selfitis, narcissism and attachment patterns.
- To understand the relation between selfitis, narcissism and attachment patterns of the males and females under study.

2. HYPOTHESES

- There is a significant impact of gender on the concepts of selfitis, narcissism and attachment patterns.
- There is a significant correlation between selfitis, narcissism and attachment patterns of the males and females under study.

3. METHODOLOGY

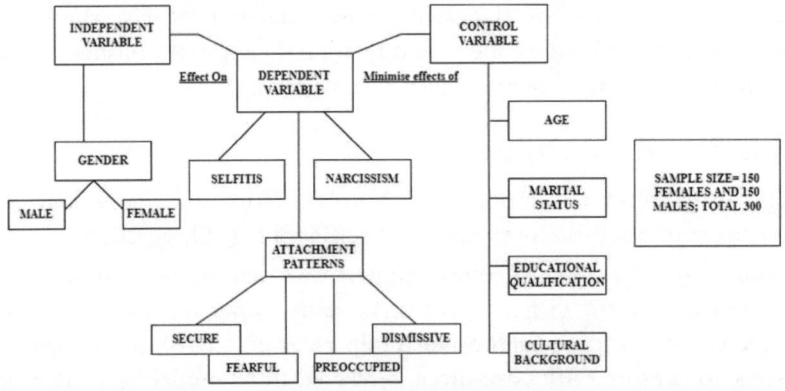

Schematic Representation of Study Design

The study was conducted through a Google form where all the questions were plotted along with specific instructions for the convenience of the subjects. Selfitis Behaviour Scale, Narcissistic Personality Inventory-40, and Attachment Styles Questionnaire were used for the respective dimensions. A total of 300 participants, i.e., 150 males, and 150 females, aged between 18 and 27 years provided the necessary information. Snowball sampling method was used. The collected data were all treated quantitatively. The data were analysed with the help of descriptive statistics, MANOVA, ANOVA and correlation tests.

4. RESULTS

Table 1. Between-subjects factors.

		N
GENDERM1F2	1.00	150
	2.00	150

Table 2. Descriptive statistics.

	GenderM1F2	Mean	Std. Deviation	N
Selfitis total	1.00	49.4000	13.77285	150
	2.00	45.9933	13.94596	150
	Total	47.6967	13.94128	300
NPI total	1.00	14.8200	6.46734	150
	2.00	17.5067	6.23343	150
	Total	16.1633	6.48203	300
Attachment secure	1.00	24.2533	3.64017	150
	2.00	23.1467	4.41685	150
	Total	23.7000	4.07825	300
Attachment fearful	1.00	17.3867	3.95238	150
	2.00	16.2000	3.66188	150
	Total	16.7933	3.84968	300
Attachment preoccupied	1.00	24.1267	4.50249	150
	2.00	22.5000	4.28318	150
	Total	23.3133	4.46186	300
Attachment dismissive	1.00	18.7533	2.92854	150
	2.00	18.2667	2.86302	150
	Total	18.5100	2.90137	300

Table 3. Multivariate tests.

GenderM1F2	Effect	Value	F	Hypothesis df	Error df	Sig.	Partial eta squared
	Wilks' Lambda	.887	6.221	6.000	293.000	.000	.113

Table 4. Tests of between-subjects effects.

Source	Dependent Variable	Type III Sum of Squares	df	Mean Square	F	Sig.	Partial Eta Squared
Gender M1F2	Selfitis total	870.403	1	870.403	4.531	.034	.015
	NPI total	541.363	1	541.363	13.420	.000	.043
	Attachment secure	91.853	1	91.853	5.608	.019	.018
	Attachment fearful	105.613	1	105.613	7.276	.007	.024
	Attachment preoccupied	198.453	1	198.453	10.278	.001	.033
	Attachment dismissive	17.763	1	17.763	2.118	.147	.007

Table 5. Descriptive statistics: selfitis total & narcissism total.

	Mean	Std. Deviation	N
Selfitis total	47.6967	13.94128	300
NPI total	16.1633	6.48203	300

Table 6. Correlations: selfitis total & narcissism total.

		Selfitis Total	NPI Total
Selfitis total	Pearson correlation	1	.217**
	Sig. (2-tailed)		.000
	N	300	300
NPI total	Pearson correlation	.217**	1
	Sig. (2-tailed)	.000	
	N	300	300

**Correlation is significant at the 0.01 level (2-tailed).

Table 7. Descriptive statistics: selfitis total & attachment styles.

	Mean	Std. Deviation	N
Selfitis total	47.6967	13.94128	300
Attachment secure	23.7000	4.07825	300
Attachment fearful	16.7933	3.84968	300
Attachment preoccupied	23.3133	4.46186	300
Attachment dismissive	18.5100	2.90137	300

Table 8. Correlations: selfitis total and attachment styles.

Selfitis total		Attachment Secure	Attachment Fearful	Attachment Preoccupied	Attachment Dismissive
	Pearson correlation	.036	.243**	.255**	−.090
	Sig. (2-tailed)	.535	.000	.000	.119
	N	300	300	300	300

**Correlation is significant at the 0.01 level (2-tailed).

Table 9. Descriptive statistics: narcissism total & attachment styles.

	Mean	Std. Deviation	N
Attachment secure	23.7000	4.07825	300
Attachment fearful	16.7933	3.84968	300
Attachment preoccupied	23.3133	4.46186	300
Attachment dismissive	18.5100	2.90137	300
NPI total	16.1633	6.48203	300

Table 10. Correlations: narcissism total & attachment styles.

NPI Total		Attachment Secure	Attachment Fearful	Attachment Preoccupied	Attachment Dismissive
	Pearson correlation	.011	−.066	−.169**	−.104
	Sig. (2-tailed)	.845	.258	.003	.071
	N	300	300	300	300

**Correlation is significant at the 0.01 level (2-tailed).

5. DISCUSSION

Selfitis can be defined as a cacoethes for clicking photos of oneself and uploading on social media sites, which gives a boost to self-esteem and improves the sense of intimacy. It is an activity that often helps individuals embrace their individuality and improves self-confidence, particularly for those who lack social support (Katz and Crocker, 2015)

Narcissism refers to a paramount degree of self-involvement by having an inordinate interest over one's own needs and physical appearance that often, makes the individual disregard the needs of those around them.

The distinct way in which an individual connects and associates with other people is known as their attachment style.

Table 1 among the result tables highlights the gender division and the age group of the participants involved in this study. The age group of 18–27 years was chosen to understand the effect that social media usage has on the cognition of today's youth. A total of 300 individuals, among whom, 150 were males and 150 were females, gave their responses for this study.

Table 2 demonstrates the mean and standard deviation of all the variables. From the table, it can be seen that males obtained a mean of 49.4 and females obtained a mean of 45.99 in selfitis. It can be concluded that males are more prone to selfitis when compared to their female counterparts. The higher mean among the males for selfitis can be attributed to certain factors like fear of missing out, seeking attention, getting social validation, social attraction, loneliness and narcissistic personality (Manago et al., 2008). The absence of recreational facilities, financial inadequacy and perceived lack of social support plays an important role in borderline selfitis (Albury, 2015). However, according to previous literature, females score higher on selfitis (Lal and Singh, 2022). Selfitis can be scored as a composite score as well as according to its six factors. For this study, the composite score was taken into account. This can be a possible explanation for the deviating result.

Regarding NPI, females scored a higher mean of 17.51, whereas males scored a mean of 14.82. For the males and females under this study, females have shown higher narcissistic traits when compared to male population. This is contradictory to the existing literature which displayed that women are more invested in interpersonal relationships than men (Gutmann, 1965). They often succumb to societal pressures and behave accordingly (Raskin, 1980), which results in higher levels of depression (Wright et al., 1989). Their idea of self-expression is also different than males (Carlson, 1970). Even the idea, that the outlook of women is changing with time and they are expressing their desires and needs more in the current timeframe as well the societal expectations are undergoing changes, can also be a cause of the deviating result. The contradictory results can be attributed to the fact the narcissism can be scored as a composite score as well as according to its seven components and for this study, the composite score was taken into account.

The males scored a mean of 24.25, whereas the female scored a mean of 23.15 in the secure style of attachment style. In the fearful style of attachment, the males scored a mean of 17.39, whereas the females scored a mean of 16.2. In the case of the preoccupied style of attachment, males obtained a higher score of 24.12, whereas females secured a mean of 22.5. In the dismissive style of attachment, the scores are almost similar with males and females obtaining 18.75 and 18.27, respectively. Regarding the attachment styles, both men and women have shown all the four types of attachment styles and the mean scores are indicative of the fact that gender is not the point of difference in this domain, except in the preoccupied and fearful style of attachment (this is also demonstrated from the findings of Table 4). According to Bakan (1966) and Carlson (1970), men tend to show self-preoccupation as they have a preference for gaining mastery, indifferent aloofness and dismissing others. Females are unconditional in attachment and have more concern towards social affiliation making them more vulnerable in relationships, which makes them open to interpersonal relationships. Females are often seen to be more fearful and less secure, preoccupied and dismissive in their attachment styles. However, the mean scores of fearful style of attachment are contradictory to this finding as the females have scored lower than the males.

From Table 3, it can be concluded that the independent variable (gender – male and female) has an effect on the dependent variables – selfitis, narcissism, secure style of attachment, fearful style of attachment, preoccupied style of attachment and dismissive style of attachment. MANOVA shows a significant difference in gender on the dependent variables. Since Wilks' Lambda is significant at 0.01 level, univariate ANOVAs were done to determine the gender differences in the case of each of the dependent variables separately.

From previous literature, we have seen that gender does have an effect on narcissism (Bakan, 1966; Carlson, 1970; Gutmann, 1965; Wright et al., 1989) and attachment patterns (Bakan, 1966; Bilsker and Marcia, 1991; Carlson, 1970; Downing, 2008; Gutmann, 1965; Schwalbe and Stapples, 1991). A significant amount of studies have not been done to analyse and state whether there exists any effect of gender on selfitis. This study proves that there exists no relationship between gender and selfitis, but the effect of gender on narcissism and preoccupied and fearful style of attachment has been proven as the results are significant at 0.01 level (Table 4). So hypothesis 1 can be partially accepted. This finding can be supported by the mean scores from Table 2. For the domain of narcissism, females scored a higher mean of 17.51, whereas males scored a mean of 14.82. For the domain of preoccupied style of attachment males obtained a higher score of 24.12, whereas females secured a mean of 22.5. For the domain of fearful style of attachment, males obtained a higher score of 17.39, whereas females secured a mean of 16.20.

Table 5 states the mean value of the total of selfitis which equals 47.70, whereas the mean value of narcissism equals 16.16.

From Table 6, it can be inferred that Narcissism and Selfitis are significantly and positively correlated with each other. As suggested by earlier works of literature, narcissism has an influence on selfitis (Buffardi and Campbell, 2008; Charoensukmongkol, 2016; Ehlin, 2014; Halpern *et al.*, 2016; McCain *et al.* 2016; Murray, 2015) and the results are in accordance with the existing literature. To highlight the individual characteristics and to gain importance of oneself from others, individuals feel motivated to take selfies. Attention-seeking, loneliness and self-centred behaviour significantly affect selfie-taking behaviour. Posting on social media leads to interaction with the external world and gives individuals a sense of social validation, which in turn may boost their confidence and self-esteem. This obsession with oneself caters to the concept of narcissism as well.

From Table 7, it can be seen that the mean selfitis total is 47.70. The mean for the secure style of attachment is 23.7, the mean for the fearful style of attachment is 16.80, the mean for the preoccupied style of attachment is 23.31, and the mean for the dismissive style of attachment is 18.51.

From Table 8, it can be seen that the secure style of attachment, the fearful style of attachment, and the preoccupied style of attachment have a positive linear correlation with the selfitis total, whereas the dismissive style of attachment has a negative linear correlation with the selfitis total.

However, only the correlation coefficient between the fearful style of attachment and the selfitis total as well as between the preoccupied style of attachment and selfitis total is significant at 0.01 level (2-tailed). It can be said that people who have a preoccupied or secure or fearful style of attachment will have higher chances of developing selfitis, whereas people who have a dismissive style of attachment will have lower chances of developing selfitis as it is negatively correlated. It can be said that people who have a fearful style of attachment suffer from self-doubt and fear of abandonment (Bartholomew and Horowitz, 1991), hence they seek for external validation. In order to complete that need, individuals post selfies on social media sites to get likes and comments from others. Similarly, people having a preoccupied attachment style tend to suffer from negative self-image, which again leads them to seek for acceptance and validation from others. These people believe that in order to stay relevant they should be constantly updating nuances of their lives on social media. This can also lead to unsolicited jealousy and give rise to other issues such as body image issues, anxiety, depression, etc. From Table 2, it can be seen that the average score of males are higher in all the three domains of preoccupied style, fearful style as well as in selfitis. This proves that in spite of having a lower mean score in narcissism, they seek external validation due to faulty attachment patterns.

The results have stated that secure style of attachment and selfitis do not have a strong correlation, which can be attributed to the fact that people having secure style of attachment believe in themselves as they have a positive self-concept and those who have a dismissive attachment style fail to make bonds with others, in spite of having a positive self-concept; this can also be the reason for an insignificant result between the two variables.

However, since no significant work has been done prior to this to formulate a relationship between attachment styles and selfitis, it can be said that this study is one of the pioneer studies in that aspect.

From Table 9, it can be seen that the mean of the secure style of attachment is 23.7, the mean for the fearful style of attachment is 16.80, the mean for the preoccupied style of attachment is 23.31 and the mean of the dismissive style of attachment is 18.51. The mean of Narcissism is 16.16.

From Table 10, it can be seen that the secure style of attachment has a positive linear correlation with the narcissism total, whereas the fearful style of attachment, the preoccupied style of attachment and the dismissive style of attachment have a negative linear correlation with the narcissism total. People with secure style attachment may display more narcissistic traits but since the correlation is not significant, it can be based on individual differences.

However, only the correlation coefficient between the preoccupied style of attachment and narcissism total is significant at 0.01 level (2-tailed). It can be said that people who have a preoccupied style of attachment would have an effect on narcissism. Previous literature has certain evidence regarding the influence on how individuals take on attachment patterns based on the narcissistic traits present in themselves as well as in their respective partners (Philipson, 1985; Reis *et al.*, 2021). Males have scored higher in preoccupied style, but scored lower in narcissism (Table 2). This is in accordance with the negative correlation between the two variables.

According to Albury (2015), it can be said that although the concept of selfitis started as a fad, it could emerge in a more dangerous format among the youth. Narcissism and attachment patterns do have an influence on selfitis. Gender does play a role in developing the narcissism and attachment styles of individuals. Experiences often shape the way individuals perceive the world and therefore have an influence on both narcissism and attachment patterns. Selfitis is a relatively new concept and hence further research is required for its proper analysis.

6. CONCLUSION

From the above empirical research, it can be deduced that the gender does not have any effect on selfitis but influences the concept of narcissism as well as attachment styles. It can be inferred that selfitis, narcissism and certain attachment patterns are significantly correlated in relation to the male and female population under study.

On a concluding note, the study provided a discreet knowledge about the concept of selfitis, narcissism and attachment patterns among the mass population of the young adults. It gave us an opportunity to broaden our horizon in the field and undertaking the survey was also a chance of applying our education in applicable grounds.

This study was an effort to understand these concepts in an effective manner. The study findings are helpful in understanding the reason behind the social media addiction, prevalent among the youth. The research facilitates further research related to the management of addiction in terms of social media usage and paves the way which facilitates discussion on several other issues related to selfitis in context with attachment patterns and narcissism.

7. LIMITATIONS OF THE STUDY

- Participants selected were young (age range fixed from 18 to 27 years). Including younger and older subjects would have added important information regarding the concepts.
- All the measures used were self-report inventories. Participants often have a tendency to fake their responses due to pre-conceived biases, like social desirability and social validation. This could not be eliminated by using these measures alone. Due to a constraint of time, other qualitative methods could not be implemented.
- As culture has an influence in the initiation of selfitis, narcissism, attachment patterns as well as on gender, this study needed to be conducted in a cross-cultural setting, but this research could not incorporate that aspect, owing to the lack of participants of other cultures, as only snowball method of sampling was used.
- This research has included the total score for narcissism and selfitis. Including the components of this scale and analysing the data, may have given a wide scope of discussion and have promoted the importance of the components in the development of narcissism and selfitis in individuals.
- The concept of selfitis is an emerging domain of study. Therefore, this study is in its initial stage and further research is required to examine other parts of this concept.

8. APPLICABILITY

- The variables of selfitis, narcissism and attachment patterns have not been studied widely in previous works of literature, hence this study adds to the repertoire of scientific research.
- The significant implication of this study lies in understanding the impact of gender on selfitis, narcissism, and attachment patterns of the youth.

- It also explores the relation between selfitis, narcissism and attachment patterns.
- The study findings are useful in understanding the reason behind the social media addiction, prevalent among the youth.
- It paves a way for further research to be conducted on the measures for handling social media addiction.

9. GUIDELINES FOR FUTURE RESEARCH

- Expanding the sample to various cultures and ages would give a more detailed understanding of the concepts.
- Increasing the sample size would enhance the scope of generalised applicability.
- As this research includes self-report measures for collecting data, there can be a chance that participants have manipulated the data to gain social validation. Therefore, future research should include qualitative forms of data collection, like interviews, to give a more holistic view of the concept and eliminate such manipulative techniques.
- It has been noted that the youth suffers from several other issues related to selfitis, such as body image issues, self-esteem issues, drug addiction, isolation, depression, anxiety, cyber-bullying, etc. This research facilitates the discussion of all these issues in context with attachment patterns and narcissism.

REFERENCES

Ainsworth, M.D.S., M.C. Blehar, E. Waters, and S. Wall (1978). *Patterns of Attachment: A Psychological Study of the Strange Situation*. Hillsdale, NJ: Erlbaum.

Ainsworth, M.S. (1979). "Infant-mother attachment," *American Psychologist* 34(10): 932–937.

Albury, K. (2015). "Selfies, selfies, sexts and sneaky hats: young people's understandings of gendered practices of self representation," *Int. J. Communication* 9:1734–1745.

Back, M.D., A.C.P. Küfner, M. Dufner, T.M. Gerlach, and J.F. Rauthmann (2013). "Narcissistic admiration and rivalry: disentangling the bright and dark sides of narcissism," *Journal of Personality and Social Psychology* 105: 1013–1037.

Bakan, D. (1966). *The Duality of Human Existence*. Chicago: Rand McNally.

Balakrishnan, J., and M.D. Griffiths (2017). "An exploratory study of 'selfitis' and the development of the selfitis behavior scale," *International Journal of Mental Health Addiction* 36(1): 3–6. doi: 10.1007/s11469-017-9844-x.

Balakrishnan, J., and M.D. Griffiths (2018). "An exploratory study of 'selfitis' and the development of the selfitis behaviour scale," *Int. J. Ment. Health Addict* 16(3): 722–736. doi: 10.1007/s11469-017-9844-x.

Bartholomew, K., and L.M. Horowitz (1991). "Attachment styles among young adults: a test of a four-category model," *Journal of Personality and Social Psychology* 61(2): 226–244.

Bevan, J.L. (2016). "Perceptions of selfie takers versus selfie stick users: exploring personality and social attraction differences," *Computers in Human Behavior* 75: 494–500.

Bilsker, D., and J.E. Marcia (1991). "Adaptive regression and ego identity," *Journal of Adolescence* 14: 75–84.

Bowlby, J. (1973). *Attachment and Loss: Separation, Anxiety and Anger*, Vol. 2. New York: Basic Books.

Buffardi, L. E., and W.K. Campbell (2008). "Narcissism and social networking web sites," *Personality and Social Psychology Bulletin* 34: 1303–1314.

Carlson, R. (1970). "Sex difference in ego functioning: exploratory studies of agency and communion," *Educational Testing Service*.

Charoensukmongkol, P. (2016). "Exploring personal characteristics associated with selfie- liking," *Cyberpsychology: Journal of Psychosocial Research and Cyberspace* 10. doi: 10.5817/CP2016-2-7.

Chauhan, S., and S. Yachu (2022). "Mental health in India: impact of social media on young Indians," *The Indian Express*.

Dokur, M., E. Petekkaya, and M. Karadag (2018). "Media-based clinical research on selfie-related injuries and deaths," *Turkish Journal of Trauma and Emergency Surgery* 24(2): 129–135.

Downing, V.L. (2008). "Attachment style, relationship satisfaction, intimacy, loneliness, gender role belief and the expression of authentic self in romantic relationship," *Dissertation Abstracts International* 69(9): 30–58.

Ehlin, L. (2014). "The subversive selfie: redefining the mediated subject," *Clothing Cultures* 2(1): 73–89.

Foster, J.D., L.K. Shiverdecker, and I.N. Turner (2016). "What does the narcissistic personality inventorymeasure across the total score continuum?" *Current Psychology*. https://doi.org/10.1007/s12144-016-9407-5. Advanced online publication.

Fox, J. and M.C. Rooney (2015). "The Dark Triad and trait self objectification as predictors of men's use and self-presentation behaviors on social networking sites," *Personality and Individual Differences* 76: 161–165.

Frosh, P. (2015). "The gestural image: the selfie, photography theory, and kinesthetic sociability," *International Journal of Communication* 9: 1607–1628.

Grijalva, E., D.A. Newman, L. Tay, M.B. Donnellan, P.D. Harms, R.W. Robins, and T. Yan (2015). "Gender differences in narcissism: a meta-analytic review," *Psychological Bulletin* 141: 261–310.

Gutmann, D. (1965). "Women and the conception of ego strength," *Merrill Palmer Quarterly* 11: 229–240.

Halpern, D., S. Valenzuela, and J.E. Katz (2016). "'Selfie-ists' or 'Narciselfiers'? A cross-lagged panel analysis of selfie taking and narcissism," *Personality and Individual Differences* 97: 98–101.

Hess, A. (2015). "Selfies‌ the selfie assemblage," *International Journal of Communication* 9: 1629–1646.

Katz, J.E, and E.T. Crocker (2015)." Selfies: selfies and photo messaging as visual conversation: Reports from the United States, United Kingdom and China," *International Journal of Communication* 9: 186172. Available at: https://ijoc.org/index.php/ijoc/article/view/3180/1405.

Ma, J.W., Y. Yang, and J.A. Wilson (2017). "A window to the ideal self: a study of UK Twitter and Chinese Sina Weibo selfie-takers and the implications for marketers," *Journal of Business Research* 74: 139–142.

Manago, A.M., M.B. Graham, P.M. Greenfield, and G. Salimkhan (2008). "Self-presentation and gender on MySpace," *J. Appl. Dev. Psychol.* 29: 446–458.

Mikulincer, M., and P.R. Shaver (2003). "The attachment behavioral system in adulthood: activation, psychodynamics, and interpersonal processes," *Advances in Experimental Social Psychology* 35: 53–152. doi: 10.1016/S0065-2601(03)01002-5.

McCain, J.L., Z.G. Borg, A.H. Rothenberg, K.M. Churillo, P. Weiler, and W. K. Campbell (2016). "Personality and selfies: narcissism and the Dark Triad," *Computers in Human Behavior* 64: 126–133.

McLeod, S.A. (2018). "Mary ainsworth."

Moon, J.H., E. Lee, J.A. Lee, T.R. Choi, and Y. Sung (2016). "The role of narcissism in self- promotion on Instagram," *Personality and Individual Differences* 101: 22–25.

Murray, D.C. (2015). "Notes to self: the visual culture of selfies in the age of social media," *Consumption Markets & Culture* 18(6): 490–516.

Philipson, I. (1985). "Gender and Narcissism," *Sage Journals*.

Polat, R. (2017). "Dijital hastalık olarak nomofobi [Nomophobıa as digital disease]," *e-JNM* 1(2): 164–172.

Rao, P.A. (2020). "Exploring correlates of selfitis and narcissism among Indian adolescent". *Clin Psychiatry* 6(1): 67.

Raskin, R., and H. Terry (1988). "A principal component analysis of the narcissistic personality inventory and further evidence of its construct validity," *Journal of Personality and Social Psychology* 54(5): 890–902.

Reis, S., E. Huxley, B. Eng Yong Feng, and B.F. Grenyer (2021). "Pathological narcissism and emotional responses to rejection: the impact of adult attachment," *Frontiers in Psychology* 12: 679168.

Rettberg, J.W. (2014). *Seeing Ourselves Through Technology: How we Use Selfies, Blogs and Wearable Devices to See and Shape Ourselves*. New York: Palgrave Macmillan.

Roberts, J. and M. Koliska (2017). "Comparing the use of space in selfies on Chinese Weibo and Twitter," *Global Media and China* 2(2): 1–16.

Safna, H.M.F. (2017). "Negative impact of selfies on youth," *International Journal of Computer Science and Information Technology Research* 5(3): 68–73.

Schwalbe, M.L., and C.L. Staples (1991). "Gender differences in source of self esteem," *Social, Psychology Quarterly* 54(2): 158–168.

Semenyna, S. (2018). "Narcissism," *Research Gate*.

Sorokowski, P., A. Sorokowska, A. Oleszkiewicz, T. Frackowiak, A. Huk, and K. Pisanski (2015). "Selfie posting behaviors are associated with narcissism among men," *Personality and Individual Differences* 85: 123–127.

Wright, F., J. O'Leary, and J. Balkin (1989). "Shame, guilt, narcissism, and depression: Correlates and sex differences," *Psychoanalytic Psychology* 6(2): 217–230. https://doi.org/10.1037/0736-9735.6.2.217

23. Role of Digital Marketing Strategies Supporting B2B Firms

Amarnath Jaithiwad

MBA Student, Woxsen University, Hyderabad

Krishna Saraf

MBA Student, Woxsen University, Hyderabad

Sandhya T

MBA Student, Woxsen University, Hyderabad

Dr. Subhadarshini Khatua

Assistant Professor, Woxsen University Hyderabad

ABSTRACT: The study aims to find out the available literature on the use of digital marketing in a business-to-business (B2B) context. It identifies gaps in the current research knowledge and proposes a research agenda for scholars and practitioners. In today's world, every company may pursue marketing objectives online. Reaching the target audience may depend on selecting the best Internet marketing tools or combining a few. Therefore, it is difficult to select the most acceptable instruments for competing with other firms without clear criteria for evaluating the influence of Internet marketing tools on company enterprises. This looks article into how Internet marketing tools affect commercial businesses. The duties of analysing the idea of Internet marketing and disclosing the standards for impact evaluation have been established to attain this goal. In marketing, brands have received a lot of attention and have grown to be integral to contemporary life. They are all around us and have influenced many facets of our lives, including the social, cultural, religious, athletic and economic ones.

KEYWORDS: Digital marketing, B2B marketing, Martech, cognitive computing, business development, online marketing tools

1. INTRODUCTION

Online, Internet and web marketing are other terms for digital marketing. It is where we promote the brands that connect with potential customers using different forms of digital media. The process of promoting a business's goods or services online using apps, websites, social media and search engines is known as digital marketing. The term digital marketing has grown in popularity

Chapter 23 DOI- 10.4324/9781003397175-23

throughout time, particularly in a number of countries. Any form of marketing that involves electronic devices is considered digital marketing. The promotion of products or services via the use of digital technology most notably the Internet, but also mobile devices, display ads and other digital media is referred to as digital marketing.

In 2015, many of my clients spent hundreds of millions of dollars on billboard advertising. Today, most companies have moved to online marketing. That is because Google and Facebook generate more revenue than any traditional media company. After all, they control more eyeballs. That is why digital marketing matters in this era, and it is the future. It is affordable and useful also. Digital marketing also provides access to more data and analytics about your customers. You need to understand and know how to leverage digital marketing for any business to succeed.

B2B is called transactions between two businesses are referred to as 'business-to-business' transactions rather than transactions between a corporation and a single client. While most retail transactions are business-to-consumer (B2C), wholesale sales are typically business-to-business (B2B). It is a sort of electronic commerce in which products, services or information are exchanged between businesses rather than between enterprises and customers. Business-to-business commerce is crucial since every firm must acquire goods and services from other firms in order to create, run and develop. Business-to-business transactions involve the sale of goods or services from one company to another. Typically, the vendor's goods and services are used by a department or group. B2B operates in every type of industry like food, education, health care, automobile and many more.

Today because of information on various goods and services that was previously unavailable is now available to B2B clients. This kind of data helps B2B customers make wise decisions. Online presence is crucial for businesses-to-business (B2B) interactions, and this may be achieved through social media, blogs, websites and online business forums. The study examines all digital media that is now accessible worldwide and evaluates how it affects B2B marketing and sales. The number of digital marketing companies that employ digital marketing to connect suppliers with online purchasers has rapidly increased, as we have observed.

Many tools are used in digital marketing and B2B search engine optimisation (SEO), search engine marketing (SEM), content marketing, influencer marketing, content automation, campaign marketing, data-driven marketing, e-commerce marketing, social media marketing, social media optimisation, e-mail direct marketing, display advertising, e-books, and optical disks and games have all become more popular as technology has advanced.

Digital marketing and B2B aid you in reaching a larger audience than you could with traditional tactics and focusing on potential clients who are most likely to acquire your commodity or service. There are endless opportunities in digital marketing such as Site development, SEO expertise, email marketing, copywriting, content production, social media marketing, advertising and search engine marketing (SEM) are just a few of the other services available. In recent years, the area of digital marketing has evolved considerably. Along with social media, the industry seems to be developing in the next few years. There will be several opportunities in this industry. To stay ahead of the curve in their field, digital marketers must stay current on the latest advances. In the near future, the opportunities will increase in it.

We know that in the 21st century, the traditional methods of marketing have been significantly impacted by digital marketing agencies around the world. The realm of digital marketing provides a platform for prospective buyers to experience brand ownership. Customers who frequently interact with brand material have the chance to feel a connection to the company. The effects of digital marketing have permanently altered how businesses function and interact with their clients. Digital marketing tactics had an impact on profit margins and a company's capacity for expansion. The world is rapidly changing as new desires, concerns, services and trends emerge. Customers are kept up to date on industry advancements thanks to the numerous digital marketing solutions available. It allowed brands to reach their target audience at a much lower cost with a better conversion rate. So, we can say that digital marketing has created a lot of impact on our daily lives.

2. LITERATURE REVIEW

It is found that some B2B companies use digital marketing tools but most of them are unable to take full benefit (Pandey *et al.*, 2020). Their research paper highlights that a few digital marketing sales and communications management areas have witnessed growth and development through electronic marketing orientation, critical success factors, decision support systems, digital marketing and many more. This has helped their organisations to align their digital marketing activities to changes in the market dynamics. It has been observed that in the past two decades, digitalisation has been growing and growing not only through consumer marketing but also through industrial marketing (Herhausen *et al.*, 2019). Industry marketing managers and scholars seek insight into understanding how their knowledge and their practice of digital marketing have helped them to structure and configure the company/organisations. It has been found that Cognitive computing is being used throughout the fourth industrial revolution for personalisation, scalability and improved accuracy (Kizgin *et al.*, 2022). B2B organisations are deciding to adopt digital marketing tools. This will help them to make moral judgments and they can decide to embrace the moral rules through their digital marketing innovations which can be a challenge since they would be making mistakes that could damage the reputations of B2B firms. Deeter-Schmelz *et al.* (2001) said that in this Internet era, the company is getting benefits through low-cost transactional networks this will save costs and it will be improving the efficiency related to the purchase through digitalisation with the help of both buyers and sellers which will maintain long-term relationships between firms and digitalisation. Maintaining it will give growth to the firms also. Dwivedi *et al.* (2019, 2021) and Tiwary *et al.* (2021) have said that B2B firms are embracing social media marketing to help their company to achieve their business opportunities and objectives by maintaining long-term relationships between them in management (Agag, 2019; Diallo and Lambey-Checchin, 2017; Iskay, 2021; Valenzuela *et al.*, 2010). As per their research paper on digital marketing context and given ethical perceptions, they have focused on loyalty, customer trust and purchase intentions. This digital marketing has created a setting that encourages unethical behaviour via their digital marketing platforms which have raised concerns about privacy among organisations and consumers (Saddhono *et al.*, 2020). As per their research paper, must explain the business-to-business digital marketing process. In that, they said that business transactions square measurements are normally done face to face but that solely one link is an extinction of the whole chain of shopper activities and the behaviour before the sale and once of the sales. This perspective has engaged the staff at the side of the firm on the whole of completing different communication touch points. This is also known as the buyers' journey for typical business – the business of the shopper purchase route which may graduate from being unaware of the retardant. Marques *et al.* (2019) have said that the B2B sector has many new challenges determining by connecting the ecosystem. In this research paper for the traditional marketing strategies, they use CRM that works with AI. However, in this research paper, they are understanding and focusing on the application of digital marketing technology which is scarce. Herhausen *et al.* (2020) in their research paper, they have been discussing and studied the gap in the literature review. In that, they identify the use of AI-based CRM which is specialised in bringing a B2B marketing ecosystem. Blanco-Gonzalez (2019) and Palos-Sanchez and Saura (2019) in their research paper apply AI in CRM in B2B digital marketing. So, it can improve data processing and the new AI can identify the new pattern by analysing the old data they have in digital environments (Xie and Lou, 2021). According to their research paper, their content marketing techniques have distributed and created relevance to attract, relate and acquire define their target audience by creating the objective of building a consumer brand and earning money for clients. Rowley and Holliman (2014), Shah and Halligan (2010) in their research paper, their marketing strategy should be changing the mindset through the communication process. Delivering your marketing communications through advertisement messages and by adopting the approach of the inbound strategy, where customers actively look for the business because they have given them interesting and useful material by entertaining and spreading awareness of the brand. Centobelli *et al.* (2020) and Rasool *et al.* (2021) in their research paper, did conceptual background work to guide the digital marketing practitioner by recommending through their empirical research so that they can build the content on marketing strategy on digitalisation.

3. RESEARCH GAP

S. No.	Year	Author	Paper	Findings	Gap
1	2022	Rajat Kumar Behera, Prdip Kumar Bala, Nripendra P Rana Hatice Kizgin	Cognitive computing based ethical principles for improving organisational reputation: A B2B digital marketing perspective	To show the implementation of CC ethical practice with the purpose of enhancing an organisation's reputation via moral dilemmas grounded on digital marketing, we developed an integrated theoretical framework known as 'CC enabled B2B ethical digital marketing'. We emphasised the following four ethical principles of cognitive computing: governance, purpose, fairness, and transparency to connect with organisational problems in B2B digital marketing.	Like all studies, this one has some restrictions as well. The main data for this research was gathered from B2B companies that operate in India, starting with a small collection of data that was sc/considered for it. While B2B digital marketplaces are still in their infancy in India, they are progressively maturing in other parts of the globe.

S. No.	Year	Author	Paper	Findings	Gap
2	2020	Ari Warokka	Digital marketing support and business development using online marketing tools: an experimental analysis	Since the beginning of the Internet Age, the quantity of studies about online business has increased dramatically. The most original and creative concept of the twenty-first century is digital marketing. Aside from its exhaustive list, there are only a few traditional forms of advertising that don't seem to fall under the umbrella of digital marketing. Any product marketing through online campaigns has become very costly and uneconomical thanks to digital advertising. By using a directory online and a database of related products and services, marketing is carried out in this manner to reach consumers in ways that are relevant, significant, distinct, and profitable.	Have a good grasp of the key components of organisational commerce plan now. That's all there is to it for the fundamentals of digital marketing; you'll need an expert square measure toughened and sensitive in each aspect of digital commerce, such as social media selling, email commerce, and so on. Though this is an ever-growing industry, the number one issue for many businesses is that it is multi-faceted. Just a short note: we've attempted to provide you with a big picture here, but this isn't a comprehensive outline of everything associated with digital trade.

| 3 | 2021 | Subrahmanian Muthuraman | Sustainable B2B marketing during a pandemic crisis: an overview of sustainable solutions and marketing practices | During the challenging times of a pandemic crisis, the principles of sustainable marketing can lead to the provision of sustainable solutions that support B2B companies in keeping their company stability and growth and concurrently contribute to environmental and societal development. The sustainability marketing strategy has been incorporated ethically and studied for the first time in Lim's research. The technical pillar and the ethical pillar both suggest a synthesis of deontology, consequentialism, and virtue ethics to support the sustainability goal. | To guarantee better pandemic and post-pandemic resilience, B2B companies need a mindset change. This article offers a paradigm for sustainable B2B marketing, taking into account the efficacy of sustainable marketing to assist organisations in keeping their company growth and adding to the environmental and social development problem. This essay also presents several workable pandemic crisis response strategies based on the ethical, social, environmental, and technical foundations of sustainable marketing. |

S. No.	Year	Author	Paper	Findings	Gap
4	2022	Ana Rita Lopes, IPAM Porto Beatriz Casais	Digital content marketing: conceptual review and recommendations for practitioners	Shows how to organise an inbound marketing plan for use with digital media, utilising the proper digital content and media at the proper stages in the client's path to address the relevant audience. The managerial implications of this philosophical evaluation are ensured by the proposed practical directions. The conceptual review employs a design science methodology to aid a practitioner who participates in conceptualisation.	Instead of focusing only on quickly growing sales, it aspires to create a connection that will last for a long time and benefit both parties. Brands can contribute value and make a lot of money by disseminating relevant content. From obtaining, retaining, connecting and attracting consumers. By considering both user-generated material and customer communications, the company can establish a long-lasting connection with its audiences. Practitioners of digital marketing can make use of the suggestions made in this paper, which addresses the philosophical foundations of content marketing, to create a content marketing strategy. The book analysis technique recommended by Centobelli et al. was used.

| 5 | 2021 | Pentti Korpela,Ari Alamaki | Digital transformation and value-based selling activities: seller and buyer perspectives | Through a more ongoing, proactive approach that heavily integrates digital value co-creation, companies are moving their B2B sales towards value-based marketing because of the digitalisation of the sales business. Similar to this, their consumers now call for more preemptive communication about novel value propositions, but the majority of B2B clients aren't big fans of social media. Since non-sellers frequently play a major role in promoting new value propositions, the sales ecosystem should be taken into account when handling digital value co-creation efforts. | Value-based selling pushes selling firms to concentrate more on instructional digital content marketing and non-seller engagement via both marketing and sales activities, therefore more research on digital value co-creation activities in the sales environment is required. |

S. No.	Year	Author	Paper	Findings	Gap
6	2021	Shahrzad Yaghtin	Sustainable B2B marketing during a pandemic crisis: an overview of sustainable solutions and marketing practices	Due to the severe societal and fiscal problems caused by the COVID-19 disaster, a broader view of marketing management is crucial in the modern environment. During the trying times of a pandemic disaster, applying marketing can result in the deliverance of long-term solutions that assist B2B businesses in maintaining their company stability and development while also improving the environment and society. An overview of study results on environmentally friendly choices that B2B businesses can consider during the COVID-19 pandemic based on the five essential elements of sustainable marketing.	To value-based pandemic and post-pandemic resilience, B2B companies need a mindset change. This article offers a paradigm for sustainable B2B marketing, taking into account the efficacy of sustainable marketing to assist organisations in keeping their company growth and adding to the environmental and social development problem. In addition, this paper presents several workable pandemic crisis solutions based on the ethical, social, environmental, and technical foundations of sustainable marketing.

| 7 | 2021 | Dwiana Rahmadiati Putri | Digital marketing strategy to increase brand awareness and customer purchase intention | Ailesh Green Consulting is facing issues with brand awareness and customer purchase intention, according to Ailesh Power the Chief of Business Operations, and potential clients who have expressed disinterest in using their services. The only digital marketing tool in use is their website, and personal selling through digital marketing has not been utilised. Their online media platform has not been effectively optimised and only provides basic information about their services and clients. Word-of-mouth advertising remains their primary means of promotion. | Changing the headline of an otherwise identical article or bit of content. The life cycle study of your organisation is a novel heading. The issue consider Ailesh Green Consulting is currently facing. Understanding of life cycle assessment by the customer. Respondents to the poll said they wouldn't use Ailesh Green Consulting's services again, according to the survey's findings. Nine out of ten respondents stated that they have no plans to use Ailesh Green Consulting to produce LCA documents for their business. The majority of them claimed that this choice cannot be made on an individual basis because it needs to be thoroughly investigated by a team selected by the business, such as the HSE section as a whole. |

S. No.	Year	Author	Paper	Findings	Gap
8	2020	S. Mahalingam. B. Ashokkumar	An overview of digital marketing practices in India	This study looked at the various types of digital marks and several potential uses for them. It also covered a wide range of adoption-related ideas and principles. E-marketplace and its uses in the adoption of digital marketing would be among the additional subjects addressed.	If a business adopts digital marketing incrementally, either by doing so without organisational changes or by implementing some organisational changes only partly, which can only lead to significant product losses, the benefits of digital marketing may not be completely realised. The potential of the Internet is recognised by new business strategies in the field of digital marketing, which offer consumers more control over their purchases whether they are made online or offline. Because of the Internet, pricing is more transparent than ever. Customers can easily assess the prices of the same or comparable goods and services provided by companies around the globe with a few quick clicks or smartphone swipes.

| 9 | 2021 | Jose Ramo n Saura, Domingo riberio-soriano | Setting B2B digital marketing in artificial intelligence-based CRMs: A review and directions for future research | If a business adopts digital marketing incrementally, either by doing so without organisational changes or by implementing some organisational changes only partly, which can only lead to significant product losses, the benefits of digital marketing may not be completely realised. The potential of the Internet is recognised by new business strategies in the field of digital marketing, which offer consumers more control over their purchases whether they are made online or offline. Because of the Internet, pricing is more transparent than ever. Customers can easily assess the prices of the same or comparable goods and services provided by companies around the globe with a few quick clicks or smartphone swipes. | The studies' sample sizes, the databases used to perform the comprehensive review of the literature, and the researchers' in-depth evaluation of the studies they selected are the sources of the research's limitations. It can also be argued that the MAC process has trouble understanding its output visibly. Future studies should continue the study paths proposed in the future to continue laying the groundwork for the use AI-powered CRMs for B2B digital marketing as the technology corresponding to their advancement is constantly increasing. |

S. No.	Year	Author	Paper	Findings	Gap
10	2020	Pandey. N., Nayal, P. and Rathore, A. S	**Digital marketing for B2B organisations: structured literature review and future research directions**	The majority of businesses use digital marketing for their operations, but most of them are unable to take full benefit of it because they haven't done enough study on the subject. The framework provides a deeper look into new subjects. The analysis reveals decision support systems, crucial success factors, an emphasis on electronic marketing, etc. Some aspects, like sales management and digital marketing communication, may experience constant improvement. It also demonstrates the expanding research fields for aspiring researchers and finds research requirements.	A thorough book review on B2B digital marketing has been conducted. The different subjects have been chosen after a thorough review of the body of literature. Semi-structured interviews with B2B marketing experts were also conducted to further concentrate the developing digital marketing topics.

3.1. Conceptual Modelling

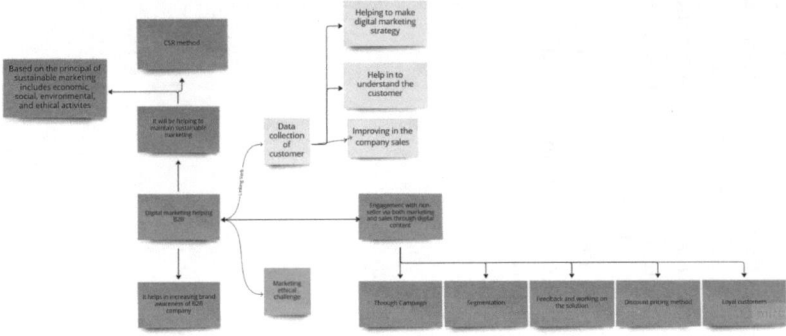

In this conceptual modelling, we are connecting the dots with why B2B needs the support of digital marketing. First, digital marketing will be storing their existing customers which is corporate people data, and the new one also. This will be helping them to maintain relationships between themselves. This existing data will be helping B2B firms to improve their company sales, this will be helping them to understand the customer's needs and problems, this will be helping to make a digital marketing strategy to create brand awareness and product awareness for their targeted customers. Second, digital marketing can bring engagement through many digitalised ways like through campaigns, segmentation, feedback taken from chatbots and taking out the solution for that problem, by targeting and following the trend you can use discount pricing method and with the digital tools you can maintain loyal customer by maintaining a relationship with them through digital and CRM tools. Third, this is not a disadvantage but digital marketing face marketing ethical challenge. But today AI is fully developed whatever tools the firms are using privacy concerns are not there. Most of the customers are very carefully related to privacy. So, looking at that privacy concerns are very less. Fourth, this digital marketing is one of the most useful tools to create brand awareness. Because of this tool, most of the firms have spread awareness of their brand. Fifth, this digital marketing tool supports the carbon footprint. They follow the principle of sustainable marketing including economic, social, ethical activities, and environmental. Sixth, digital marketing tools can spread awareness about the firm's CSR activities. They can show it as a blog with the help of digital marketing tools.

4. METHODOLOGY

This research paper aims to analyse how digital marketing and its tools play an important role in b2b firms. To reach the aim, we prepared a structured questionnaire, and out of 100 people, only 70 responded to the questionnaire. In this research paper, we used both primary and secondary data like primary data we have taken the questionnaire and secondary data is from Google and articles. This research used the descriptive method to get the statistical result from respondents.

4.1. Data Analysis and Interpretation

Digital marketing tools important for B2B firms: Of the 70 respondents who were asked whether digital marketing tools are important for B2B firms, 66 answered in the affirmative, while only 4 respondents said no. This means that more than 90% of the respondents believe that digital marketing tools are important for B2B firms. Therefore, we can conclude that there is a strong consensus among the respondents that digital marketing tools are indeed crucial for B2B firms.

Data collection of the customer helps B2B firms: Out of the 70 respondents who were asked whether data collection of customers helps B2B firms, 68 responded in the affirmative, while only 2 respondents disagreed. This means that most of the respondents believe that data collection of customers is indeed beneficial for B2B firms. Therefore, we can conclude that there is a strong consensus among the respondents that data collection of customers is a useful tool for B2B firms.

By doing advertisements in the way of digital marketing, will it help in the growth of sales of B2B firms: of the 70 respondents who were asked whether advertisement through digital marketing helps in the growth of sales of B2B firms, 34 respondents agreed, while 15 strongly agreed, making a total of 49 respondents in agreement. On the other hand, 3 respondents strongly disagreed, while 4 disagreed, and 14 were neutral. Therefore, we can conclude that the majority of the respondents agree that advertisement through digital marketing can help in the growth of sales of B2B firms. The responses indicate that digital marketing can be an effective tool for B2B firms looking to boost their sales.

The advantages of data collection in digital marketing: of the 70 respondents who were asked about the advantages of data collection in digital marketing, 62 respondents agreed that it helps in understanding customers' needs. Additionally, 46 respondents believed that data collection helps in making digital marketing strategies. Only one respondent each thought that data collection can improve precision in targeting customers and help the company perform better overall compared to competitors. Therefore, we can conclude that the majority of respondents agree that data collection in digital marketing can help in understanding customers' needs and in developing effective marketing strategies. These advantages can ultimately lead to increased customer satisfaction and improved business outcomes for the company.

Digital marketing will be helping to bring engagement to increase sales: of the 70 respondents who were asked about how digital marketing can help bring engagement and increase sales, 46 respondents believed that campaigns are an effective tool for this purpose. Additionally, 36 respondents believed that segmentation can help bring engagement and increase sales. Feedback was also considered important, with 40 respondents believing it can help improve engagement and increase sales. Discount pricing methods were also seen as useful by 29 respondents. Lastly, only 20 respondents believed that retention can help bring engagement and increase sales. Therefore, we can conclude that according to most of the respondents, campaigns, segmentation and feedback are important tools that can help bring engagement and increase sales in digital marketing.

Digital marketing help in increasing brand awareness of B2B firm: of the 70 respondents who were asked whether digital marketing helps in increasing brand awareness of B2B firms, 67 respondents answered in the affirmative, while only 3 respondents said no. This means that most of the respondents believe that digital marketing does indeed help in increasing brand awareness of B2B firms. Therefore, we can conclude that there is a strong consensus among the respondents that digital marketing is an effective tool for increasing brand awareness for B2B firms.

Ethics plays an important role in marketing for B2B companies: of the 70 respondents who were asked whether ethics play an important role in marketing for B2B companies, 64 respondents answered in the affirmative, while only 6 respondents said no. This indicates that most of the respondents believe that ethics do indeed play an important role in marketing for B2B companies. Therefore, we can conclude that there is a strong consensus among the respondents that B2B companies should prioritise ethical considerations in their marketing practices. Adhering to ethical principles can help these companies build a positive reputation and establish trust with their clients, ultimately leading to improved business outcomes.

CSR important in B2B firms: of the 70 respondents who were asked whether Corporate Social Responsibility (CSR) is important in B2B firms, 63 respondents answered in the affirmative, while only 7 respondents said no. This indicates that the vast majority of the respondents believe that CSR is indeed important in B2B firms. Therefore, we can conclude that there is a strong consensus among the respondents that B2B firms should prioritise their social responsibility and contribute to the betterment of society. Incorporating CSR initiatives in their business practices can help these firms build a positive image, enhance their reputation and create a sense of goodwill among their stakeholders. This can ultimately lead to improved business outcomes and long-term success.

Social media branding strategy aligned with my overall branding strategy: of the 70 respondents who were asked whether their social media branding strategy is aligned with their overall branding strategy, 34 respondents answered in the affirmative, while 9 respondents strongly agreed, 25 respondents were neutral and only 2 respondents disagreed. This indicates that a majority of the respondents believe that their social media branding strategy is aligned with their overall branding strategy. Therefore, we can conclude that there is a consensus among the respondents that social media branding should be consistent with the overall branding strategy of a company. Having a cohesive branding strategy across all channels can help companies establish a strong brand identity, increase brand recognition and build trust with their customers.

The top social media platforms for B2B marketing: of the 70 respondents who were asked about the top social media platforms for B2B marketing, 52 respondents answered Instagram, 49 respondents answered Facebook and 45 respondents answered LinkedIn. These three social media platforms emerged as the most popular choices for B2B marketing among the respondents. Other social media platforms, such as Twitter and WhatsApp, were chosen by 26 and 20 respondents respectively, while only 4 respondents selected 'Others'. This indicates that Instagram, Facebook and LinkedIn are the top social media platforms for B2B marketing, and companies should prioritise their marketing efforts on these platforms to effectively reach their target audience, build brand awareness and generate leads.

5. CONCLUSION

Now with the rise of technology, digital marketing research and practice are getting better. Technology development creates various benefits while also presenting marketers with unusual obstacles. Marketers utilise Digital Portfolios as a platform to advertise a professional brand by accurately describing the goods. The CSR tool plays important role in B2B firms. B2B buyers prefer cold emails and calls, according to research, but digital marketing offers plenty of options for connecting with the correct audience through feedback and campaigns thanks to its efficient search engine and relationships. Today's consumers demand information that could help them solve problems with a product. Because of the organisation's or brand's significant online presence, people might form perceptions of the company or product.

REFERENCES

Alamäki, A., and P. Korpela (2021). "Digital transformation and value-based selling activities: seller and buyer perspectives."

Hawaldar, Iqbal Thonse, Mithun S. Ullal, Adel Sarea, Rajesha T. Mathukutti, and Nympha Joseph (2022). "The study on digital marketing influences on sales for B2B start-ups in South Asia."

Herhausen, Dennis, Dario Miočević, Robert E. Morgan, and Mirella H.P. Kleijnen (2020). "The digital marketing capabilities gap."

Hofacker, C., I. Golgeci, K.G. Pillai, and D.M. Gligor (2020). "Digital marketing and business-to-business relationships: a close look at the interface and a roadmap for the future."

Lopes, Ana Rita, IPAM Porto Beatriz Casais, and University of Minho (2022). "Digital content marketing: conceptual review and recommendations for practitioners."

Mahalingam, S., and B. Ashok Kumar (2020). "An overview of digital marketing practices in India."

Moreno-Camacho, C.A., J.R. Montoya-Torres, A. Jaegler, and N. Gondran (2019). "Sustainability metrics for real case applications of the supply chain network design problem: a systematic literature review."

Muthuraman, Subrahmanian, Mohammed Al Haziazi, Rengarajan Veerasamy, and Nasser Al Yahyaei (2021). "SME – Key Drivers for Economic Development in the Sultanate of Oman."

Pandey, N., P. Nayal, and A.S. Rathore (2020), "Digital marketing for B2B organizations: structured literature review and future research directions."

Putri, Dwiana Rahmadiati (2021). "Digital marketing strategy to increase brand awareness and customer purchase intention (Case Study: Ailesh Green Consulting)."

Saura, Jose Ramon, Domingo Ribeiro-Soriano, and Daniel Palacios-Marqués (2021). "Setting B2B digital marketing in artificial intelligence-based CRMs: a review and directions for future research."

Siamagka, Nikoletta-Theofania, George Christodoulides, Nina Michaelidou, and Aikaterini Valvi (2015). "Determinants of social media adoption by B2B organizations."

Urte STURIENE (2019). "Internet marketing tools."

Warokka, Ari, Herman Sjahruddin, Sriyanto Sriyanto, Endang Noerhartati, and Kundharu Saddhono (2020). "Digital marketing support and business development using online marketing tools: an experimental analysis."